APPLIED FORECASTING METHODS

APPLIED FORECASTING METHODS

NICK T. THOMOPOULOS

Harold Leonard Stuart School of Management and Finance
Illinois Institute of Technology

PRENTICE-HALL, INC., *Englewood Cliffs, New Jersey 07632*

Library of Congress Cataloging in Publication Data

Thomopoulos, Nicholas T
Applied forecasting methods.

Bibliography: p.
Includes index.
1. Forecasting. I. Title.
HD30.27.T56 338.5′442 79-266
0-13-040139-0

Editorial/production supervision and interior design
by Barbara A. Cassel
Cover design by Jorge Hernandez
Manufacturing buyer: Gordon Osbourne

Printed in the United States of America

10 9 8 7 6 5 4 3 2

PRENTICE-HALL INTERNATIONAL, INC., *London*
PRENTICE-HALL OF AUSTRALIA PTY. LIMITED, *Sydney*
PRENTICE-HALL OF CANADA, LTD., *Toronto*
PRENTICE-HALL OF INDIA PRIVATE LIMITED, *New Delhi*
PRENTICE-HALL OF JAPAN, INC., *Tokyo*
PRENTICE-HALL OF SOUTHEAST ASIA PTE. LTD., *Singapore*
WHITEHALL BOOKS LIMITED, *Wellington, New Zealand*

This book is dedicated to

Elaine, Marie, Melina, and Diana

CONTENTS

Appendices

PREFACE

This book is an introduction to forecasting for practitioners and students in business administration, management science, and industrial engineering. It covers the principles and techniques of forecasting the future demands of an item for applications in the industrial, retail, and public sectors of the economy; the forecasts are projections that are based on the flow pattern of past demands. The book also gives the fundamentals on how forecasts are used in controlling customer service. These are highly important tools for management and line personnel in their daily course of activities and decision making.

The book has evolved during the past several years from classroom lectures on forecasting at the Illinois Institute of Technology. It is based also on the author's consulting experience in forecasting for inventory and production applications throughout the United States, Europe, and Japan.

The bulk of the book is nontechnical and is written for readers with a modest amount of mathematical and statistical training. Each section of the book begins with a description of the technique at hand and gives the major results and implications needed in its application. One or more numerical examples are presented to give insight and to smooth the way for the reader. To close the section, a short mathematical basis is sometimes provided for further clarification, which many readers may choose to skip since the findings are essentially proofs and are not needed in applying the techniques. Homework problems are included with each chapter, along with pertinent references.

The author is grateful to the many persons who have made numerous contributions and suggestions. They include Allen Endres, Marinos Frangiadis, James Hall, Willard Huson, Kazuo Iwata, Keith Mahal, Yoshio Mizoroki, Pricha Pantumsinchai, and Reino Warren. Special thanks to Donna Cahill and Catherine Hiles for the tedious task of typing and retyping the manuscript. I am grateful to my wife Elaine for her assistance in checking the drafts and final proofs; to my children, Marie, Melina, and Diana, for their patience; and to my mother, Marie Thomopoulos, for helping with the children. Thanks also to Paul Becker, Barbara Cassel, and Hank Kennedy of Prentice-Hall for their guidance.

NICK T. THOMOPOULOS

1

INTRODUCTION

In today's complex world, management is faced with a never-ending flow of business planning and decision making. A forecast of some kind is oftentimes used as a basis to meet these needs, whereby the more reliable the forecasts are, the better is the outcome from the planning and decisions. Forecasting is not a new problem—it has plagued management for centuries. In the recent past, however, with the advent of computers, it has become possible to use forecasting methods that were previously impossible to explore. The purpose of this book is to introduce the various types of forecasting techniques that have been reported in the literature or used in industry with relative acceptance.

Forecasts are no longer a luxury; they are a necessity since they allow management to cope with the ever-changing shifts in demands for their products and resources. A company with an oversupply in inventory incurs undue costs caused by stocking, deterioration, or obsolesence of the items. With an undersupply, bad will and lost sales may result. Reliable forecasts are essential for a company to survive and grow.

In a manufacturing environment, management must forecast the future demands for its products and on this basis provide for the materials, labor, and capacity to fulfill these needs. These resources are planned and scheduled well before the demands for the products are placed on the firm.

Forecasting is the heart of an inventory control system. A firm with hundreds or thousands of items must anticipate in advance demands that will occur against these items. This is needed to have the proper inventory available to fill customers' demands as they come in. Management must plan several months in advance for this inventory, since procurement lead times from suppliers generally runs from 2 to 6 months. With each item, forecasts are needed for the months in the planning horizon. The forecasts are used to determine whether or not an order to the supplier is needed now, and if so how large the order should be.

Forecasting techniques can be categorized into three groups. The first is called *qualitative*, where all information and judgment relating to an item are used to forecast the items demands. This technique is often used when little or no demand history is available. The forecasts may be based on marketing research studies, the Delphi method, or similar methods.

The second group is called *causal*, where a cause-and-effect type of relation is sought. Here, the forecaster seeks a relation between an item's demands and other factors, such as business industrial and national indices. The relationship is used to forecast the future demands of the item.

The third group is called *time-series analysis*, where a statistical analysis on past demands is used to generate the forecasts. A basic assumption is that the underlying trends of the past will continue into the future.

This book is primarily concerned with forecasting as it relates to time-series analysis. In this context the time series represents the demands recorded over past time intervals. The forecasts are estimates of the demands over future time intervals and are generated using the flow of demands from the past.

1-1 PLAN OF THE BOOK

This book is written for practitioners and students who desire an understanding in forecasting without the need to delve deeply into the mathematical theory of forecasting. The forecasting methods are described, with emphasis placed on how to use the techniques. Examples are cited along the way to guide the reader. All the forecasting methods have some mathematical basis associated with them. For purposes of continual flow, if the mathematical basis is associated with a section of a chapter, it is presented at the end of the section. These (mathematical basis) sections can be skipped over by most readers, since they primarily contain proofs and are not needed to carry out the methods. They also require a higher level of mathematical training than is needed in the body of the book.

The book contains 15 chapters, of which this introduction is the first. A short summary of each chapter follows.

Chapter 2 provides a review of certain statistical concepts that are useful in the development and understanding of forecasting methods. These include topics in summations, probability, means, variances, least squares, and matrices. The reader does not have to master the material of this chapter to move forward, but may instead refer back to this chapter as the need arises.

Chapter 3 reviews certain aspects of the demand history that the forecaster should recognize before he begins to generate a forecast. First is a review of the demand patterns with which an item's demand history may be classified. Some items have demands that fluctuate around a fixed level while others move with an increasing or decreasing trend. Some have seasonal or other influences as well. It is important for the forecaster to recognize the pattern in order to choose the appropriate forecast model. Second is a review of filtering methods. Filtering is used to screen the demand entries prior to forecasting in order to seek out and modify any entries that are far removed from the normal flow of all other entries. The aim is to provide reliable data to generate the forecasts.

Chapter 4 describes three forecasting models that are applicable when the demand pattern is horizontal. This is where the demands per time period fluctuate in a random manner about a fixed value, and an average demand of the past is used as the forecast of future demands.

Chapter 5 presents four forecasting models that are used when the demand level is assumed to be moving upward or downward in a straight-line manner. A linear fit of the past demands is found and is projected forward to forecast the future demands.

Chapter 6 shows two forecasting models that can be used when the demand pattern is quadratic. In these situations, the level of the demand is rising or falling in a nonlinear fashion over the neighborhood of the recent past and the future planning horizon.

Chapter 7 describes various forecasting methods that are useful over a wider range of demand patterns. The models include regression, discounted regression, and the adaptive smoothing methods. In all cases the method of least squares is used to generate the forecasts. The discounted regression method allows the forecaster to give higher weights to the more current demand entries. The adaptive smoothing model also uses discounting and provides a procedure of updating the forecasts in a relatively simple manner.

The adaptive smoothing method is carried forward and used in Chapter 8 to forecast trigonometric-type demand patterns. In these patterns the demands are assumed to flow with various sine-wave and trending influences. These patterns are often interchangeable with trend-seasonal demand patterns.

Chapter 9 is devoted to seasonal and trend-seasonal forecasting models. The models are useful for items with seasonal and trend influences. One seasonal model and two trend seasonal models are described (the multiplicative and additive trend seasonal models).

Chapter 10 describes various adaptive control and blending methods. These methods are applied by using some of the forecasting models presented in the earlier chapters. The goal of adaptive control techniques is to seek out the best parameter(s) of a forecasting model. The parameter(s) may or may not be altered at each time period, depending on the flow of the forecast errors. Blending techniques combine results from two independent forecasts to generate a new forecast.

Chapter 11 is devoted to the Box–Jenkins forecasting method. This method follows a systematic procedure in seeking an appropriate forecast model from which the forecasts are based. This method generally gives better forecasts than the standard techniques but requires more data and calculations. For these reasons the Box–Jenkins method is more applicable to key time series of relative importance and not to the vast majority of items in an inventory.

Chapter 12 presents various methods in forecasting that are used in special situations. Forecasting techniques are shown for lumpy items, for items without regularly posted demands, and for items with some knowledge of future demands available in advance. In addition, methods of forecasting the all time requirements of an item are shown. The cumulative sum technique is described as it is used to detect a change in the level of an item. Finally, methods of reducing the drop in fast decreasing forecasts is given, and a method of measuring the rate of growth in the forecasts is shown.

Chapter 13 describes various forecasting methods that are applicable when the problem setting contains multiple dimensions. Ranking methods and vector smoothing are used to estimate the probability distribution of future demands for an item. The vector smoothing method is modified to generate forecasts of demands for the mix of items in a product line, and for the demands of an item by region. Blending techniques are used to forecast the mix of items in a product line as well as the mix of items by characteristics, such as size and color. The percent done method is shown as it is used to estimate the total demand of an item over a season. Finally, the percent of aggregate demand method is presented. This method is used to forecast each item in a line.

Chapter 14 pertains to forecasting errors and tracking signals. Methods of measuring the standard deviation and mean absolution deviation of the forecasting error are shown. These measures are used in a variety of ways to help control the forecasts and the quantity of stock that is provided in the inventory. The tracking signal is a quality-control-type device that allows the forecaster to detect the adequacy of the currently employed forecast model and to help guide the forecasting system along the way. Two methods of measuring the tracking signal are given.

Chapter 15 is concerned with customer service. The goal is to provide the minimum safety stock to reach an acceptable service level. Methods of providing for this goal are shown for the normal and the truncated normal distributions, and for individual and a group of items. The sensitivity of the service level and safety stock are shown as they are affected by order sizes, lead times, and forecast errors.

1-2 NOTATION

The following notation is used throughout this book. A time period represents a point in time and the time periods $t = 1, 2, 3, \ldots$ designate equally spaced points in time. The current time period is denoted as T, so the full history of time periods is $t = 1, 2, \ldots, T$. The future time periods are listed as $t = T + \tau$, where τ represents the τth future time period.

The amount of demand in time period t is x_t and the history of demand is $(x_1, x_2, \ldots, x_T)$. Note that x_T identifies the most current demand entry.

The forecast at time T for the τth future demand is designated as $\hat{x}_T(\tau)$. This is the τ-period-ahead forecast for the (as yet) unknown observation $x_{T+\tau}$.

Often, the cumulative τ-period forecast is needed. This is the forecast for the sum of the next τ time periods. This forecast is listed as $\hat{X}_T(\tau)$, where

$$\hat{X}_T(\tau) = \hat{x}_T(1) + \hat{x}_T(2) + \ldots + \hat{x}_T(\tau)$$

The τ-period-ahead forecast error is denoted as $e_{(\tau)}$, where this error is obtained by

$$e_{(\tau)} = x_{T+\tau} - \hat{x}_T(\tau)$$

in some applications and by

$$e_{(\tau)} = \hat{x}_T(\tau) - x_{T+\tau}$$

in others. The τ-period cumulative forecast error is given by

$$E_\tau = e_{(1)} + e_{(2)} + \ldots + e_{(\tau)}$$

REFERENCES FOR FURTHER STUDY

CHAMBERS, J. C., S. K. MULLICK, AND D. D. SMITH, "How to Choose the Right Forecasting Technique." *Harvard Business Review*, July–August 1971, pp. 45–74.

CHAMBERS, J. C., S. K. MULLICK, AND D. D. SMITH, *An Executive's Guide to Forecasting*. New York: John Wiley & Sons, Inc., 1974.

DAUTEN, C. A., AND L. M. VALENTINE, *Business Cycles and Forecasting*. Cincinnati, Ohio: South-Western Publishing Company, 1974.

GROSS, C. W., AND R. T. PETERSON, *Business Forecasting*. Boston: Houghton Mifflin Company, 1976.

2

STATISTICAL CONCEPTS

This chapter gives an introduction to certain concepts in statistics and related topics that are useful in forecasting. These include topics in summations, probability, random variables, means, mean absolute deviation, variance, covariance, correlation, linear combinations, the normal distribution, confidence intervals, the central limit theorem, least squares, and matrices. The reader does not have to master the topics of this chapter before moving forward, but instead may refer back to this chapter as the need arises.

2-1 SUMMATIONS

A common operation in forecasting is to find the sum of a set of terms. Let N designate the number of such terms and $x_1, x_2, \ldots, x_N$ the particular values for each of the N terms. The summation becomes

$$\sum_{i=1}^{N} x_i = x_1 + x_2 + \ldots + x_N$$

The capital Greek letter Σ (sigma) is interpreted as the "sum of" and $\sum_{i=1}^{N} x_i$ represents the sum of x_i for i ranging from 1 to N. The letter i is called the *summation index*. The summation may also be carried out over transformations of the x_i, such as

$$\sum_{i=1}^{N} x_i^2 = x_1^2 + x_2^2 + \ldots + x_N^2$$

Example 2.1

If $x_1 = 5$, $x_2 = 7$, $x_3 = 1$, and $x_4 = 3$, then

$$\sum_{i=1}^{4} x_i = 5 + 7 + 1 + 3 = 16$$

$$\sum_{i=1}^{4} x_i^2 = 5^2 + 7^2 + 1^2 + 3^2 = 84$$

The summations often include one or more constants. Letting a represent a constant, then

$$\sum_{i=1}^{N} a = Na$$

$$\sum_{i=1}^{N} ax_i = a \sum_{i=1}^{N} x_i$$

$$\sum_{i=1}^{N} (x_i - a)^2 = (x_1 - a)^2 + (x_2 - a)^2 + \ldots + (x_N - a)^2$$

A common situation with two constants (a, b) is

$$\sum_{i=1}^{N} (ax_i + b) = a \sum_{i=1}^{N} x_i + Nb$$

Example 2.2

Suppose that $x_1 = 5$, $x_2 = 7$, $x_3 = 1$, $x_4 = 3$, $a = 2$, and $b = 5$. Some examples are:

$$\sum_{i=1}^{4} 2 = 4(2) = 8$$

$$\sum_{i=1}^{4} 2x_i = 2(5) + 2(7) + 2(1) + 2(3) = 2(16) = 32$$

$$\sum_{i=1}^{4} (x_i - 2)^2 = (5 - 2)^2 + (7 - 2)^2 + (1 - 2)^2 + (3 - 2)^2 = 36$$

$$\sum_{i=1}^{4} (2x_i + 5) = 2 \times 16 + 4 \times 5 = 52$$

The summations can also be carried out with two or more variables (say x and y). Some examples used in forecasting are:

$$\sum_{i=1}^{N} (x_i + y_i) = \sum_{i=1}^{N} x_i + \sum_{i=1}^{N} y_i$$

$$\sum_{i=1}^{N} x_i y_i = x_1 y_1 + x_2 y_2 + \ldots + x_N y_N$$

Example 2.3

Using $x_1 = 5, x_2 = 7, x_3 = 1, x_4 = 3, y_1 = -2, y_2 = 0, y_3 = 4$ and $y_4 = -1$, then

$$\sum_{i=1}^{4} (x_i + y_i) = \sum_{i=1}^{4} x_i + \sum_{i=1}^{4} y_i = 16 + 1 = 17$$

$$\sum_{i=1}^{4} x_i y_i = 5 \times -2 + 7 \times 0 + 1 \times 4 + 3 \times -1 = -9$$

2-2 PROBABILITY

Statistics and probability are so closely related that it is difficult to discuss statistics without an understanding of probability. Probability theory is used in many ways to interpret statistical result. In this section some of the fundamental elements of probability theory are given.

Perhaps an appropriate start is to give the following four definitions:

experiment—a set procedure of obtaining a result from some process (e.g., record all demands for an item over a given month's time duration).

trial—actual performance of the experiment (e.g., record all demands in June).

outcome—the results of the trial (e.g., the sum of all demands in June is 140 pieces).

event—a particular outcome (e.g., the sum of all demands in June is greater than 100).

Probability may be described as follows: when an event may occur x times in N trials, the probability of the event occurring (in a single trial) is x/N.

When the result above is generated from a theoretical basis, it is called the *theoretical probability*. When it is estimated from historical data, it is an *empirical probability*. For example, in the toss of a die, the theoretical probability for an odd number to occur is 1/2; whereas if the demand for an item has been greater than 100 in 4 of the past 24 months, the empirical probability that the demand will exceed 100 in the coming month is $\frac{4}{24} = 0.167$.

For an event A, the probability that the event A will occur in any given trial is denoted as $P(A)$. It is always true that $0 \leq P(A) \leq 1$. When $P(A) = 0$, the event A will never occur; when $P(A) = 1$, it always occurs; and when $P(A) = \frac{1}{2}$ (say), the event A has a one-half chance of occurring on any given trial.

The *complementary event* of A is denoted as $\bar{A}$. This is the event that A will not occur and the probability of event $\bar{A}$ is

$$P(\bar{A}) = 1 - P(A)$$

Note that in any given trial, either A or $\bar{A}$ will occur and hence

$$P(A) + P(\bar{A}) = 1$$

Two events are said to be *mutually exclusive* if they cannot occur simultaneously (e.g., A = event that the demand will be greater than 100, and B = event that the demand will fall between 50 and 100). Letting $P(A)$ and $P(B)$ represent the probability that A and B occur, respectively, the probability that A or B occurs in a single trial is

$$P(A \text{ or } B) = P(A) + P(B)$$

When the events are not mutually exclusive (A = event demand is between 100 and 200, and B = event demand is greater than 150), the *joint probability* is used. This is denoted as $P(AB)$[1] and represents the probability that events A and B both occur in a single trial. In this situation the probability that A or B occurs in a single trial is

$$P(A \text{ or } B) = P(A) + P(B) - P(AB)$$

For example, if $P(A) = 0.20$, $P(B) = 0.25$ and $P(AB) = 0.10$, then

$$P(A \text{ or } B) = 0.20 + 0.25 - 0.10 = 0.35$$

gives the probability that the demand will be greater than or equal to 100. Note that when the events (A, B) are mutually exclusive, $P(AB) = 0$.

Two or more events are said to be *independent* when the probability of any of them occurring is not influenced by the occurrence of the other(s). Otherwise, the events are dependent. When events A and B are independent,

$$P(AB) = P(A)P(B)$$

For example, if A is the event of a 3 occurring in a roll of a die $[P(A) = 1/6]$ and B is the event of a head in a toss of a coin $[P(B) = 1/2]$, then A and B are independent because the outcome of one has no influence on the other. Further,

$$P(AB) = \frac{1}{6} \times \frac{1}{2} = \frac{1}{12}$$

is the probability that a three (from the die) and a head (from the coin) will jointly occur in an experiment where a die and coin are tossed together.

When the events A and B are *dependent* (not independent), the joint probability of A and B is related by way of a *conditional probability*. The

[1] $P(AB) = P(A \text{ and } B)$.

conditional probability is denoted as

$$P(A \mid B) = \text{probability of event } A \text{ given event } B \text{ has occurred}$$

The vertical line separating A and B is interpreted as "given." All events to the left of the line are the events of interest and those to the right are the events that are given (or known). When $P(B) \neq 0$, then

$$P(A \mid B) = \frac{P(AB)}{P(B)}$$

$$P(AB) = P(A \mid B)P(B)$$

Recall that when A and B are independent, then $P(AB) = P(A) \cdot P(B)$. This yields

$$P(A \mid B) = \frac{P(AB)}{P(B)} = \frac{P(A)P(B)}{P(B)} = P(A)$$

Example 2.4

Consider an experiment where a die is rolled only if the outcome from the toss of a coin is a head. Let H = event that a head occurs $[P(H) = 1/2]$ and A = event that a three will occur in a roll of the die (should it be rolled at all). Note that

$$P(A \mid H) = \frac{1}{6}$$

$$P(A \mid \bar{H}) = 0$$

the joint probability of A and H is

$$P(AH) = P(A \mid H)P(H) = \left(\frac{1}{6}\right)\left(\frac{1}{2}\right) = \frac{1}{12}$$

Also, $P(A\bar{H}) = 0$.

Example 2.5

Consider customers of long-sleeved shirts in a retail store. Let A = event that the collar size is 16 and B = event that the arm length is 33. Assume it is known that for an arbitrary customer, $P(A) = 0.10$ and $P(B \mid A) = 0.50$. With this information the probability that a customer wishes a (16×33) shirt is easily found. This is

$$P(AB) = P(B \mid A)P(A) = 0.50 \times 0.10 = 0.05$$

Example 2.6

Suppose that A = event that demand is greater than 150, and B = event demand is between 100 and 200. Assume also that $P(B) = 0.20$ and $P(AB) = 0.10$ are known. Then

$$P(A \mid B) = \frac{P(AB)}{P(B)} = \frac{0.10}{0.20} = 0.5$$

yields the conditional probability that the demand is greater than 150 given the demand lies within 100 and 200.

The K events $(A_1, A_2, \ldots, A_K)$ are said to be *exhaustive* if any possible outcome from a trial must include at least one of the events. If they are *mutually exclusive and exhaustive*, the outcome from an arbitrary trial must be one of the K events. In the latter situation,

$$\sum_{i=1}^{K} P(A_i) = 1$$

Example 2.7

Suppose that

$$A_1 = \text{event demand} \leq 50$$
$$A_2 = \text{event demand is within (51 and 100)}$$
$$A_3 = \text{event demand} > 100$$

The events A_1, A_2, and A_3 are exhaustive and mutually exclusive since one of the three events must occur in a given trial. Hence,

$$P(A_1) + P(A_2) + P(A_3) = 1$$

2-3 RANDOM VARIABLES

A random variable can be discrete or continuous, depending on the values with which it may assume. A description of each type follows.

Suppose that a variable x can incur the values $x_1, x_2, \ldots, x_K$ with probabilities $P_1, P_2, \ldots, P_K$, respectively. That is,

$$P(x = x_i) = P_i \qquad \text{for } i = 1, 2, \ldots, K$$

Because x can assume certain values with specific probabilities, x is a *discrete random variable*. When

$$\sum_{i=1}^{K} P_i = 1$$

then P_i $(i = 1, 2, \ldots, K)$ is the probability distribution of x_i.

Example 2.8

Suppose that a truck dealer has four trucks in stock. Let x represent the number of these trucks that will be sold during the coming month. Hence, $x = 0, 1, 2, 3$, or 4. Assume that the probabilities associated with the values of x are $P(x = 0) = 0.10$, $P(x = 1)$, $= 0.20$, $P(x = 2) = 0.30$, $P(x = 3) = 0.30$, and $P(x = 4) = 0.10$. In this situation x is a discrete random variable and $P(x)$ is the probability distribution of x.

Note also that

$$\sum_{x=0}^{4} P(x) = 1$$

For a *continuous random variable*, the relative frequency of each value of x is a polygon shaped like a continuous curve. The frequency at the value x is represented by a nonnegative value $f(x)$. The total area under this curve is 1 and $f(x)$ is called the probability density function of x.

Example 2.9

Consider a car wash with 2 gallons of liquid wax available for the spray booth. Let x represent the amount of wax that is needed for a given day. Suppose that the relative frequency of x increases linearly from $x = 0$ to $x = 2$, as depicted in Figure 2-1.

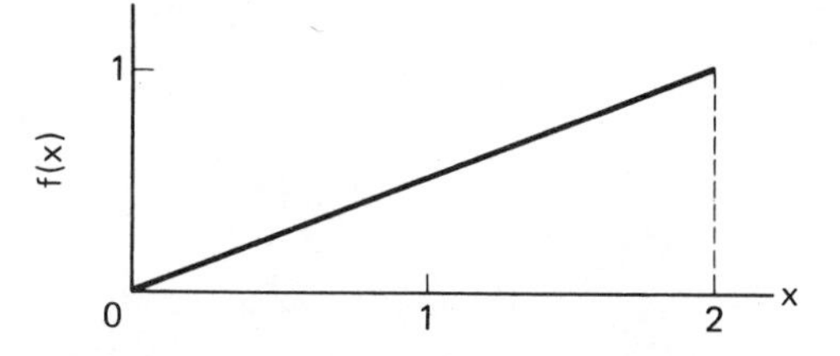

Figure 2-1. The probability density in the car wash example.

The probability density becomes

$$f(x) = \begin{cases} \frac{1}{2}x & \text{if } 0 \leq x \leq 2 \\ 0 & \text{otherwise} \end{cases}$$

In this situation x is a continuous random variable that can lie anywhere between 0 and 2 and $f(x)$ is the probability density function of x. Note (in Figure 2-1) that the total area under the curve is 1.

2-4 THE MEAN

The mean of a random variable is the weighted average of all possible values of x. When the probability distribution (for x discrete) or the probability density (for x continuous) is known, the true value of the mean may be measured. This is called the *expected value* of x and is denoted by $E(x)$, whereby

$$E(x) = \begin{cases} \sum_x xP(x) & \text{for } x \text{ discrete} \\ \int_x xf(x)\,dx & \text{for } x \text{ continuous} \end{cases}$$

It is common to denote the true value of the mean by the Greek letter μ (mu). Hence, $E(x) = \mu$.

Example 2.10

Suppose that x is a discrete random variable with $P(x = 0) = 0.1$, $P(x = 1) = 0.2$, $P(x = 2) = 0.3$, $P(x = 3) = 0.3$, and $P(x = 4) = 0.1$. The expected value of x is

$$\begin{aligned} E(x) &= 0 \cdot P(x = 0) + 1 \cdot P(x = 1) + 2 \cdot P(x = 2) + 3 \cdot P(x = 3) \\ &\quad + 4 \cdot P(x = 4) \\ &= 0(0.1) + 1(0.2) + 2(0.3) + 3(0.3) + 4(0.1) = 2.1 \end{aligned}$$

Example 2.11

If x is continuous with

$$f(x) = \begin{cases} \frac{1}{2}x & \text{for } 0 \le x \le 2 \\ 0 & \text{otherwise} \end{cases}$$

then

$$E(x) = \int_0^2 x\frac{x}{2}\,dx = \int_0^2 \frac{x^2}{2}\,dx = \frac{4}{3}$$

The following relationships concerning $E(x)$ are always true, where a and b are fixed constants.

$$E(a) = a$$

$$E(ax) = aE(x)$$

$$E(ax + b) = aE(x) + b$$

When $E(x)$ is not known, but a set of N observations of x ($x_1, x_2, \ldots, x_N$) are available, the mean is estimated from

$$\bar{x} = \frac{x_1 + x_2 + \ldots + x_N}{N} = \frac{\sum_{i=1}^{N} x_i}{N}$$

Here $\bar{x}$ is the estimate of μ and is called the *sample mean.*

Example 2.12

Suppose that five samples of x give the values (3, 5, 4, 9, 1). The sample mean is

$$\bar{x} = \frac{3 + 5 + 4 + 9 + 1}{5} = \frac{22}{5} = 4.4$$

2-5 THE MEAN ABSOLUTE DEVIATION

The mean absolute deviation from a set of N numbers $x_1, x_2, \ldots, x_N$ is the average difference between each x_i and $\bar{x}$ regardless of the sign. This is

$$\text{mean absolute deviation} = \frac{\sum_{i=1}^{N} |x_i - \bar{x}|}{N}$$

Example 2.13

Consider the five numbers (3, 5, 2, 9, 6), where $\bar{x} = 5$. In this situation

$$\begin{aligned}\text{mean absolute deviation} &= \frac{|3-5| + |5-5| + |2-5| + |9-5| + |6-5|}{5} \\ &= \frac{|-2| + |0| + |-3| + |4| + |1|}{5} \\ &= \frac{2+0+3+4+1}{5} = 2\end{aligned}$$

2-6 THE VARIANCE AND STANDARD DEVIATION

The variance of random variable x gives a measure of the spread. The true variance of x is commonly denoted by either $V(x)$ or σ^2. When μ is the true mean of x, then

$$\sigma^2 = E(x - \mu)^2 = E(x^2) - \mu^2$$

where

$$E(x^2) = \begin{cases} \sum_x x^2 P(x) & \text{for } x \text{ discrete} \\ \int_x x^2 f(x)\, dx & \text{for } x \text{ continuous} \end{cases}$$

Example 2.14

Suppose that x is discrete with $P(x = 0) = 0.1$, $P(x = 1) = 0.2$, $P(x = 2) = 0.3$, $P(x = 3) = 0.3$, and $P(x = 4) = 0.1$. Then $\mu = 2.1$,

$$\begin{aligned}E(x^2) &= 0^2 \cdot P(x=0) + 1^2 \cdot P(x=1) + 2^2 \cdot P(x=2) + 3^2 \cdot P(x=3) \\ &\quad + 4^2 \cdot P(x=4) \\ &= 0(0.1) + 1(0.2) + 4(0.3) + 9(0.3) + 16(0.1) = 5.7\end{aligned}$$

and

$$\sigma^2 = 5.7 - 4.41 = 1.29$$

Example 2.15

If x is continuous with

$$f(x) = \begin{cases} \frac{1}{2}x & \text{for } 0 \leq x \leq 2 \\ 0 & \text{otherwise} \end{cases}$$

then $\mu = \frac{4}{3}$,

$$E(x^2) = \int_0^2 x^2 \frac{x}{2}\, dx = \int_0^2 \frac{x^3}{2}\, dx = 2$$

and

$$\sigma^2 = 2 - \left(\frac{4}{3}\right)^2 = \frac{2}{9} = 0.222$$

The standard deviation is the square root of the variance and is denoted by σ. In Example 2.14, $\sigma = \sqrt{1.29} = 1.14$, and in Example 2.15, $\sigma = \sqrt{0.22} = 0.47$.

The variance can be found when transformations of x using constant a and b are involved. These are

$$V(a) = 0$$

$$V(ax) = a^2 V(x)$$

$$V(ax + b) = a^2 V(x)$$

Example 2.16

Suppose that $y = 2x$, $z = 3x - 4$, and $V(x) = 5$; then

$$V(y) = 4V(x) = 4(5) = 20$$

$$V(z) = 9V(x) = 9(5) = 45$$

An estimate of the variance of x can be found when N observations of x $(x_1, x_2, \ldots, x_N)$ are available. The variance is denoted by S^2 and is obtained with the relation

$$S^2 = \frac{\sum_{i=1}^{N} (x_i - \bar{x})^2}{N - 1}$$

The estimate of the standard deviation is S (the square root of S^2).

Example 2.17

Consider the five numbers (3, 5, 2, 9, 6), where $\bar{x} = 5$. Here

$$S^2 = \frac{(3-5)^2 + (5-5)^2 + (2-5)^2 + (9-5)^2 + (6-5)^2}{4}$$

$$= \frac{(-2)^2 + (0)^2 + (-3)^2 + (4)^2 + (1)^2}{4}$$

$$= 7.5$$

and

$$S = \sqrt{7.5} = 2.74$$

Often, the standard deviation of the mean is of interest and is commonly also referred to as the *standard error of the mean*. When the mean is based on N observations, the standard error of the mean is

$$\sigma_{\bar{x}} = \frac{\sigma}{\sqrt{N}} \qquad \text{(when } \sigma \text{ is known)}$$

and

$$S_{\bar{x}} = \frac{S}{\sqrt{N}} \qquad \text{(when } \sigma \text{ is not known and is estimated by } S\text{)}$$

Example 2.18

Consider the five numbers (3, 5, 2, 9, 6), where $S = 2.74$. The standard error of the mean is

$$S_{\bar{x}} = \frac{2.74}{\sqrt{5}} = 1.23$$

2-7 COVARIANCE

The covariance between two paired variables (x, y) represents a measure of the association between x and y. This is commonly denoted by either $C(xy)$ or σ_{xy} when developed on a theoretical basis. The covariance is

$$C(xy) = E(x - \mu_x)(y - \mu_y) = E(xy) - \mu_x\mu_y$$

where μ_x is the mean of x and μ_y is the mean of y.

Notice that when x and y are independent,

$$E(xy) = E(x)E(y) = \mu_x \mu_y$$

and

$$C(xy) = 0$$

An estimate of the covariance can be found when N paired observations of x and y are available; i.e., $(x_1, y_1), (x_2, y_2), \ldots, (x_N, y_N)$. The estimate is denoted by S_{xy} and is derived by

$$S_{xy} = \frac{\sum_{i=1}^{N} (x_i - \bar{x})(y_i - \bar{y})}{N - 1}$$

$$= \frac{\sum_{i=1}^{N} x_i y_i - N\bar{x}\bar{y}}{N - 1}$$

Example 2.19

Consider the following paired observations of x and y.

x	3	10	8	7	2
y	2	6	5	4	3

These values yield

$$\bar{x} = \frac{3 + 10 + 8 + 7 + 2}{5} = 6$$

$$\bar{y} = \frac{2 + 6 + 5 + 4 + 3}{5} = 4$$

$$\sum_{i=1}^{5} x_i y_i = 3 \times 2 + 10 \times 6 + 8 \times 5 + 7 \times 4 + 2 \times 3 = 140$$

and

$$S_{xy} = \frac{140 - 5 \times 6 \times 4}{4} = 5$$

The covariance of certain transformations of x and y are easily found. For example, when $x' = ax$ and $y' = by$, where a and b are constants, then

$$C(x', y') = abC(x, y)$$

Example 2.20

Suppose that $C(x, y) = 8$, $x' = 10x$, and $y' = 2y$; then

$$C(x'y') = 10 \times 2C(xy) = 20(8) = 160$$

2-8 CORRELATION

The correlation is a standardized measure of the association between two paired variables (x, y). The true correlation between x and y is denoted by the Greek letter ρ (rho). This is found by the relation

$$\rho = \frac{\sigma_{xy}}{\sigma_x \sigma_y}$$

where σ_x and σ_y are the standard deviations of x and y, respectively. ρ always lies within the interval $(-1, 1)$. When x and y are independent, then $\rho = 0$. If x and y both tend to increase simultaneously, then $\rho > 0$, and if x increases when y tends to decrease, then $\rho < 0$. A perfect relation between x and y occurs when $\rho = 1$ or $\rho = -1$.

The correlation between x and y can be estimated from N paired observations $(x_1, y_1), (x_2, y_2), \ldots, (x_N, y_N)$. This is denoted by r and is measured from

$$r = \frac{S_{xy}}{S_x S_y}$$

where S_x is the standard deviation of x and S_y is the corresponding value for y.

Example 2.21

Consider the five paired observations in Example 2.19, where $S_{xy} = 5$, $\bar{x} = 6$, and $\bar{y} = 4$. Here

$$S_x^2 = \frac{(3-6)^2 + (10-6)^2 + (8-6)^2 + (7-6)^2 + (2-6)^2}{4}$$

$$= \frac{46}{4} = 11.5$$

$$S_x = 3.39$$

$$S_y^2 = \frac{(2-4)^2 + (6-4)^2 + (5-4)^2 + (4-4)^2 + (3-4)^2}{4}$$

$$= \frac{10}{4} = 2.5$$

$$S_y = 1.58$$

and

$$r = \frac{5.00}{3.39 \times 1.58} = 0.93$$

Therefore, the set of data is highly correlated.

2-9 LINEAR COMBINATIONS

Consider K variables $(x_1, x_2, \ldots, x_K)$ with means $\mu_1, \mu_2, \ldots, \mu_K$, respectively. Now suppose that y is related to the K variables by the following linear combination:

$$y = a_1x_1 + a_2x_2 + \ldots + a_Kx_K$$

where $(a_1, a_2, \ldots, a_K)$ are constants. The expected value of y is

$$E(y) = a_1\mu_1 + a_2\mu_2 + \ldots + a_K\mu_K$$

Let σ_i^2 represent the variance of x_i and σ_{ij} the covariance between x_i and x_j. Now, the variance of y becomes

$$V(y) = \sum_{i=1}^{K} a_i^2 \sigma_i^2 + 2\sum_{i=1}^{K}\sum_{j>i}^{K} a_i a_j \sigma_{ij}$$

Should all the x_i be independent, then $\sigma_{ij} = 0$ for all i and j and

$$V(y) = \sum_{i=1}^{K} a_i^2 \sigma_i^2$$

Example 2.22

Consider x and y, where $\mu_x = 10$, $\mu_y = 5$, and $w = 2x + 3y$. Then

$$\begin{aligned}\mu_w &= 2\mu_x + 3\mu_y \\ &= 2 \times 10 + 3 \times 5 = 35\end{aligned}$$

Example 2.23

In Example 2.22, if $\sigma_x^2 = 20$, $\sigma_y^2 = 10$, and $\sigma_{xy} = -5$, then

$$\begin{aligned}V(w) &= 4(20) + 9(10) + 2 \times 2 \times 3(-5) \\ &= 110\end{aligned}$$

Example 2.24

In Example 2.22, if $\sigma_x^2 = 20$, $\sigma_y^2 = 10$, and x and y are independent ($\sigma_{xy} = 0$), then

$$V(w) = 4(20) + 9(10) = 170$$

2-10 THE NORMAL DISTRIBUTION

One of the most useful probability distributions in forecasting is the *normal probability distribution*. The normal distribution is often called the *normal curve* and represents in reality a probability density since it pertains to a continuous variable x. Letting μ represent the mean and σ^2 the variance of x then

$$f(x) = \frac{1}{\sqrt{2\pi}\,\sigma} \exp\left[-\frac{1}{2}\left(\frac{x-\mu}{\sigma}\right)^2\right]$$

A standardized normal distribution is obtained when $z = (x - \mu)/\sigma$, which gives $\mu_z = 0$ and $\sigma_z = 1$. The probability density is

$$f(z) = \frac{1}{\sqrt{2\pi}} e^{-z^2/2}$$

A graph of the standardized normal distribution is given in Figure 2-2.

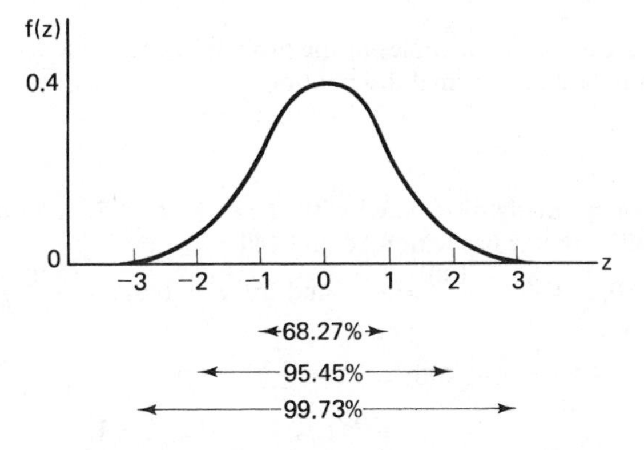

Figure 2-2. The standardized normal distribution.

The total area under the curve is 1 and the percent of the area contained for

z within ± 1 is 68.27%

z within ± 2 is 95.45%

z within ± 3 is 99.73%

Hence, the probability is 0.6827 that z will lie within ± 1, it is 0.9545 that z will lie within ± 2, and so on.

Appendix B contains a table showing the total area above any value of z for the range of -3 to $+3$. Figure 2-3 gives some results that can be found by use of the appendix. In the figure the shaded areas represent the segments of the curve that are of interest.

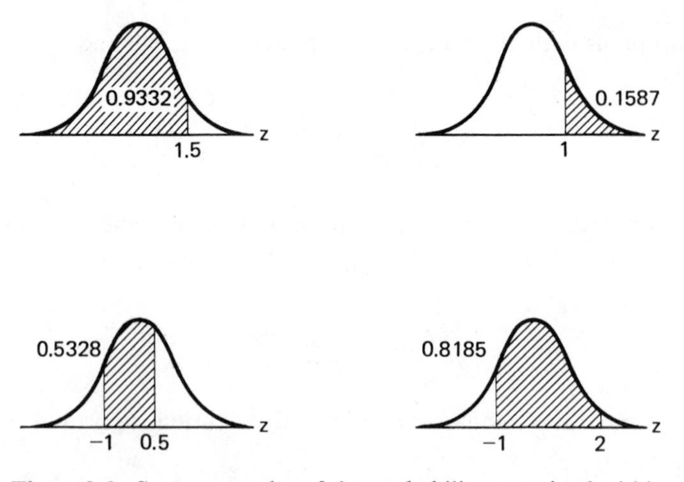

Figure 2-3. Some examples of the probability contained within the standardized normal distribution.

Example 2.25

Suppose that x is normally distributed with $\mu = 100, \sigma = 20$, and it is desired to find the probability that x lies within 80 and 140.

For $x = 80$, $z = \dfrac{80 - 100}{20} = -1$, and for $x = 140$, $z = \dfrac{140 - 100}{20} = 2$. So

$$\begin{aligned} P(80 \leq x \leq 140) &= P(-1 \leq z \leq 2) \\ &= P(z \leq 2) - P(z \leq -1) \\ &= 0.9772 - 0.1587 = 0.8185 \end{aligned}$$

2-11 CONFIDENCE INTERVALS

Forecasters are generally interested in two types of estimates, point estimates and interval estimates. A *point estimate* gives a single value for the item being estimated. For example, $\bar{x}$ is a point estimate of μ. An *interval estimate* specifies a range of values wherein the parameter being estimated is said to lie with a stated degree of confidence.

A *confidence interval* is such an interval estimate. The probability associated with the interval is called the *confidence coefficient*, and the limits of the interval are called *confidence limits*. Hence, if the statement is made that the true mean (μ) lies between 10 and 30 with 95% confidence, the confidence interval is 10 to 30, the confidence limits are 10 and 30, and the confidence coefficient is 95%.

Example 2.26

Suppose that 36 observations of x yield $\bar{x} = 100$ and $S = 24$. Also, assume that x is normally distributed. The 95% confidence interval of μ is

$$\bar{x} \pm 1.96 S_{\bar{x}}$$

where

$$S_{\bar{x}} = \frac{S}{\sqrt{N}} = \frac{24}{6} = 4$$

Hence

$$\bar{x} \pm 1.96 S_{\bar{x}} = 100 \pm 7.84 \quad \text{or} \quad (92.16 \text{ to } 107.84)$$

2-12 CENTRAL LIMIT THEOREM

One of the most noteworthy theorems in statistics is the central limit theorem. The *central limit theorem* states: If n samples are taken from a population with mean μ and standard deviation σ, the distribution of the sample mean ($\bar{x}$) will approach a normal distribution with mean μ and standard deviation $\sigma_{\bar{x}} = \sigma/\sqrt{n}$ as n increases.

The theorem holds (almost always) for discrete and continuous random variables and for any shape of the probability distribution.

Example 2.27

Consider a service part that is stocked in a parts depot. An order placed on the part may be for one or more pieces. The size of each order on the part follows an arbitrary probability distribution and has an unknown mean μ and standard deviation σ. A sample of $n = 36$ order sizes yields a mean $\bar{x} = 20$ and a standard deviation of $S = 6$. Now using the central limit theorem, the distribution of $\bar{x}$ is approximately normal with mean μ and standard deviation $\sigma_{\bar{x}} = \sigma/\sqrt{n}$. Since μ and σ are unknown, they are estimated by $\bar{x} = 20$ and $S_{\bar{x}} = S/\sqrt{n} = 6/\sqrt{36} = 1$, respectively. So the 95% confidence interval on the true value of μ is approximately

$$\bar{x} \pm 2S_{\bar{x}} = 20 \pm 2 \quad \text{or} \quad (18 \text{ to } 22)$$

2-13 LEAST-SQUARES METHOD

Many of the forecasting models presented in this book are developed using the least-squares method. The past demand entries $(x_1, x_2, \ldots, x_T)$ are associated with specific time periods $(1, 2, \ldots, T)$, respectively, and may be plotted against time as shown in Figure 2-4. The forecaster may seek an equation of a particular type which relates x_t to t. For each time period t, the equation yields a fitted value, denoted as $\hat{x}_t$.

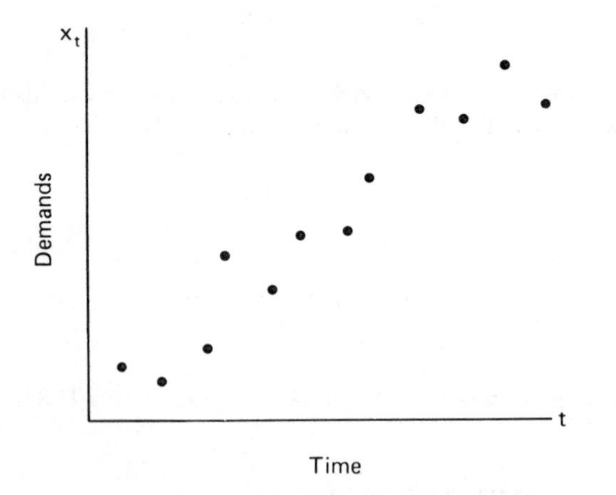

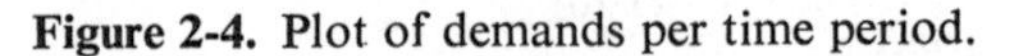

Figure 2-4. Plot of demands per time period.

The method of least squares is used to find the equation (or fit) through the demands which gives the minimum sum of squared differences between each entry x_t and its fitted value $\hat{x}_t$. That is, the sum

$$S(e) = \sum_{t=1}^{T} (x_t - \hat{x}_t)^2 = \sum_{t=1}^{T} e_t^2$$

is minimized, where

$$e_t = x_t - \hat{x}_t$$

To fit a straight line through the demands, the fitted line is of the form

$$\hat{x}_t = a + bt$$

where the coefficients a and b yield the minimum value of $S(e)$. The least-squares solution for a and b becomes

$$b = \frac{T \sum tx_t - \sum x_t \sum t}{T \sum t^2 - (\sum t)^2}$$

$$a = \frac{\sum x_t - b \sum t}{T}$$

where all summations range from $t = 1$ to $t = T$.

Example 2.28

Consider the following ten demand entries:

t	1	2	3	4	5	6	7	8	9	10
x_t	2	3	2	5	6	4	8	10	7	11

Now the least-squares straight-line fit that relates x_t to t is obtained using the table below:

t	x_t	t^2	tx_t
1	2	1	2
2	3	4	6
3	2	9	6
4	5	16	20
5	6	25	30
6	4	36	24
7	8	49	56
8	10	64	80
9	7	81	63
10	11	100	110
$\sum t = 55$	$\sum x_t = 58$	$\sum t^2 = 385$	$\sum tx_t = 397$

Since $T = 10$, then

$$b = \frac{10(397) - 58(55)}{10(385) - (55)^2} = \frac{780}{825} = 0.95$$

$$a = \frac{58 - 0.95(55)}{10} = 0.575$$

The results yield

$$\hat{x}_t = 0.575 + 0.95t$$

and the plot of the fit is given in Figure 2-5.

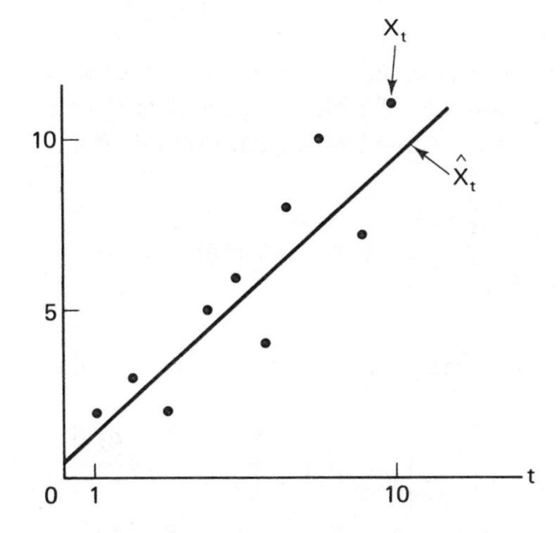

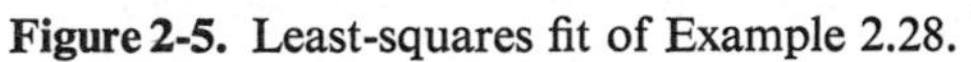
Figure 2-5. Least-squares fit of Example 2.28.

2-14 MATRICES

A matrix is a rectangular array of numbers enclosed by a pair of brackets and is subjected to certain rules of operation. Examples of matrices are

$$\begin{bmatrix} 1 & 3 & 4 \\ 2 & 1 & 0 \end{bmatrix} \quad \text{and} \quad \begin{bmatrix} 3 & 10 & 5 & 1 \\ 6 & -8 & 4 & -3 \\ 5 & 8 & 4 & 7 \end{bmatrix}$$

The *order* of the matrix is denoted by (r, c), where r represents the number of rows and c the number of columns. The matrix above on the left is of order (2, 3), and the one on the right is of order (3, 4).

A *square matrix* is one where $r = c$, such as

$$\begin{bmatrix} 1 & -1 \\ 2 & 4 \end{bmatrix} \quad \text{or} \quad \begin{bmatrix} 4 & 3 & 7 \\ 6 & 1 & 4 \\ 5 & 10 & 2 \end{bmatrix}$$

A *vector* is a matrix where either r or c is 1. When $r = 1$, the vector is called a *row vector*, e.g.,

$$[1 \quad 4 \quad 2]$$

and when $c = 1$, it is a *column vector*, e.g.,

$$\begin{bmatrix} 7 \\ 4 \end{bmatrix}$$

When a matrix A of order (r, c) has all its rows and columns interchanged, the new matrix is of order (c, r) and is the transpose of A. The *transpose* of matrix A is denoted as A'. An example is:

$$\begin{bmatrix} 1 & 4 & 5 \\ 2 & 1 & 6 \end{bmatrix} \quad \text{is the transpose of} \quad \begin{bmatrix} 1 & 2 \\ 4 & 1 \\ 5 & 6 \end{bmatrix}$$

Matrices of the same order may be *summed* as follows:

$$\begin{bmatrix} a_1 & a_2 \\ a_3 & a_4 \\ a_5 & a_6 \end{bmatrix} + \begin{bmatrix} b_1 & b_2 \\ b_3 & b_4 \\ b_5 & b_6 \end{bmatrix} = \begin{bmatrix} a_1 + b_1 & a_2 + b_2 \\ a_3 + b_3 & a_4 + b_4 \\ a_5 + b_5 & a_6 + b_6 \end{bmatrix}$$

Example 2.29

$$\begin{bmatrix} 1 \\ 2 \end{bmatrix} + \begin{bmatrix} 4 \\ 2 \end{bmatrix} - \begin{bmatrix} -6 \\ 1 \end{bmatrix} = \begin{bmatrix} 11 \\ 3 \end{bmatrix}$$

The matrices A and B can be *multiplied* ($C = AB$) when the number of columns of A is the same as the number of rows in B. An example is shown where A is of order (2, 3) and B is of order (3, 1). The results give

$$\begin{bmatrix} a_1 & a_2 & a_3 \\ a_4 & a_5 & a_6 \end{bmatrix} \begin{bmatrix} b_1 \\ b_2 \\ b_3 \end{bmatrix} = \begin{bmatrix} c_1 \\ c_2 \end{bmatrix}$$

where

$$c_1 = a_1b_1 + a_2b_2 + a_3b_3$$

$$c_2 = a_4b_1 + a_5b_2 + a_6b_3$$

In general, the element in the ith row and jth column of C is obtained by multiplying the ith row of A by the jth column of B. Note that $C = AB$ is not necessarily equal to $D = BA$ (even if D exists).

Example 2.30

$$\begin{bmatrix} 2 & 5 \\ -1 & 0 \end{bmatrix} \begin{bmatrix} 1 & 3 & 6 \\ 4 & 1 & 1 \end{bmatrix} = \begin{bmatrix} 2 \cdot 1 + 5 \cdot 4 & 2 \cdot 3 + 5 \cdot 1 & 2 \cdot 6 + 5 \cdot 1 \\ -1 \cdot 1 + 0 \cdot 4 & -1 \cdot 3 + 0 \cdot 1 & -1 \cdot 6 + 0 \cdot 1 \end{bmatrix}$$

$$= \begin{bmatrix} 22 & 11 & 17 \\ -1 & -3 & -6 \end{bmatrix}$$

An identity matrix (I) is always a square matrix with all elements in the main diagonal equal to 1 and all other elements equal to zero, e.g.,

$$\begin{bmatrix} 1 & 0 \\ 0 & 1 \end{bmatrix} \quad \text{or} \quad \begin{bmatrix} 1 & 0 & 0 & 0 \\ 0 & 1 & 0 & 0 \\ 0 & 0 & 1 & 0 \\ 0 & 0 & 0 & 1 \end{bmatrix}$$

If A and B are square matrices of the same order and $AB = I$, then B is the *inverse* of A and A is the inverse of B. It is common to denote the inverse of the matrix A as A^{-1}. Hence,

$$I = AA^{-1} = A^{-1}A$$

Example 2.31

Since

$$\begin{bmatrix} 2 & 3 \\ 1 & 4 \end{bmatrix}\begin{bmatrix} \frac{4}{5} & -\frac{3}{5} \\ -\frac{1}{5} & \frac{2}{5} \end{bmatrix} = \begin{bmatrix} 1 & 0 \\ 0 & 1 \end{bmatrix}$$

then the matrix

$$\begin{bmatrix} \frac{4}{5} & -\frac{3}{5} \\ -\frac{1}{5} & \frac{2}{5} \end{bmatrix}$$

is the inverse to

$$\begin{bmatrix} 2 & 3 \\ 1 & 4 \end{bmatrix}$$

Example 2.32

The inverse of

$$\begin{bmatrix} 1 & 2 & 3 \\ 1 & 3 & 3 \\ 1 & 2 & 4 \end{bmatrix}$$

is

$$\begin{bmatrix} 6 & -2 & -3 \\ -1 & 1 & 0 \\ -1 & 0 & 1 \end{bmatrix}$$

since

$$\begin{bmatrix} 1 & 2 & 3 \\ 1 & 3 & 3 \\ 1 & 2 & 4 \end{bmatrix}\begin{bmatrix} 6 & -2 & -3 \\ -1 & 1 & 0 \\ -1 & 0 & 1 \end{bmatrix} = \begin{bmatrix} 1 & 0 & 0 \\ 0 & 1 & 0 \\ 0 & 0 & 1 \end{bmatrix}$$

PROBLEMS

2-1. If $x_1 = 5$, $x_2 = -3$, $x_3 = 0$, $x_4 = 2$, and $x_5 = 1$, find the following summations:

a. $\sum_{i=1}^{5} x_i$

b. $\sum_{i=1}^{5} x_i^2$

c. $\sum_{i=1}^{5} (3x_i + 2)^2$

d. $\sum_{i=1}^{5} (3x_i + 2)$

e. $\sum_{i=1}^{2} 4x_i - \sum_{i=3}^{5} 2x_i$

f. $\sum_{i=1}^{5} 7$

2-2. Using $(x_1, y_1) = (3, 5)$, $(x_2, y_2) = (4, 9)$, $(x_3, y_3) = (0, 1)$, and $(x_4, y_4) = (6, 3)$, find:

a. $\sum_{i=1}^{4} (2x_i + 3y_i)$

b. $\sum_{i=1}^{4} (2x_i y_i)$

c. $\sum_{i=1}^{4} (x_i - y_i)^2$

d. $\sum_{i=1}^{4} (2x_i + 3y_i)^2$

2-3. In the roll of a die, let A be the event that a 1 or a 2 will occur, and let B represent the event of a 2, 3, or 4 occurring. Fing $P(A)$, $P(B)$, $P(AB)$, $P(A \text{ or } B)$, $P(A|B)$, $P(B|A)$, $P(\hat{A})$, $P(\hat{B})$, and $P(\hat{A} \text{ or } \hat{B})$.

2-4. In Problem 2-3, are the events A and B mutually exclusive? Are they exhaustive? Why?

2-5. In the toss of a die, consider the following events:

$$A = 1 \text{ or } 2$$
$$B = 2, 3, \text{ or } 4$$
$$C = 4, 5, \text{ or } 6$$
$$D = 3$$

a. Are any combinations of the events above mutually exclusive?
b. Are any exhaustive?

2-6. Suppose that A, B, and C are independent events with $P(A) = 0.2$, $P(B) = 0.3$, and $P(C) = 0.6$. Find $P(AB)$, $P(AC)$, $P(BC)$, $P(ABC)$, and $P(A \text{ or } B \text{ or } C)$.

2-7. Consider a discrete random variable x with $P(x = 0) = 0.1$, $P(x = 1) = 0.3$, $P(x = 2) = 0.3$, $P(x = 3) = 0.2$, and $P(x = 4) = 0.1$. Find:
a. $P(x \leq 2)$, $P(x > 1)$, $P(1 \leq x \leq 3)$, and $P(x = 1 | x \leq 2)$.
b. $E(x)$, $E(x^2)$, and $V(x)$.
c. If $y = 2x + 1$, find $E(y)$, $E(y^2)$, and $V(y)$.

2-8. Assume that the demands for an item over the past nine time periods are (8, 2, 6, 7, 1, 9, 5, 0, 7). Find:
a. The sample mean ($\bar{x}$).
b. The mean absolute deviation.
c. The sample variance (S^2).
d. The standard deviation (S).
e. The standard error of the mean ($S_{\bar{x}}$).

2-9. Suppose that the demands for items X and Y over the past 10 time periods are as follows:

Time period (t)	1	2	3	4	5	6	7	8	9	10
X demands (x_t)	6	5	4	1	9	3	2	0	6	4
Y demands (y_t)	7	8	7	0	6	7	5	3	9	8

a. Find the means of x and y ($\bar{x}$, $\bar{y}$).
b. Find the standard deviation of x and y (S_x, S_y).
c. Find the covariance between x and y (S_{xy}).
d. Find the correlation between x and y (r).

2-10. Consider two independent variables x and y with $\mu_x = 5$, $\mu_y = 10$, $\sigma_x = 1$, and $\sigma_y = 2$.
a. If $w = x + y$, find $E(w)$ and $V(w)$.
b. If $w = x - y$, find $E(w)$ and $V(w)$.
c. If $w = -4x + 2y$, find $E(w)$ and $V(w)$.

2-11. Suppose that x and y are two variables that are not independent with $\mu_x = 5$, $\mu_y = 10$, $\sigma_x = 3$, $\sigma_y = 2$, and $\sigma_{xy} = 4$.
a. If $w = x + y$, find $E(w)$ and $V(w)$.
b. If $w = x - y$, find $E(w)$ and $V(w)$.
c. If $w = -4x + 2y$, find $E(w)$ and $V(w)$.

2-12. Suppose that z is a standard normal variate ($\mu_z = 0$, $\sigma_z = 1$). Use Appendix B to find the following probabilities:
a. $P(z \leq 1)$ b. $P(z > -2)$
c. $P(1 \leq z \leq 2)$ d. $P(-2 \leq z \leq -1)$

2-13. Suppose that x is a normally distributed variable with $\mu_x = 10$ and $\sigma_x = 2$. Find:
a. $P(x \leq 12)$ b. $P(x > 6)$
c. $P(8 \leq x \leq 10)$ d. $P(6 \leq x \leq 12)$

2-14. A sample of 36 observations yields $\bar{x} = 20$ and $S = 5$. Assuming that x is normally distributed, find
a. The 95% confidence interval of μ.
b. The 90% confidence interval of μ.

2-15. Suppose that a sample of size 64 yields $\bar{x} = 100$ and $S = 16$. Using the central limit theorem, find:
a. The 95% confidence interval of μ.
b. The probability that $\mu \leq 102$.

2-16. Consider the eight most recent demands for item x:

Time period (t)	1	2	3	4	5	6	7	8
x demands (x_t)	2	5	4	7	10	9	12	15

a. Find the least-squares fit of the form $\hat{x}_t = a + bt$.
b. Using the results above, calculate $\hat{x}_t$ for $t = 1, \ldots, 8$.
c. Now find e_t ($t = 1, \ldots, 8$) and $S(e)$.

2-17. Consider the matrices

$$A = \begin{bmatrix} 1 & 2 \\ 3 & 4 \end{bmatrix} \quad \text{and} \quad B = \begin{bmatrix} 4 & 0 \\ 1 & 3 \end{bmatrix}$$

a. Find the transpose of A and of B.
b. If $C = A + B$, find C.
c. If $C = A - B$, find C.
d. If $C = AB$, find C.

2-18. For the matrices

$$A = \begin{bmatrix} 1 \\ 2 \\ 3 \end{bmatrix} \quad \text{and} \quad B = \begin{bmatrix} 2 \\ -1 \\ 1 \end{bmatrix}$$

find:

a. $C = A + B$.
b. The transpose of A (say A').
c. If $D = A'B$, find D.
d. If $E = BA'$, find E.

2-19. Show that $A = \begin{bmatrix} 4 & 6 \\ 2 & 8 \end{bmatrix}$ is the inverse to

$$B = \begin{bmatrix} 0.4 & -0.3 \\ -0.1 & 0.2 \end{bmatrix}$$

REFERENCES FOR FURTHER STUDY

AYRES, F., *Matrices*. New York: McGraw-Hill Book Company, 1960.

LEVIN, R. I., *Statistics for Management*. Englewood Cliffs, N.J.: Prentice-Hall, Inc., 1978.

WONNACOTT, T. H., AND R. J. WONNACOTT, *Introductory Statistics*. New York: John Wiley & Sons, Inc., 1977.

3

DEMAND PATTERNS AND FILTERING

In time-series analysis, the principal data that are used in generating forecasts is the history of the past demand entries $(x_1, \ldots, x_T)$. The forecasting process begins by developing a statistical type of fit through the demands of the past. The fit that is sought may vary depending on the flow pattern of the history of demands, herein called the *demand pattern*. The fit is then projected forward to estimate the demands for the time periods of the future. These estimates are essentially the forecasts of future demands.

The objective of this chapter is twofold. The first is to identify the more common type of demand patterns within which the history of demands may fall. This is of interest to the forecaster, since the forecast model selected to generate the forecasts should be compatible with the demand pattern of the data. Selecting the wrong forecast model may yield misleading results.

The second objective is to describe various methods of filtering the demand history. This entails the need to seek out and adjust any entries that are far removed from the normal flow of all other entries. The purpose is to attempt to avoid any entries that may have been erroneously transmitted or may be due to some unnatural phenomenon that is nonrecurring. Should these entries not be adjusted, the forecasts could be distorted for many time periods to come.

3-1 DEMAND PATTERNS

The purpose of this section is to describe some of the key demand patterns that are encountered in time-series analysis. The demand patterns are classified according to the flow of the averages of the demands. In all situations, the average at time period t is denoted as μ_t and is often called the *level of the demand* at time period t.

Horizontal Demand Patterns

When the level is the same for all time periods, then

$$\mu_t = \mu$$

for all t and the demand pattern is *horizontal*. An example is shown in Figure 3-1.

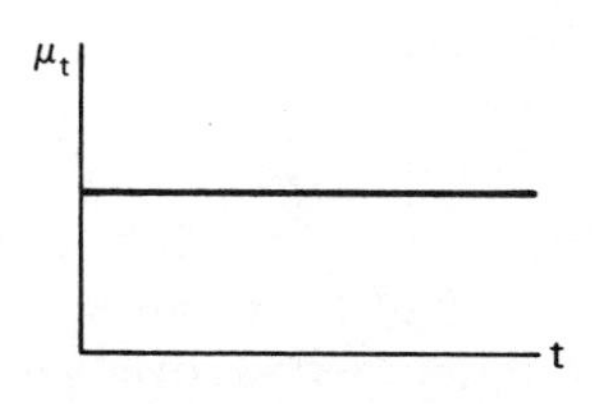

Figure 3-1. Example of the horizontal demand pattern.

Trend Demand Patterns

A *trend* demand pattern is encountered when the level is linearly changing over time. The level at time t is defined by

$$\mu_t = a + bt$$

where a is the level at $t = 0$ and b is the slope. The trend is called *positive* when the level is rising, and *negative* when the level is falling. Examples are shown in Figure 3-2.

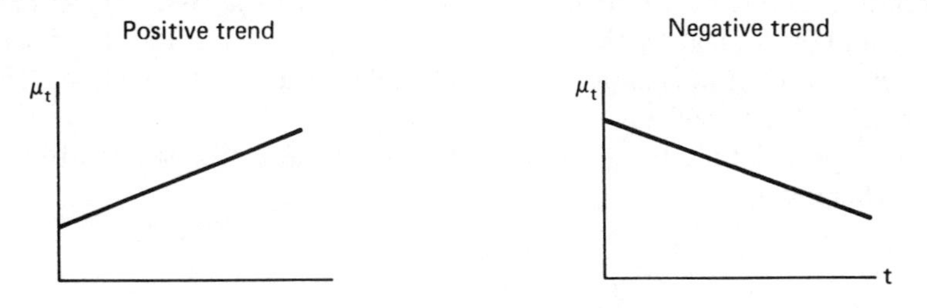

Figure 3-2. Examples of trend demand pattern.

Quadratic Demand Patterns

In the *quadratic* demand patterns the level is rising or falling in a non-linear manner. The level at time t is in the form

$$\mu_t = a + bt + ct^2$$

Generally, such a pattern does not hold for an item over the long run but may be representative over the local time horizon. Various examples are shown in Figure 3-3.

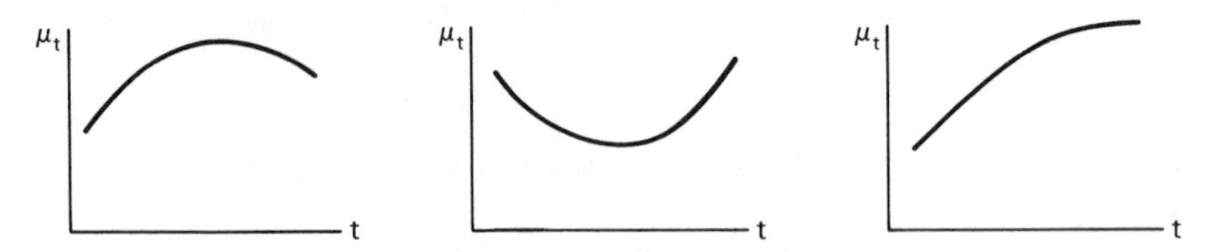

Figure 3-3. Examples of quadratic demand patterns.

Seasonal Demand Patterns

SEASONAL CYCLE. In order to describe a seasonal demand pattern, it is first helpful to define the *length of the seasonal cycle*. This is denoted as M and represents the number of time periods over which the season completes a full cycle. Should the cycle be repeated once each year, $M = 12$ when the time periods represent monthly occurrences. For weekly time periods, $M = 52$; and for quarterly time periods, $M = 4$.

The level at time period t is defined by

$$\mu_t = \mu\rho_t$$

where μ is the long-run average level and ρ_t ($\rho_t \geq 0$) is the *seasonal ratio* at time t. Because there is no change in the average level from cycle to cycle, this pattern is often called the *horizontal seasonal demand pattern*.

The average of the seasonal ratios over the course of a cycle is 1, and in this manner the average level over the cycle is μ. The seasonal ratios give the relative increase or decrease of μ_t with respect to μ. When $\rho_t = 1$, then $\mu_t = \mu$ and the level at t is the same as the average level over the entire season. When $\rho_t > 1$, then μ_t is larger than μ by a relative amount. In a like fashion, μ_t is smaller than μ when $\rho_t < 1$.

It is noted that the seasonal ratios do repeat themselves after each cycle, whereby

$$\rho_{t+M} = \rho_t$$

and

$$\mu_{t+M} = \mu_t$$

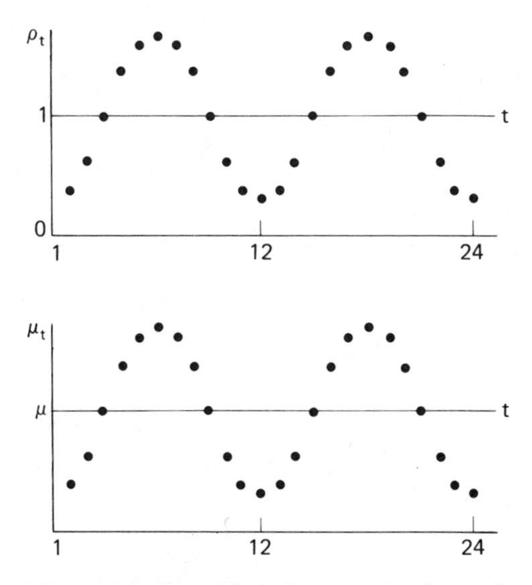

Figure 3-4. Examples of seasonal ratios and the corresponding seasonal demand pattern.

An example of the seasonal ratios and the corresponding seasonal demand pattern is given in Figure 3-4 for a situation where $M = 12$.

Trend Seasonal Demand Patterns

The *trend seasonal demand pattern* is encountered when both trend and seasonal influences are active. Two patterns of this type are possible, the multiplicative and the additive.

In the *multiplicative* case, the level at time t is defined by

$$\mu_t = (a + bt)\rho_t$$

Here, $(a + bt)$ gives the trend influence and ρ_t is the seasonal influence. As before, the average value of ρ_t over a cycle is 1 and $\rho_{t+M} = \rho_t$. An example with $M = 12$ is cited in Figure 3-5. Here the seasonal and trend components are shown individually and in combination to yield μ_t. Note how μ_t fluctuates wider as the trend influence reaches the higher values.

In the *additive* case, the level becomes

$$\mu_t = (a + bt) + \delta_t$$

Again, $(a + bt)$ gives the trend influence. However, now δ_t represents the seasonal influence and is called the *seasonal increment*. The average value of δ_t over a seasonal length is zero. When $\delta_t > 0$, the level at time t is higher than the trend influence by the amount δ_t. With $\delta_t < 0$, the level at time t is

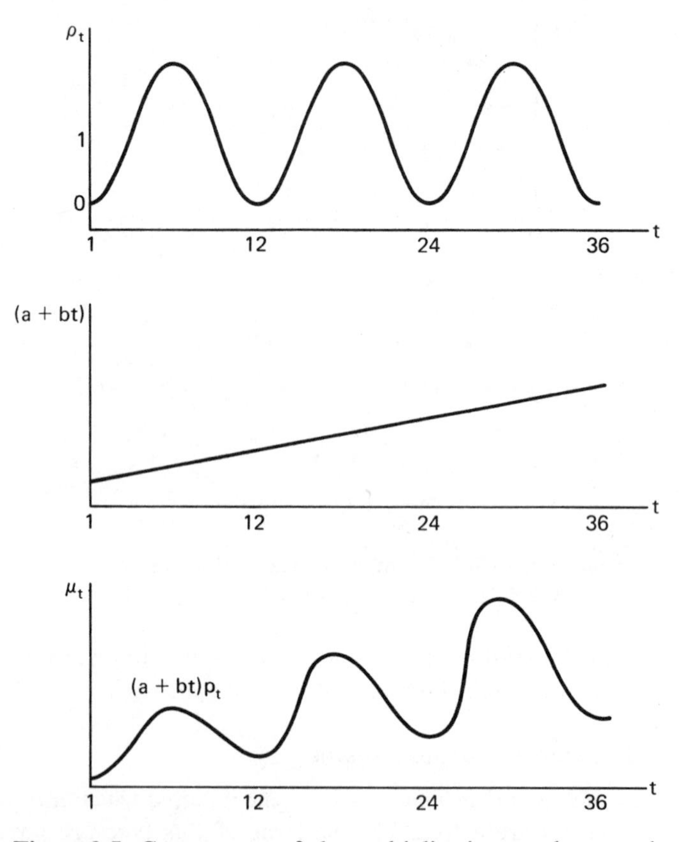

Figure 3-5. Components of the multiplicative trend seasonal demand pattern.

below the underlying trend. Figure 3-6 shows how the components vary individually and in combination to give μ_t. In this situation $M = 12$ is assumed. It is noted that the seasonal fluctuations in μ_t are the same from season to season. This is because $\delta_t = \delta_{t+M}$.

Trigonometric Demand Patterns

A class of demand patterns that fluctuates up and down in a periodic manner are the *trigonometric* demand patterns. The level is defined using a series of sine and cosine terms. Because the level flows in a wave manner, the demand pattern is often used in place of the seasonal and trend seasonal patterns cited earlier. Again the cycle length is defined as M and represents the number of time periods included in one seasonal cycle. The demand pattern takes on a different shape, depending on the number of terms

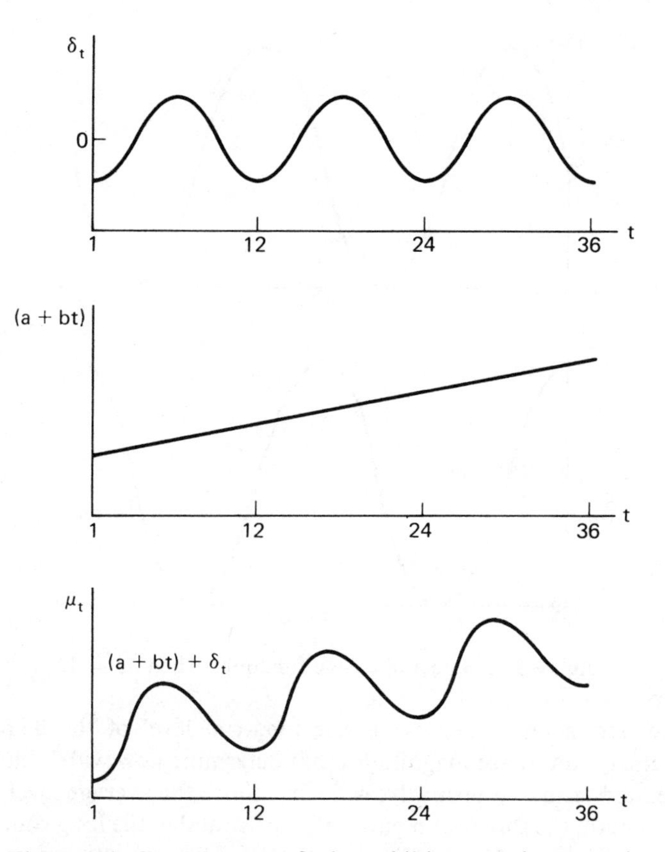

Figure 3-6. Components of the additive trend seasonal demand pattern.

included in defining the level μ_t. Three cases are given here: the three-, four-, and six-term patterns.

In the *three-term* case, the level is defined by

$$\mu_t = a_1 + a_2 \sin \omega t + a_3 \cos \omega t$$

where

$$\omega = \frac{2\pi}{M}$$

Using ω in this manner allows the sine ($\sin \omega t$) and cosine ($\cos \omega t$) components to complete one full cycle everytime t spans M time periods.

The average value of $\sin \omega t$ over a cycle length (M) is zero and the range from its minimum to maximum values is -1 to $+1$. The same is true for $\cos \omega t$. Examples are given in Figure 3-7, where $M = 12$.

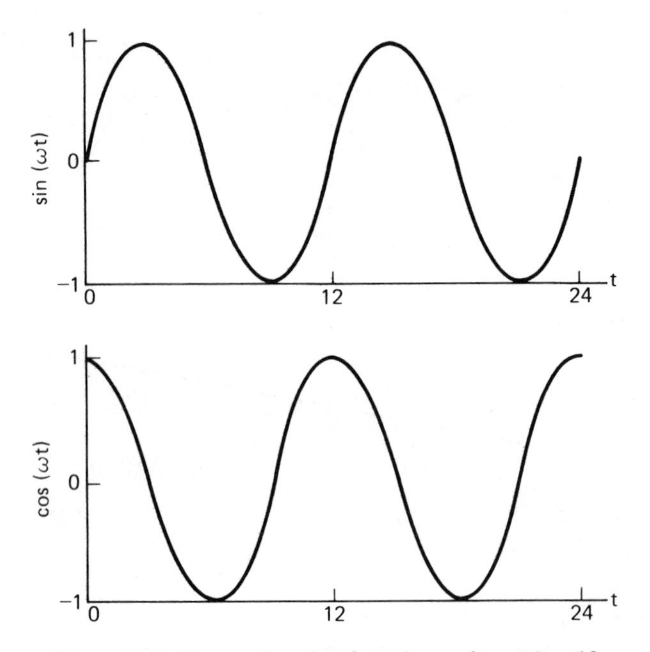

Figure 3-7. Sine and cosine functions when $M = 12$.

The parameter a_1 represents the long-run level of the demand. The parameters a_2 and a_3 are magnitudes that determine how widely the sine and cosine related terms, respectively, will vary. Since the average level is a_1 over the cycle length, the three-term pattern is horizontal in the long run.

An example is given in Figure 3-8, where the various components are individually depicted. The example assumes that $M = 12$.

The *four-term* pattern is essentially the same as the three-term pattern except that a trend influence is included. The level is defined by

$$\mu_t = a_1 + a_2 t + a_3 \sin \omega t + a_4 \cos \omega t$$

Figure 3-9 illustrates how the components combine to form the demand pattern. Again, $M = 12$ is assumed.

The level in the *six-term* pattern includes two more terms than that of the four-term pattern, i.e.,

$$\mu_t = a_1 + a_2 t + a_3 \sin \omega t + a_4 \cos \omega t + a_5 \sin 2\omega t + a_6 \cos 2\omega t$$

The component sin $2\omega t$ completes two full cycles in one seasonal cycle length (M), during which the average value is zero and the minimum and maximum values are -1 and $+1$. These characteristics pertain also to cos $2\omega t$, as seen in Figure 3-10. An example of the six-term pattern is given in Figure 3-11.

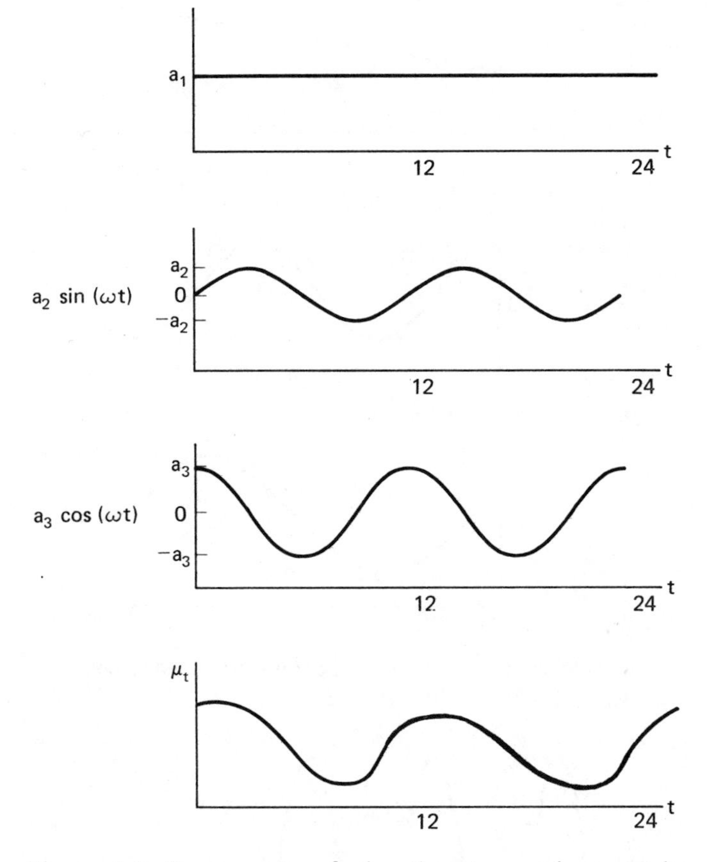

Figure 3-8. Components of the three-term trigonometric demand pattern with $M = 12$.

3-2 FILTERING

Since in time-series analysis, the forecasts of future demands is based on the flow of the past demands, the forecaster obviously wants the history of demands to be as accurate as possible. Occasionally, one observation is widely different from all other observations. The problem that confronts the forecaster is whether to keep the suspect observation in computing the forecasts or whether to discard it as being a faulty measurement or an occurrence from an unusual circumstance that is not likely to repeat. For example, these may be due to data-transmission errors, unexpected strikes, or weather conditions. The faulty observation, if used in the computations, may distort the forecasts for many time periods to come.

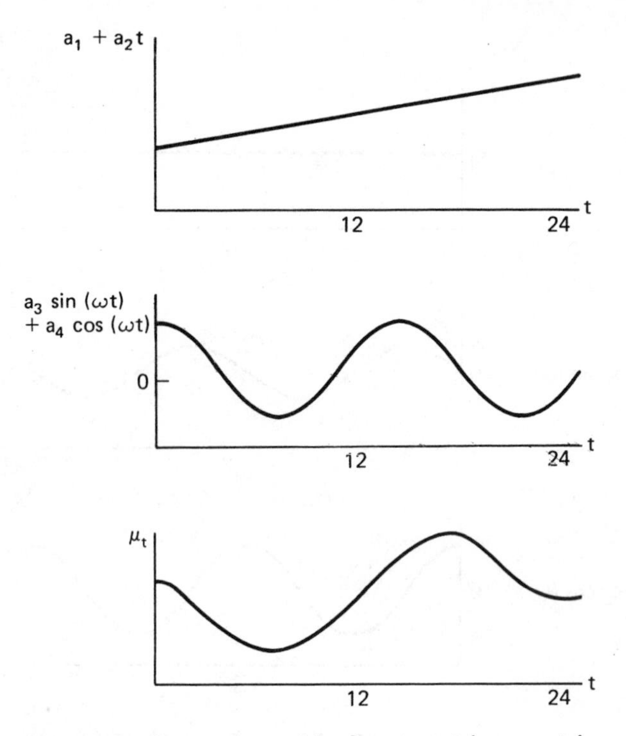

Figure 3-9. Components of the four-term trigonometric demand pattern with $M = 12$.

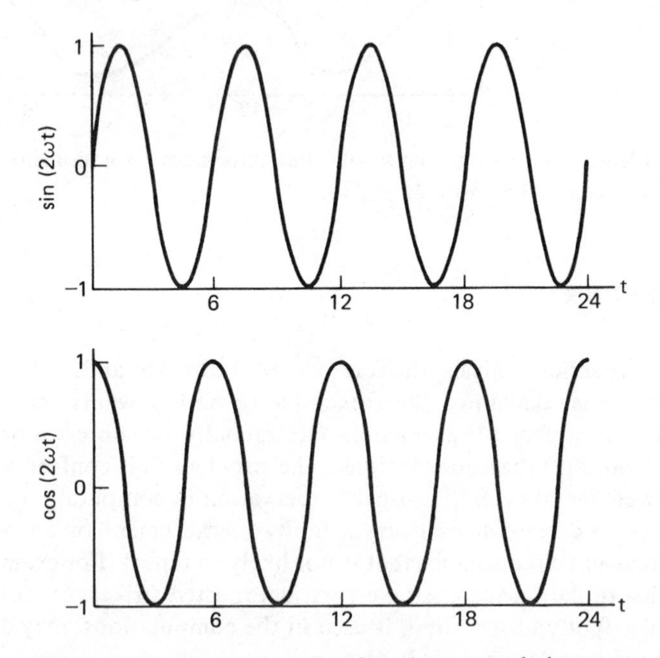

Figure 3-10. Sine and cosine functions completing one cycle every six time periods.

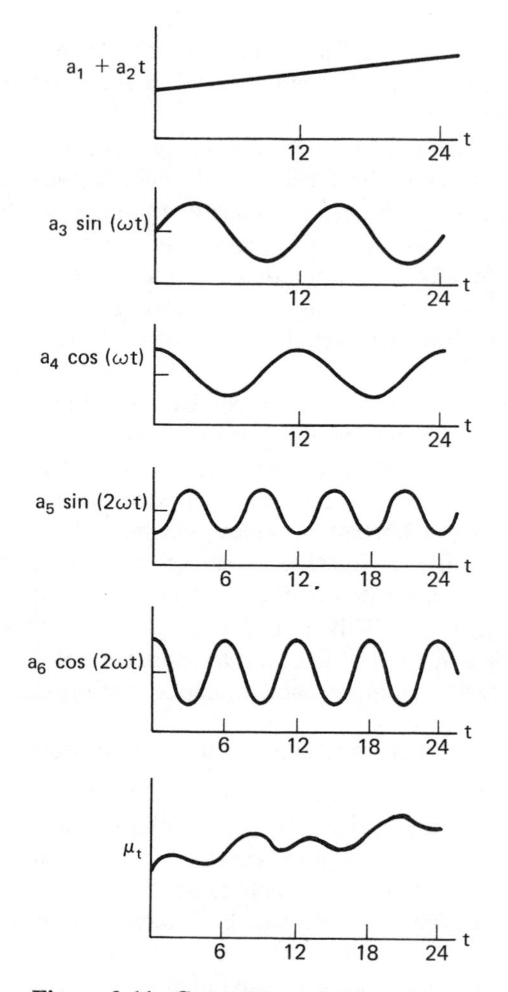

Figure 3-11. Components of the six-term trigonometric demand pattern.

The suspected faulty observation is often called an *outlier*. In an endeavor to identify and adjust such outliers, various methods are possible. This is performed through a filtering process, which examines each observation to determine its representativeness with all other observations. When not representative, the observation is identified as an outlier. This outlier may be discarded completely or modified before use in computations.

There is no guarantee that any filtering process will truly detect all outliers, nor that it will not overreact and falsely identify a valid demand as an outlier. These are difficultities that are possible whenever filtering is used.

In the ideal situation, the outlier entry is brought to the attention of the forecaster, who then decides whether or not to include the entry in generating the forecasts. It is only the forecaster who can actually change the data record. This situation is readily practical when the number of items to be forecast is of a moderate size.

Unfortunately, however, there are many situations where there is a large volume of items to be forecast and such personal attention is not possible. Also, the forecasting step is usually an initial phase of a computerized inventory control system and forecasts are needed without interruption in order to move forward. In these situations, the outlier is identified and may need to be replaced in calculating the forecast (but not in the permanent record of the item). The methods of this section apply to these circumstances primarily.

This section shows how filtering may be applied in four different situations. The first is where the demand pattern is assumed to be horizontal and the demands at each time period are filtered. Second is when each order received is filtered. By filtering each order, it is not necessary to assume that the demand pattern is horizontal, since the method is applicable to any demand pattern. Third is a situation where the demands for each time period are filtered using as information the total demands and total number of orders in any time period. This procedure is also applicable to any demand pattern. Finally is a method of filtering the most current demand entry using the forecast as a basis. Again, the method applies for any demand pattern.

Filtering Demands from a Horizontal Demand Pattern

Suppose that the demand pattern is horizontal and the individual demand entries are assumed to be generated from a normal probability distribution. As of the current time period T, the demand entries that are available are $x_1, x_2, \ldots, x_T$. These entries will be used to forecast the future demands. However, before doing so, the forecaster may wish to filter the entries to screen out any outliers.

Although several statistically based methods of identifying outliers are available [1],* the procedures generally require sorting the data from low to high values as a first step. This need is time-consuming and may not be feasible in large inventory systems.

One possibility is to use tolerance limits [2] where a value of K is found so that one can assert with confidence P_1 that the proportion of observations contained within $\bar{X} \pm KS_x$ is at least P_2. In the above, $\bar{X}$ is the mean and S_x is the standard deviation estimated from the data. Table 3-1 lists K values for $P_1 = (0.95, 0.99)$, $P_2 = (0.99, 0.999)$, and for selected sample sizes. For example, with a sample size of 40, $P_1 = 0.95$, and $P_2 = 0.99$, $K = 3.21$ and the tolerance limits are $\bar{X} \pm 3.21S_x$. This means that with 95% confidence, the proportion of the population contained within $(\bar{X} - 3.21S_x)$

---

*Numbers in square brackets refer to references at the end of this chapter.

Table 3-1. Tolerance Factors (K) for Normal Distributions

	$P_1 = 0.95$		$P_1 = 0.99$	
	P_2		P_2	
Number of Observations	0.99	0.999	0.99	0.999
10	4.43	5.65	5.59	7.13
15	3.88	4.95	4.60	5.88
20	3.61	4.61	4.16	5.31
25	3.46	4.41	3.90	4.99
30	3.35	4.28	3.73	4.77
35	3.27	4.18	3.61	4.61
40	3.21	4.10	3.52	4.49
45	3.16	4.04	3.44	4.40
50	3.13	3.99	3.38	4.32
60	3.07	3.92	3.29	4.21
70	3.02	3.86	3.22	4.12
80	2.99	3.81	3.17	4.05
90	2.96	3.78	3.13	4.00
100	2.93	3.74	3.10	3.95
∞	2.58	3.29	2.58	3.29

Source: Adapted by permission from *Techniques of Statistical Analysis* by C. Eisenhart, M. W. Hastay, and W. A. Wallis, Copyright 1947, McGraw-Hill Book Company, Inc.

and $(\bar{X} + 3.21S_x)$ is at least 0.99. Any entry outside these limits is questionable.

The method may be applied to the series $(x_1, x_2, \ldots, x_T)$ by following the four steps listed below:

1. Using $x_1, x_2, \ldots, x_T$, the sample mean ($\bar{x}$) and standard deviation (S_x) are calculated as shown in Chapter 2.
2. With T observations, Table 3-1 is used to find the appropriate value of K for given values of P_1 and P_2.
3. Low and high limits on x are now derived. These are

$$x_{\text{low}} = \bar{x} - KS_x$$

$$x_{\text{high}} = \bar{x} + KS_x$$

   Should x_{low} be below zero, then x_{low} is reset to be exactly zero. This is necessary to avoid any demand entries that are not positive.
4. Any entry x_t which lies outside these limits is rejected.

A fifth step pertains to the adjusted value of the rejected (or questionable) entry. In the ideal situation, the forecaster investigates the item and each rejected entry to determine what adjusted value, if any, to use.

On the other hand, it is often not possible to give such personal attention to each item. This occurs when a large number of items are being processed in a computer system and a decision is needed within the framework of the computer system without interruption. In these situations the forecaster must select an adjustment scheme which automatically changes the questionable values in a manner that seems suitable to the types of items in the inventory. One possibility is to adjust each questionable entry with its closest limit x_{low} or x_{high}.

Example 3.1

Consider the following 36 demand entries: 8, 4, 12, 7, 3, 11, 15, 6, 7, 4, 0, 8, 2, 6, 8, 60, 4, 3, 7, 8, 0, 15, 2, 17, 6, 10, 2, 7, 6, 3, 5, 4, 8, 1, 2, and 9. Filter the demands using $P_1 = 0.99$ and $P_2 = 0.99$.

Carrying out the four steps cited above, the following results are found:

1. $\bar{x} = 7.78$, $S_x = 9.86$.
2. Using $T = 36$ in Table 3-1, the table entry yields $K \approx 3.6$.
3. The limits are

$$x_{\text{high}} = 7.78 + 3.6 \times 9.86 = 43.26$$

$$x_{\text{low}} = 0$$

since $7.78 - 3.6 \times 9.86 = -27.71$.

4. The entry $x_{16} = 60$ is rejected since it is greater than x_{high}.

Filtering Each Order

When the demand (x_t) in time period t is comprised of n_t orders, then

$$x_t = d_{1t} + d_{2t} + \ldots + d_{n_t t}$$

where d_{it} represents the number of pieces requested in the ith order. If each of the d_{it} is known for the history of demands, it is possible to filter each order before using the demand entries $x_1, x_2, \ldots, x_T$ in forecasting. In this way the filtering process is applicable for any type of demand pattern, assuming that d entries are from the same population.

In this section it is assumed that the order sizes (d) are normally distributed with a fixed mean and standard deviation. Further, the number of orders (n) per time period are assumed to fluctuate in accordance with an arbitrary demand pattern. It is the order sizes that are filtered and not the number of orders.

The following steps are followed to filter the orders:

1. Using all the orders for the T time periods, find the sample mean $(\bar{d})$ and the standard deviation (S_d).

2. Letting the total number of observations be denoted by N, where $N = \sum_{t=1}^{T} n_t$, then N is used in Table 3-1 to find the appropriate value of K.
3. Low and high limits for each order size are found. These are

$$d_{\text{low}} = \bar{d} - KS_d$$
$$d_{\text{high}} = \bar{d} + KS_d$$

If d_{low} is negative, it is adjusted up to zero.
4. Any entry d_{it} which lies outside the limits is rejected. As stated earlier, any rejected entry is replaced using a scheme selected by the forecaster.

Example 3.2

Consider an item with the following entry of orders over the past 12 time periods:

t	d_{it}	t	d_{it}
1	3, 6	7	6, 4, 3, 2, 4
2		8	5, 6, 2, 1
3	4, 7, 3	9	4, 4, 3
4	2, 2, 1, 5	10	2, 3
5	6, 5, 4	11	3, 4
6	8, 5, 30, 2, 1	12	2

Note, for instance, that when $t = 1$, then $x_1 = 9$, $n_1 = 2$, $d_{11} = 3$, and $d_{21} = 6$. Filter the orders using $P_1 = 0.99$ and $P_2 = 0.99$.

Following are the results from the steps listed above:

1. $\bar{d} = 4.47$, $S_d = 4.84$.
2. Since the number of entries is $N = 34$, this value is used in Table 3-1 to obtain $K \approx 3.62$.
3. The limits on d become

$$d_{\text{high}} = 4.47 + 3.62 \times 4.84 = 21.99$$
$$d_{\text{low}} = 0$$

since $4.47 - 3.62 \times 4.84 = -13.05$ is below zero.
4. The entry $d_{3,6} = 30$ is rejected since this entry exceeds the high limit of $d_{\text{high}} = 21.99$.

Filtering Demands from an Arbitrary Demand Pattern

The prior method of filtering is only possible to apply when all the orders in the demand history of the item are available to the forecaster. Generally, all such data are not available because of computer storage limitations. A method is presented here that uses fewer data and is also applicable for any demand pattern. The data required in this situation are the history of the number of orders $(n_1, n_2, \ldots, n_T)$ and the corresponding demands $(x_1, x_2, \ldots, x_T)$, where T represents the most current time period.

The following steps are carried out to detect any outlier demands:

1. The mean and standard deviation of the order sizes are obtained as follows:

$$\bar{d} = \frac{\sum_{t=1}^{T} x_t}{\sum_{t=1}^{T} n_t}$$

$$S_d = \sqrt{\sum_{t=1}^{T} (x_t - n_t\bar{d})^2 \frac{N}{N^2 - \sum n_t^2}}$$

where

$$N = \sum_{t=1}^{T} n_t$$

2. Table 3-1 is used with a value T^*, where T^* is the number of time periods with $n_t > 0$. The corresponding value of K is identified.
3. High and low limits for each time period are now approximated by

$$x_{\text{low}\,t} = n_t\bar{d} - K\sqrt{n_t}\,S_d$$

$$x_{\text{high}\,t} = n_t\bar{d} + K\sqrt{n_t}\,S_d$$

As before, $x_{\text{low}\,t}$ is reset to zero should it be negative using the above relation.

4. Any x_t that lies outside the limits is rejected. As before, the rejected values are replaced using a scheme selected by the forecaster.

Example 3.3

Consider the most recent 12 months of demand activity for an item.

t	n_t	x_t	t	n_t	x_t
1	2	9	7	5	19
2	0	0	8	4	14
3	3	14	9	3	11
4	4	10	10	2	5
5	3	15	11	2	7
6	5	46	12	1	2

Filter the demands (x_t) using $P_1 = 0.99$ and $P_2 = 0.99$.

The steps given above are followed with the following results:

1. $N = 2 + 0 + \ldots + 1 = 34$

   $\sum_{t=1}^{12} n_t^2 = 2^2 + 0^2 + \ldots + 1^2 = 122$

   $\bar{d} = \frac{152}{34} = 4.47$

   $S_d = \sqrt{[(9 - 8.94)^2 + \ldots + (2 - 4.47)^2]\frac{34}{34^2 - 122}} = 4.72$
2. Using Table 3-1 with $T^* = 11$, $K \approx 5.40$.
3. The limits become

$$x_{\text{low}\,t} = 4.47n_t - 5.40\sqrt{n_t}\,4.72$$

$$x_{\text{high}\,t} = 4.47n_t + 5.40\sqrt{n_t}\,4.72$$

For the 12 time periods, these are:

t	$x_{\text{low},t}$	$x_{\text{high},t}$	t	$x_{\text{low},t}$	$x_{\text{high},t}$
1	0	44.9	7	0	79.4
2	0	.0	8	0	68.9
3	0	57.6	9	0	57.6
4	0	68.9	10	0	44.9
5	0	57.6	11	0	44.9
6	0	79.4	12	0	30.0

4. Since all the demand entries x_t are within their corresponding limits, none of the entries are rejected.

Mathematical Basis. Assuming that the d_{it} are independent and normally distributed with mean μ and standard deviation σ, then $x_t = d_{1t} + d_{2t} + \ldots + d_{n_t t}$ is normally distributed and has $\mu_x = n_t\mu$ and $\sigma_x^2 = n_t\sigma^2$. The expected value of $\bar{d}$ is

$$E(\bar{d}) = E\left(\frac{\sum_{t=1}^{T} x_t}{N}\right) = \frac{\sum_{t=1}^{T} n_t\mu}{N} = \mu$$

for

$$N = \sum_{t=1}^{T} n_t$$

Also, since

$$E(x_t - n_t\bar{d})^2 = E(x_t^2) + E[(n_t\bar{d})^2] - 2E(x_t n_t \bar{d})$$
$$= \sigma^2 n_t \left(\frac{N - n_t}{N}\right)$$

where

$$E(x_t^2) = n_t(n_t\mu^2 + \sigma^2)$$
$$E[(n_t\bar{d})^2] = n_t^2\left(\mu^2 + \frac{\sigma^2}{N}\right)$$
$$E(x_t n_t \bar{d}) = n_t^2\left(\mu^2 + \frac{\sigma^2}{N}\right)$$

then

$$E(S_{\bar{d}}^2) = E\left[\sum_{t=1}^{T} (x_t - n_t\bar{d})^2 \left(\frac{N}{N^2 - \sum_{t=1}^{T} n_t^2}\right)\right] = \sigma^2$$

Filtering the Current Demand Entry with Use of the Forecast

Brown [3] shows how the most current demand entry (x_T) can be filtered using the one-period-ahead forecast of this entry [$\hat{x}_{T-1}(1)$]. The entry x_T is deemed an outlier if it is so far removed from $\hat{x}_{T-1}(1)$ that the probability of this difference occurring by chance is small. The difference is measured by the relation

$$K = \left|\frac{x_T - \hat{x}_{T-1}(1)}{S_e}\right|$$

where S_e is an estimate of the standard deviation of the forecast error.

Generally, $K = 4$ is used as the critical value to detect outliers. Brown states that $K = 4$ is adequate for normal control, $K = 3.5$ may be used for tight control, and $K = 5$ for loose control. When the calculated value of K is less than the critical value, x_T is not treated as an outlier; otherwise, it is assumed to be an outlier.

Example 3.4

Suppose that $x_T = 40$, $\hat{x}_{T-1}(1) = 100$, and $S_e = 30$. Determine whether x_T is an outlier, using the criteria of $K = 4$. Since the calculations give

$$K = \left|\frac{40 - 100}{30}\right| = 2.0$$

the current demand is not an outlier.

PROBLEMS

3-1. Draw a graph of $\mu_t = a + bt$ for $t = 1, \ldots, 12$ when:
a. $a = 10$ and $b = 1$
b. $a = 10$ and $b = 3$
c. $a = 100$ and $b = -1$
d. $a = 100$ and $b = 0$

3-2. Plot the expected demand $\mu_t = a + bt + ct^2$ for $t = 1, \ldots, 12$ with:
a. $a = 100$, $b = 1$, and $c = -0.1$
b. $a = 50$, $b = -1$, and $c = 0.2$
c. $a = 50$, $b = 2$, and $c = 0.1$

3-3. Suppose that $\mu_t = \mu\rho_t$ represents the quarterly ($M = 4$) expected demand of an item with $\mu = 200$, $\rho_1 = 0.90$, $\rho_2 = 1.20$, $\rho_3 = 1.10$, and $\rho_3 = 0.80$. Plot μ_t for $t = 1, \ldots, 8$.

3-4. Consider the multiplicative trend seasonal demand pattern with $M = 4$ and $\mu_t = (a + bt)\rho_t$. Assuming that $a = 100$, $b = 2$, $\rho_1 = 0.80$, $\rho_2 = 0.90$, $\rho_3 = 1.20$, and $\rho_4 = 1.10$, plot μ_t for $t = 1, \ldots, 12$.

3-5. Suppose that the additive trend seasonal demand pattern is in effect where $M = 4$,

$$\mu_t = a + bt + \delta_t$$

and $a = 100$, $b = 2$, $\delta_1 = 20$, $\delta_2 = 30$, $\delta_3 = 0$, and $\delta_4 = -50$. Plot μ_t for $t = 1, \ldots, 12$.

3-6. Consider the three-term trigonometric demand pattern with $M = 4$, i.e.,

$$\mu_t = a_1 + a_2 \sin \omega t + a_3 \cos \omega t$$

Note here that $\omega = 90°$ and

t	0	1	2	3	4
ωt	0°	90°	180°	270°	360°
$\sin \omega t$	0.00	1.00	0.00	−1.00	0.00
$\cos \omega t$	1.00	0.00	−1.00	0.00	1.00

Now, if $a_1 = 100$, $a_2 = 20$, and $a_3 = 10$, plot μ_t for $t = 1, \ldots, 8$.

3-7. Assume that the four-term trigonometric model with $M = 4$ is in effect, i.e.,

$$\mu_t = a_1 + a_2 t + a_3 \sin \omega t + a_4 \cos \omega t$$

If $a_1 = 200$, $a_2 = -20$, $a_3 = 30$, and $a_4 = 20$, plot μ_t for $t = 1, \ldots, 8$.

3-8. Consider the following 24 demand entries: 3, 8, 4, 3, 2, 10, 5, 6, 9, 2, 0, 4, 9, 4, 200, 3, 2, 5, 8, 6, 4, 2, 1, and 3. In this situation $\bar{x} = 12.62$ and $S_x = 40$. Assuming that the horizontal demand pattern is in effect, filter the demands using the procedure shown in Example 3.1 with $P_1 = 0.99$ and $P_2 = 0.99$.

3-9. Suppose that the order sizes for each of the past 12 months are available for an item and are the following:

t	*Number of Orders*	*Order Size*	t	*Number of Orders*	*Order Size*
1	3	2, 4, 5	7	1	3
2	5	3, 6, 2, 4, 2	8	2	4, 1
3	6	4, 2, 3, 5, 6, 2	9	2	2, 6
4	4	5, 2, 1, 2	10	3	1, 2, 4
5	2	1, 50	11	4	5, 4, 2, 6
6	0		12	3	1, 5, 4

In this situation $\bar{d} = 4.60$ and $S_d = 8.07$. Now filter each order by the procedure applied in Example 3.2, using $P_1 = 0.99$ and $P_2 = 0.99$.

3-10. Consider the demands for an item over the past 12 months, where the number of orders (n_t) and the monthly demands (x_t) are the following:

t	n_t	x_t	t	n_t	x_t
1	3	11	7	1	3
2	5	17	8	2	5
3	6	22	9	2	8
4	4	10	10	3	7
5	2	51	11	4	17
6	0	0	12	3	10

Filter the demands using the method applied in Example 3.3 with $P_1 = 0.99$ and $P_2 = 0.99$.

3-11. Suppose that the one-period-ahead forecast at $t = T - 1$ is $\hat{x}_{T-1}(1) = 50$ and the standard deviation of the forecast error is estimated as $S_e = 10$. Also, assume that the demand at $t = T$ is $x_T = 100$ and the critical value of K is set at 4. Use the method in Example 3.4 to determine if the current demand is an outlier.

REFERENCES

[1] NATRELLA, M. G., *Experimental Statistics*. Washington, D.C.: National Bureau of Standards Handbook 91, 1963, Chap. 17.

[2] MILLER, I., AND J. E. FREUND, *Probability and Statistics for Engineers*. Englewood Cliffs, N.J.: Prentice-Hall, Inc., 1977, pp. 434–35.

[3] BROWN, R. G., *Materials Management Systems*. New York: John Wiley & Sons, Inc., 1977, Chap. 8.

4

HORIZONTAL MODELS

An item following the horizontal demand pattern is one where the level is $\mu_t = \mu$ for all t. In this respect the average value of the demands is constant over the long run. The individual demand entries x_t will seldom equal μ but instead will fluctuate around μ in a random fashion. The role of the horizontal forecasting model is to estimate the average demand from the entries of the past and to use this average as the forecast of demands for the future time periods.

Three such forecasting models are described in this chapter: the naive, moving average, and single smoothing models. The single smoothing model is also often called the exponential smoothing model and is perhaps the forecasting model in greatest use in industry today.

4-1 NAIVE MODEL

The naive forecasting model is where the estimates of all future demands are merely taken to be the same as the most current demand entry. Although this forecast model is easy to apply and requires minimal data, the forecasts generated are poor and of small practical value. The model is presented at this time to serve as a comparison to other, more refined forecasting models.

In order to apply the naive model, the only data required are x_T, the most current demand. With this, the forecast as of time T of the τth future demand is

$$\hat{x}_T(\tau) = x_T \qquad \tau = 1, 2, \ldots$$

The forecast of the cumulative demands for the next τ time periods is

$$\begin{aligned}\hat{X}_T(\tau) &= \hat{x}_T(1) + \hat{x}_T(2) + \ldots + \hat{x}_T(\tau) \\ &= \tau\, x_T\end{aligned}$$

Example 4.1

Suppose that at $t = 1$, the demand entry is $x_1 = 95$. Using the naive model, the forecast of the demands for the next two time periods are

$$\hat{x}_1(1) = 95$$

$$\hat{x}_1(2) = 95$$

Table 4-1. EXAMPLE OF THE NAIVE FORECASTING MODEL

t	x_t	$\hat{x}_t(\tau)$	$\hat{X}_t(2)$
1	95	95	190
2	100	100	200
3	87	87	174
4	123	123	246
5	90	90	180
6	96	96	192
7	75	75	150
8	78	78	156
9	106	106	212
10	104	104	208
11	89	89	178
12	83	83	166
13	118	118	236
14	86	86	172
15	86	86	172
16	112	112	224
17	85	85	170
18	101	101	202
19	135	135	270
20	120	120	240
21	76	76	152
22	115	115	230
23	90	90	180
24	92	92	184

and the corresponding cumulative forecast is

$$\hat{X}_1(2) = 2(95) = 180$$

Continuing with this example, the one-period-ahead forecasts for 24 time periods are listed in Table 4-1. A plot showing the demands and their associated one-period-ahead forecasts is given in Figure 4-1.

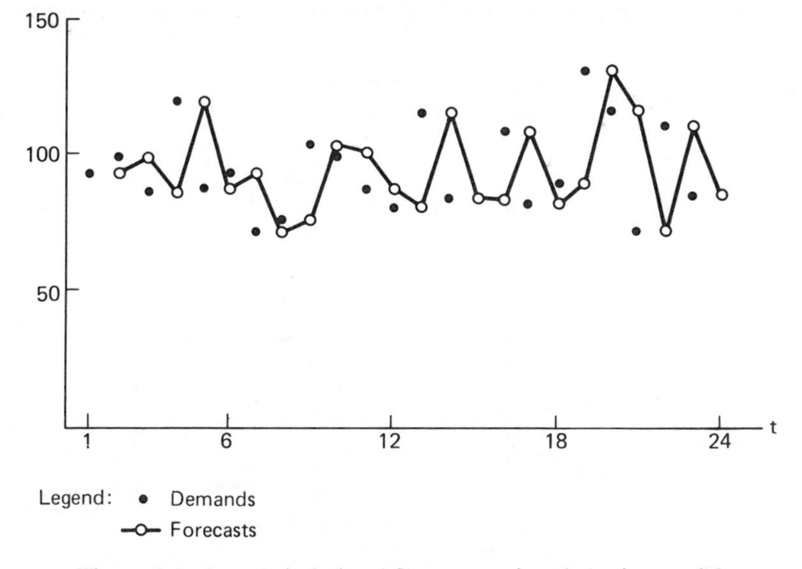

Figure 4-1. One-period-ahead forecasts using the naive model.

Forecast Errors

An important measure in forecasting is the error associated with the forecast. The forecast error of the τ-period-ahead forecast of $x_{T+\tau}$ is

$$e_{(\tau)} = x_{T+\tau} - \hat{x}_T(\tau)$$

For the naive model the expected value of $e_{(\tau)}$ is zero when the horizontal demand pattern is truly in effect. Assuming that the demands are independent with a standard deviation σ, the standard deviation of $e_{(\tau)}$ is

$$\sigma_{e_{(\tau)}} = \sqrt{2}\,\sigma$$

Since $\sqrt{2} = 1.41$, the forecast error has a standard deviation that is 1.41 times as large as the standard deviation of the demand entries.

The forecast error for the τ-period-cumulative forecast is

$$\begin{aligned} E_\tau &= (x_{T+1} + \ldots + x_{T+\tau}) - \hat{X}_T(\tau) \\ &= e_{(1)} + \ldots + e_{(\tau)} \end{aligned}$$

so, as before, the mean of E_τ is zero. The standard deviation of E_τ becomes

$$\sigma_{E_\tau} = \sqrt{\tau(\tau + 1)}\,\sigma$$

This relation shows that σ_{E_τ} increases with τ. As one would suspect, the further the forecasts are projected, the larger is the forecast error. Table 4-2 shows how σ_{E_τ} increases with respect to σ for values of τ ranging from 1 to 12.

Table 4-2. RATIOS OF σ_{E_τ}/σ FOR THE NAIVE MODEL

τ	σ_{E_τ}/σ
1	1.41
2	2.45
3	3.46
4	4.47
5	5.48
6	6.48
7	7.48
8	8.49
9	9.49
10	10.49
11	11.49
12	12.49

Example 4.2

Suppose that the naive forecast model is in use for an item whose demands are horizontal and where $\sigma = 50$. Find the standard deviation of the forecast errors $e_{(1)}$, $e_{(2)}$, and E_2.

Using Table 4.2,

$$\sigma_{e(1)} = 1.41(50) = 70.50$$

$$\sigma_{e(2)} = 1.41(50) = 70.50$$

$$\sigma_{E_2} = 2.45(50) = 122.50$$

MATHEMATICAL BASIS. When $E(x_t) = \mu$, $V(x_t) = \sigma^2$, and the x_t are independent,

$$E(\hat{x}_T(\tau)) = E(x_T) = \mu$$

$$E(e_{(\tau)}) = E(x_{T+\tau} - \hat{x}_T(\tau)) = \mu - \mu = 0$$

$$V(\hat{x}_T(\tau)) = V(x_T) = \sigma^2$$

$$V(e_{(\tau)}) = V(x_{T+\tau} - \hat{x}_T(\tau)) = 2\sigma^2$$

Also,

$$E(\hat{X}_T(\tau)) = \tau E(\hat{x}_T(1)) = \tau\mu$$

$$E(E_\tau) = E(x_{T+1} + \dots + x_{T+\tau} - \hat{X}_T(\tau))$$

$$= \tau\mu - \tau\mu = 0$$

$$V(\hat{X}_T(\tau)) = V(\tau\hat{x}_T(1)) = \tau^2\sigma^2$$

$$V(E_\tau) = V(x_{T+1} + \dots + x_{T+\tau} - \hat{X}_T(\tau))$$

$$= \tau\sigma^2 + \tau^2\sigma^2$$

$$= \tau(\tau + 1)\sigma^2$$

4-2 MOVING-AVERAGE MODEL

In the *moving-average model*, the average demand from the N most-current time periods is found and used as the forecast of demands for the future time periods. To get started, the forecaster must select a parameter N from which his forecasts will be based. Once selected, the corresponding demand entries $(x_T, x_{T-1}, \dots, x_{T-N+1})$ are gathered and their average (M_T) is obtained, i.e.,

$$M_T = \frac{x_{T-N+1} + \dots + x_{T-1} + x_T}{N}$$

M_T is called the *moving average* as of time T.

The average shown above gives the forecast for the τth future time period. This is

$$\hat{x}_T(\tau) = M_T$$

It is noted that the moving averages at $t = T$ and $t = T - 1$ are related as follows:

$$M_T = M_{T-1} + \frac{x_T - x_{T-N}}{N}$$

When N is large, this recursive relation is particularly useful to shorten the number of calculations in finding the updated moving average.

The forecast of the demands for the sum of the next τ time periods becomes

$$\hat{X}_T(\tau) = \tau M_T$$

This is called the *cumulative forecast* and is easy to calculate once M_T is known.

Example 4.3

Suppose that the forecasts $\hat{x}_T(\tau)$ and $\hat{X}_T(2)$ are required at each time period. Assume that $N = 6$ is to be used and the first six demand entries are 95, 100, 87, 123, 90, and 96. The six-period moving average at $T = 6$ becomes

$$M_6 = \frac{95 + 100 + 87 + 123 + 90 + 96}{6} = 98.50$$

Hence, $\hat{x}_6(\tau) = 98.50$ and $\hat{X}_6(2) = 2 \times 98.50 = 197.00$.

At the next time period ($t = 7$), a new demand entry (say, $x_7 = 75$) becomes available. This gives rise to a new moving average,

$$M_7 = \frac{100 + 87 + 123 + 90 + 96 + 75}{6} = 95.17$$

or, using the recursive relation,

$$M_7 = 98.50 + \frac{75 - 95}{6} = 95.17$$

The corresponding forecasts are $\hat{x}_7(\tau) = 95.17$ and $\hat{X}_7(2) = 190.34$.

Table 4-3. Example of the Moving-Average Forecast Model with $N = 6$

t	x_t	$\hat{x}_t(\tau)$	$\hat{X}_t(2)$
1	95		
2	100		
3	87		
4	123		
5	90		
6	96	98.50	197.00
7	75	95.17	190.34
8	78	91.50	183.00
9	106	94.67	189.34
10	104	91.50	183.00
11	89	91.33	182.66
12	83	89.17	178.34
13	118	96.34	192.68
14	86	97.67	195.34
15	86	94.34	188.68
16	112	95.67	191.34
17	85	95.00	190.00
18	101	98.00	196.00
19	135	100.84	201.68
20	120	106.50	213.00
21	76	104.84	209.68
22	115	105.34	210.68
23	90	106.17	212.34
24	92	104.67	209.34

This example continues in Table 4-3, where new forecasts unfold as each new demand entry becomes available. A plot of the results is given in Figure 4-2. Notice how much more stable the forecasts are than those obtained with the naive model (Figure 4-1).

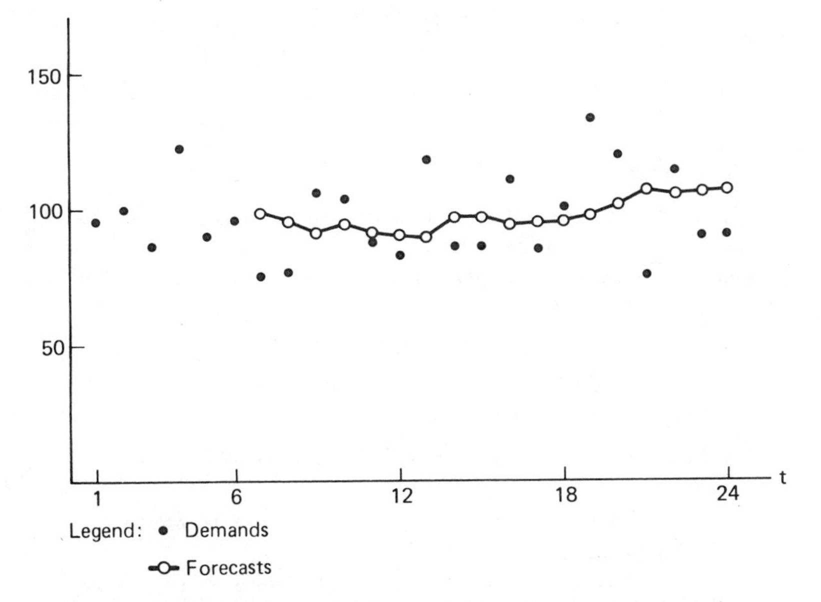

Figure 4-2. One-period-ahead forecasts using the six-period moving-average model.

Early Stages

In order to apply the moving-average model in its true sense, the forecaster has to wait until N demand entries are available before starting. This could be too long a wait since forecasts may be needed sooner. To overcome this difficulty, a scheme of the following type can be applied. At $T = 1$, then $M_1 = x_1$; at $T = 2$, then $M_2 = 0.5\,(x_1 + x_2)$, and in general for those time periods when $T < N$, the moving average is found by using the average of the T most current entries, i.e.,

$$M_T = \frac{1}{T}(x_1 + x_2 + \ldots + x_T)$$

As before, the τ-period-ahead forecasts are $\hat{x}_T(\tau) = M_T$.

Example 4.4

Using the data of Table 4-3, show what the τ period ahead forecasts would be (for $t = 1$ to 5) with the method above.

The forecasts are:

$$\hat{x}_1(\tau) = M_1 = 95$$

$$\hat{x}_2(\tau) = M_2 = \tfrac{1}{2}(95 + 100) = 97.50$$

$$\hat{x}_3(\tau) = M_3 = \tfrac{1}{3}(95 + 100 + 87) = 94.00$$

$$\hat{x}_4(\tau) = M_4 = \tfrac{1}{4}(95 + 100 + 87 + 123) = 101.25$$

$$\hat{x}_5(\tau) = M_5 = \tfrac{1}{5}(95 + 100 + 87 + 123 + 90) = 99.00$$

Average Age of Demands

It is sometimes helpful to determine the average age of the demand entries used in calculating the N-period moving average. Using the age of the most current demand (x_T) as zero, the age of the immediately prior entry (x_{T-1}) is 1, and so forth. With this scale the average age (A) is

$$A = \frac{1}{N}(0 + 1 + 2 + \ldots + N - 1) = \frac{N-1}{2}$$

Table 4-4 shows how A varies for selected values of N.

Table 4-4. Average Age of the Demands in an N-Period Moving-Average Model

N	6	12	18	24	30	36
A	2.5	5.5	8.5	11.5	14.5	17.5

Example 4.5

What is the average age of the demands when $N = 10$?

This is

$$A = \frac{10 - 1}{2} = 4.5$$

Standard Deviation of the Forecast Errors

The expected value of the τ-period-ahead forecast error ($e_{(\tau)} = x_{T+\tau} - \hat{x}_T(\tau)$) is zero and the standard deviation is

$$\sigma_{e(\tau)} = \sqrt{\frac{N+1}{N}}\,\sigma$$

where again σ is the standard deviation of the demand entries. This relation shows that the larger N becomes, the better the forecasts are. The results, however, are based on the assumption that the demand pattern is truly horizontal.

The standard deviation of the error for the τ-period cumulative forecast

$$E_\tau = (x_{T+1} + \dots + x_{T+\tau}) - \hat{X}_T(\tau)$$

is

$$\sigma_{E_\tau} = \sqrt{\frac{\tau^2 + \tau N}{N}}\,\sigma$$

Table 4-5 shows how σ_{E_τ} is related to σ for various values of N and τ.

Table 4-5. RATIOS OF σ_{E_τ}/σ FOR THE MOVING-AVERAGE MODEL

	N			
τ	6	12	18	24
1	1.08	1.04	1.03	1.02
2	1.63	1.53	1.49	1.47
3	2.12	1.94	1.87	1.84
4	2.58	2.31	2.21	2.16
5	3.03	2.66	2.53	2.46
6	3.46	3.00	2.83	2.74
7	3.89	3.33	3.12	3.01
8	4.32	3.65	3.40	3.27
9	4.73	3.97	3.67	3.52
10	5.16	4.28	3.94	3.76
11	5.58	4.59	4.21	4.01
12	6.00	4.90	4.47	4.24

Example 4.6

Suppose that cumulative forecasts for 6 months ahead are needed and the moving average model is in use. How much more accuracy is gained by using $N = 18$ instead of $N = 12$?

Since $\sigma_{E_6} = 3.00\sigma$ for $N = 12$ and $\sigma_{E_6} = 2.83\sigma$ for $N = 18$, the decrease in the standard deviation is

$$\frac{3.00 - 2.83}{3.00} = 5.7\%$$

and on the average the forecast errors will be smaller by the same percentage.

Mathematical Basis. Assuming that the x_t are independent and have a mean μ and standard deviation σ, then

$$E(\hat{x}_T(\tau)) = E\left[\frac{1}{N}(x_{T-N+1} + \ldots + x_{T-1} + x_T)\right] = \mu$$

$$E(e_{(\tau)}) = E[x_{T+\tau} - \hat{x}_T(\tau)] = \mu - \mu = 0$$

$$V(\hat{x}_T(\tau)) = V\left[\frac{1}{N}(x_{T-N+1} + \ldots + x_{T-1} + x_T)\right] = \frac{\sigma^2}{N}$$

$$V(e_{(\tau)}) = V[x_{T+\tau} - \hat{x}_T(\tau)] = \sigma^2 + \frac{\sigma^2}{N} = \frac{N+1}{N}\sigma^2$$

For $\hat{X}_T(\tau) = \tau\hat{x}_T(\tau)$,

$$E[\hat{X}_T(\tau)] = E[\tau\hat{x}_T(\tau)] = \tau\mu$$

$$E[E_\tau] = E[x_{T+1} + \ldots + x_{T+\tau} - \hat{X}_T(\tau)] = \tau\mu - \tau\mu = 0$$

$$V[\hat{X}_T(\tau)] = V[\tau\hat{x}_T(\tau)] = \frac{\tau^2\sigma^2}{N}$$

$$V[E_\tau] = V[x_{T+1} + \ldots + x_{T+\tau} - \hat{X}_T(\tau)] = \tau\sigma^2 + \frac{\tau^2\sigma^2}{N}$$

4-3 SINGLE SMOOTHING MODEL

Perhaps the most common forecasting model in use today is the *single smoothing model.* This model was developed by Holt [1] and Brown [2] and is often called *exponential smoothing*. At each time period the forecasts are updated conveniently in a recursive manner using the most current demand entry. The model assigns more weight to the more current demands, and in this way the forecasts can react more quickly to potential shifts in the level of demands. Because little data storage is needed to carry out the calculations, the model is well suited to applications where a large number of items are in the inventory.

To apply the single smoothing model, the forecaster selects a smoothing parameter α (Greek lowercase letter alpha), which must lie between zero and 1. The forecast at time T is found using α, the current demand x_T, and the prior one-period-ahead forecast $\hat{x}_{T-1}(1)$. The forecasts become

$$\hat{x}_T(\tau) = \alpha x_T + (1 - \alpha)\hat{x}_{T-1}(1) \qquad \tau = 1, 2, \ldots$$

It is notationally convenient, however, to let $S_T = \hat{x}_T(1)$ and $S_{T-1} = \hat{x}_{T-1}(1)$, whereby

$$S_T = \alpha x_T + (1 - \alpha)S_{T-1}$$

S_T is called the *smoothed average* of x_t $(t = 1, 2, \ldots, T)$ as of time T.

So at each time period, S_T is obtained, as shown above. The forecasts for the τth future time period is

$$\hat{x}_T(\tau) = S_T$$

and the cumulative forecast for the following τ time periods is

$$\hat{X}_T(\tau) = \tau S_T$$

Example 4.7

Suppose that the single smoothing model with $\alpha = 0.1$ is in use. Assume that, as of time T, $x_T = 95$ and $S_{T-1} = 80$. Find the forecasts associated with $\tau = 1$ and 2.

The results are:

$$\begin{aligned} S_T &= 0.1(95) + 0.9(80) = 81.5 \\ \hat{x}_T(1) &= 81.5 \\ \hat{x}_T(2) &= 81.5 \\ \hat{X}_T(2) &= 163.0 \end{aligned}$$

Initial Forecasts

A difficulty arises in getting started since at $t = 1$ there is a value for x_1 but none for S_0. As such, the smoothed average S_1 cannot be found with the relation shown above unless S_0 is estimated in some manner.

Two methods are shown to overcome this difficulty. These are the following:

1. Let $S_0 = x_1$ (x_1 = the first demand entry).
2. Let $S_0 = \hat{\mu}$, where $\hat{\mu}$ is an estimate of the average demand.

In the first situation, the smoothed average at $t = 1$ becomes

$$\begin{aligned} S_1 &= \alpha x_1 + (1 - \alpha)S_0 \\ &= \alpha x_1 + (1 - \alpha)x_1 = x_1 \end{aligned}$$

With the second,

$$S_1 = \alpha x_1 + (1 - \alpha)\hat{\mu}$$

For $t = 2$ and beyond, the updating procedure is carried on in the normal manner.

Example 4.8

Suppose that $\alpha = 0.1$, $T = 1$, $x_1 = 100$, and the forecaster chooses to initialize by using $S_0 = x_1$. The smoothed average at $T = 1$ is

$$S_1 = 0.1(100) + 0.9(100) = 100$$

and $\hat{x}_1(\tau) = 100$.

Example 4.9

Consider an item with no prior history and where the forecaster estimates the average demand as $\hat{\mu} = 10$. Let $\alpha = 0.1$, $T = 1$, and $x_1 = 5$. The smoothed average at $T = 1$ is

$$S_1 = 0.1(5) + 0.9(10) = 9.5$$

and the forecasts are $\hat{x}_1(\tau) = 9.5$.

Early Stages

Another problem that faces the forecaster concerns the choice of the smoothing parameter at the early stages (or time periods). With but a few demands available, the forecaster may find it advantageous to use a higher level of α. With a higher α, more weight is given to the current demand entry, which reduces the bias caused from S_0 should it be initially set too low or too high.

Several methods are possible in this endeavor. Three alternative schemes are described below.

In the first method the forecaster selects a desired value for the smoothing parameter (say α). This is the smoothing parameter that he would use when he feels the forecasts have reached their equilibrium level. At the early stages, however, he chooses a smoothing parameter which is higher than α. For convenience, the value chosen at the tth time period is here denoted as α_t. The values of α_t are obtained from

$$\alpha_t = \frac{1}{t}$$

as long as $1/t$ is greater than α. Otherwise, the desired value α is used.

For example, suppose that $\alpha = 0.1$. Now, using the relation above,

$$\begin{aligned} &\alpha_1 = 1.00, \quad \alpha_2 = 0.50, \quad \alpha_3 = 0.33, \quad \alpha_4 = 0.25 \\ &\alpha_5 = 0.20, \quad \alpha_6 = 0.17, \quad \alpha_7 = 0.15, \quad \alpha_8 = 0.12 \\ &\alpha_9 = 0.11, \quad \text{and} \quad \alpha_t = 0.10 \qquad \text{for } t \geq 10 \end{aligned}$$

Note at $t = 1$ that $\alpha_1 = 1.00$ and $S_1 = x_1$. This has the same effect as setting $S_0 = x_1$ as described earlier in seeking the initial forecast. Also, at $t = 2$,

$$S_2 = \tfrac{1}{2}x_2 + \tfrac{1}{2}S_1 = \tfrac{1}{2}(x_2 + x_1)$$

which is the average of the first two demand entries. Now at $t = 3$,

$$\begin{aligned} S_3 &= \tfrac{1}{3}x_3 + \tfrac{2}{3}S_2 \\ &= \tfrac{1}{3}x_3 + \tfrac{2}{3}[\tfrac{1}{2}(x_2 + x_1)] \\ &= \tfrac{1}{3}(x_3 + x_2 + x_1) \end{aligned}$$

which is the average of the first three demand entries. Continuing in this manner, the values of S_t become exactly the average of the first t demands. This is true until the desired value of α is reached.

Example 4.10

Suppose that $\alpha = 0.20$ is the desired value of the smoothing parameter and the forecaster wishes to use the scheme above in finding S_t for the early stages. The method is applied to the first six entries ($x_1 = 4$, $x_2 = 8$, $x_3 = 7$, $x_4 = 12$, $x_5 = 13$, and $x_6 = 10$) and gives the following results:

t	x_t	α_t	S_t
1	4	1.00	4.00
2	8	0.50	6.00
3	7	0.33	6.33
4	12	0.25	7.75
5	13	0.20	8.80
6	10	0.20	9.00

A second method of treating the early-stage forecasts is similar to the one shown above. Here the forecaster selects specific values for α_t that he deems best for his situation. For example, he may select

$$\alpha_t = \begin{cases} 1.0 & \text{for } t = 1 \\ 0.3 & \text{for } t = 2, 3, 4 \\ 0.2 & \text{for } t = 5, 6 \\ 0.1 & \text{for } t \geq 7 \end{cases}$$

Example 4.11

Use the weights above with the demand entries $x_1 = 4$, $x_2 = 8$, $x_3 = 7$, $x_4 = 12$, $x_5 = 13$, $x_6 = 10$, and $x_7 = 4$.

The smoothed averages become

t	x_t	α_t	S_t
1	4	1.0	4.00
2	8	0.3	5.20
3	7	0.3	5.74
4	12	0.3	7.62
5	13	0.2	8.70
6	10	0.2	8.96
7	4	0.1	8.46

A third method uses one value of α throughout and begins each time period (T) by measuring the average ($\bar{x}$) of all prior demands. This is

$$\bar{x} = \frac{x_1 + \ldots + x_T}{T}$$

Next, S_0 is set equal to $\bar{x}$ and the smoothed averages are updated by

$$S_t = \alpha x_t + (1 - \alpha)S_{t-1} \qquad t = 1, 2, \ldots, T$$

The τ-period-ahead forecast at time period T is $\hat{x}_T(\tau) = S_T$. This cycle continues at each time period until a preset time period, say t_0. Thereupon, the updating is carried on in the normal manner.

Example 4.12

Apply the method used above when the demands are $x_1 = 4$, $x_2 = 8$, $x_3 = 7$, $x_4 = 12$, $x_5 = 13$, and $x_6 = 10$, and the smoothing parameter is $\alpha = 0.10$.

The results are as follows for each of the first six time periods:

t	x_t	$\bar{x} = S_0$	S_1	S_2	S_3	S_4	S_5	S_6	$\hat{x}_t(\tau)$
1	4	4.00	4.00						4.00
2	8	6.00	5.80	6.02					6.02
3	7	6.33	6.10	6.29	6.36				6.36
4	12	7.75	7.38	7.44	7.39	7.85			7.85
5	13	8.80	8.32	8.29	8.16	8.54	8.99		8.99
6	10	9.00	8.50	8.45	8.31	8.67	9.11	9.20	9.20

At $t = 3$, for instance,

$$S_0 = \bar{x} = \tfrac{1}{3}(4 + 8 + 7) = 6.33$$
$$S_1 = 0.1(4) + 0.9(6.33) = 6.10$$
$$S_2 = 0.1(8) + 0.9(6.10) = 6.29$$
$$S_3 = 0.1(7) + 0.9(6.29) = 6.36$$

and $\hat{x}_3(\tau) = S_3 = 6.36$.

Example 4.13

Consider an item that is being forecasted with $\alpha = 0.1$ and where the initial smoothed average is $S_0 = x_1$. The demands and the forecasts for the first 24 time periods are listed in Table 4-6 and the results are plotted in Figure 4-3. The plot is comparable to the corresponding plot for the six-period moving-average model (Figure 4-2) and the naive model (Figure 4-1).

Table 4-6. EXAMPLE OF THE SINGLE SMOOTHING MODEL WITH $\alpha = 0.1$ AND INITIAL FORECAST x_1

t	x_t	$\hat{x}_t(\tau)$
1	95	95.00
2	100	95.50
3	87	94.65
4	123	97.49
5	90	96.74
6	96	96.66
7	75	94.50
8	78	92.85
9	106	94.16
10	104	95.15
11	89	94.53
12	83	93.38
13	118	95.84
14	86	94.86
15	86	93.97
16	112	95.77
17	85	94.70
18	101	95.33
19	135	99.29
20	120	101.36
21	76	98.83
22	115	100.45
23	90	99.40
24	92	98.66

The Smoothing Weights

The smoothing parameter α is used to assign weights to the past demand entries. Notice that

$$\begin{aligned} S_T &= \alpha x_T + (1-\alpha)S_{T-1} \\ &= \alpha x_T + (1-\alpha)\,[\alpha x_{T-1} + (1-\alpha)S_{T-2}] \\ &= \alpha x_T + \alpha(1-\alpha)x_{T-1} + (1-\alpha)^2 S_{T-2} \\ &= \alpha x_T + \alpha(1-\alpha)x_{T-1} + (1-\alpha)^2[\alpha x_{T-2} + (1-\alpha)S_{T-3}] \end{aligned}$$

and so on until the following result appears:

$$\begin{aligned} S_T &= \alpha x_T + \alpha(1-\alpha)x_{T-1} + \alpha(1-\alpha)^2 x_{T-2} + \dots \\ &\qquad + \alpha(1-\alpha)^{T-1}x_1 + (1-\alpha)^T S_0 \\ &= \alpha \sum_{j=0}^{T-1} (1-\alpha)^j x_{T-j} + (1-\alpha)^T S_0 \end{aligned}$$

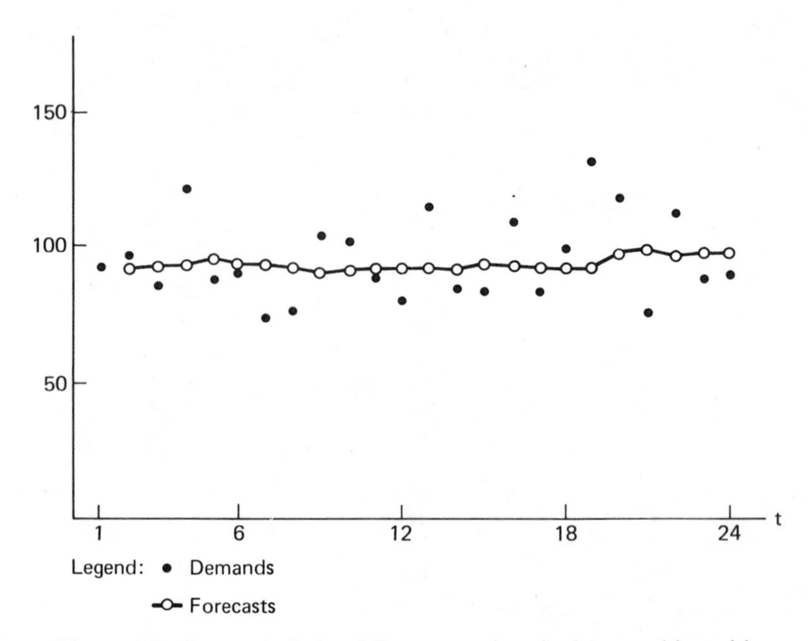

Figure 4-3. One-period-ahead forecasts using single smoothing with $\alpha = 0.1$.

Because $(0 < \alpha < 1)$, the weights assigned get smaller as the demands become older.

Table 4-7 lists the weights that are given to the past demands for various values of α. The higher the value of α, the sooner the weights approach zero. In these and in all situations, the weights follow a curve reminiscent of the exponential function (see Figure 4-4). For this reason the forecast model is often called the *exponential smoothing model.*

The sum of all the weights is

$$\alpha \sum_{j=0}^{T-1} (1-\alpha)^j + (1-\alpha)^T = 1$$

It is convenient to assume that T is large enough so that the weight given to S_0 [i.e., $(1-\alpha)^T$] approaches zero. Using $T = \infty$, then

$$\alpha \sum_{j=0}^{\infty} (1-\alpha)^j = 1$$

and the expansion of S_T becomes

$$S_T = \alpha \sum_{j=0}^{\infty} (1-\alpha)^j x_{T-j}$$

For the remainder of this book, this assumption is used.

Table 4-7. WEIGHTS USING SINGLE SMOOTHING

	$\alpha = 0.1$		$\alpha = 0.2$		$\alpha = 0.3$	
Age of Data	*Weight*	*Cumulative Weight*	*Weight*	*Cumulative Weight*	*Weight*	*Cumulative Weight*
0	0.100	0.100	0.200	0.200	0.300	0.300
1	0.090	0.190	0.160	0.360	0.210	0.510
2	0.081	0.271	0.128	0.488	0.147	0.657
3	0.073	0.344	0.102	0.590	0.103	0.760
4	0.066	0.410	0.082	0.672	0.072	0.832
5	0.059	0.469	0.066	0.738	0.050	0.882
6	0.053	0.522	0.052	0.790	0.035	0.918
7	0.048	0.570	0.042	0.832	0.025	0.942
8	0.043	0.613	0.034	0.866	0.017	0.960
9	0.039	0.651	0.027	0.893	0.012	0.972
10	0.035	0.686	0.021	0.914	0.008	0.980
11	0.031	0.718	0.017	0.931	0.006	0.986
12	0.028	0.746	0.014	0.945	0.004	0.990
13	0.025	0.771	0.011	0.956	0.003	0.993
14	0.023	0.794	0.009	0.965	0.002	0.995
15	0.021	0.815	0.007	0.972	0.001	0.997
16	0.019	0.833	0.006	0.977	0.001	0.998
17	0.017	0.850	0.005	0.982	0.001	0.998
18	0.015	0.865	0.004	0.986	0.000	0.999
19	0.014	0.878	0.003	0.988	0.000	0.999
20	0.012	0.891	0.002	0.991	0.000	0.999
21	0.011	0.902	0.002	0.993	0.000	1.000
22	0.010	0.911	0.001	0.994	0.000	1.000
23	0.009	0.920	0.001	0.995	0.000	1.000
24	0.008	0.928	0.001	0.996	0.000	1.000
25	0.007	0.935	0.001	0.997	0.000	1.000

Standard Deviation of the Forecast Errors

The standard deviation of the τ-period-ahead forecast error $[e_{(\tau)} = \hat{x}_T(\tau) - x_{T+\tau}]$ is

$$\sigma_{e(\tau)} = \sqrt{\frac{2}{2-\alpha}}\sigma$$

where σ is the standard deviation of x. Also, the standard deviation for the τ-period cumulative forecast error is

$$\sigma_{E_\tau} = \sqrt{\frac{\tau^2\alpha + \tau(2-\alpha)}{2-\alpha}}\sigma$$

Table 4-8 lists the ratio σ_{E_τ}/σ for selected values of α and for $\tau = 1$ to 12.

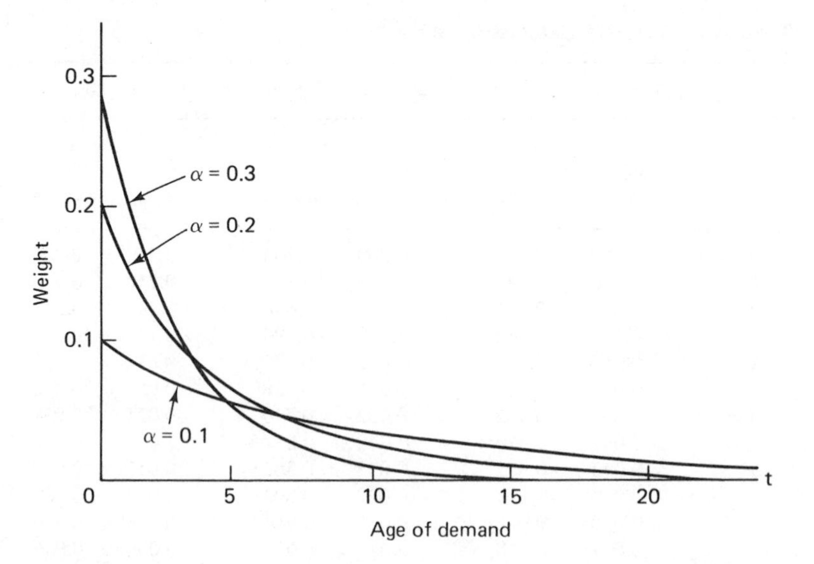

Figure 4-4. Weight assigned to demands in the single smoothing model.

Table 4-8. RATIOS OF σ_{E_τ}/σ FOR THE SINGLE SMOOTHING MODEL

	α					
τ	0.05	0.10	0.20	0.30	0.40	0.50
1	1.01	1.03	1.05	1.08	1.12	1.15
2	1.45	1.49	1.56	1.64	1.73	1.83
3	1.80	1.86	2.00	2.14	2.29	2.45
4	2.10	2.20	2.40	2.61	2.83	3.06
5	2.38	2.51	2.79	3.07	3.35	3.65
6	2.63	2.81	3.16	3.52	3.87	4.24
7	2.87	3.10	3.53	3.96	4.39	4.83
8	3.11	3.37	3.89	4.39	4.90	5.42
9	3.33	3.64	4.24	4.83	5.41	6.00
10	3.55	3.91	4.60	5.26	5.92	6.58
11	3.76	4.17	4.94	5.69	6.42	7.17
12	3.96	4.43	5.29	6.12	6.93	7.75

Example 4.14

Suppose that single smoothing with $\alpha = 0.1$ is being used with an item that has a standard deviation of $\sigma = 20$. Find the standard deviation of the cumulative forecast error when $\tau = 5$. This becomes

$$\sigma_{E_5} = 2.51(20) = 50.26$$

It is also noted that

$$\sigma_{e(\tau)} = 1.03(20) = 20.60 \qquad \text{for } \tau = 1, 2, 3, 4, 5$$

Average Age of Demands

The average age (A) of the demands can be measured as was the case in the moving-average model. Assigning an age of zero to the most current demand, 1 to the prior demand, and so on, then

$$\begin{aligned} A &= 0\alpha + 1\alpha(1-\alpha) + 2\alpha(1-\alpha)^2 + 3\alpha(1-\alpha)^3 + \ldots \\ &= \alpha \sum_{j=0}^{\infty} j(1-\alpha)^j \\ &= \frac{1-\alpha}{\alpha} \end{aligned}$$

Table 4-9 shows how A varies for selected values of α. The lower the value of α, the larger the average age. Notice when $\alpha = 0.1$, the average age of the demands is $A = 9.00$. Also, when $\alpha = 0.2$, the average age becomes $A = 4.00$.

Table 4-9. AVERAGE AGE OF DEMANDS USING THE SINGLE SMOOTHING MODEL

α	0.05	0.10	0.15	0.20	0.25	0.30	0.35	0.40	0.45	0.50
A	19.00	9.00	5.67	4.00	3.00	2.33	1.86	1.50	1.22	1.00

It is sometimes useful to find the value of N from the moving-average model that corresponds to a particular α in the single smoothing model when the average age of the demands is used as a criteria. Recall that in the moving-average model

$$A = \frac{N-1}{2}$$

and in single smoothing

$$A = \frac{1-\alpha}{\alpha}$$

Equating the two relations gives

$$\alpha = \frac{2}{N+1}$$

or

$$N = \frac{2-\alpha}{\alpha}$$

Now if the forecaster wishes to use a value of α that corresponds to a 12-month moving average (with respect to A), he/she chooses

$$\alpha = \tfrac{2}{13} = 0.15$$

Should he/she wish a value of N corresponding to $\alpha = 0.2$, he/she uses

$$N = \frac{1.8}{0.2} = 9$$

MATHEMATICAL BASIS. Assume that the demands are independent with mean μ and standard deviation σ. Now since

$$\hat{x}_T(\tau) = S_T = \alpha \sum_{j=0}^{\infty} (1 - \alpha)^j x_{T-j}$$

then

$$E(\hat{x}_T(\tau)) = \alpha \sum_{j=0}^{\infty} (1 - \alpha)^j \mu = \mu$$

$$E(e_{(\tau)}) = E(\hat{x}_T(\tau) - x_{T+\tau}) = \mu - \mu = 0$$

$$V(\hat{x}_T(\tau)) = \alpha^2 \sum_{j=0}^{\infty} (1 - \alpha)^{2j} \sigma^2 = \frac{\alpha}{2 - \alpha} \sigma^2$$

$$V(e_{(\tau)}) = V(\hat{x}_T(\tau) - x_{T+\tau}) = \frac{\alpha}{2 - \alpha} \sigma^2 + \sigma^2$$

$$= \frac{2}{2 - \alpha} \sigma^2$$

For the τ-period cumulative forecast,

$$\hat{X}_T(\tau) = \tau S_T$$

$$E(\hat{X}_T(\tau)) = \tau \mu$$

$$E(E_\tau) = E(\hat{X}_T(\tau) - x_{T+1} - \ldots - x_{T+\tau})$$

$$= \tau\mu - \tau\mu = 0$$

$$V(\hat{X}_T(\tau)) = \tau^2 V(\hat{x}_T(\tau)) = \frac{\tau^2 \alpha}{2 - \alpha} \sigma^2$$

$$V(E_\tau) = V(\hat{X}_T(\tau) - x_{T+1} - \ldots - x_{T+\tau})$$

$$= \frac{\tau^2 \alpha}{2 - \alpha} \sigma^2 + \tau \sigma^2 = \frac{\tau^2 \alpha + \tau(2 - \alpha)}{2 - \alpha} \sigma^2$$

PROBLEMS

4-1. Consider an item whose demand is horizontal with an average demand of μ and a standard deviation of σ. Suppose that $\hat{\sigma} = 10$ is an estimate of σ and μ will be estimated at each time period using the naive model. If the demand of the most current time period is 50 pieces, find:
a. The forecast of demand for each of the next four time periods.
b. The forecast for the sum of the demands over the next four time periods.
c. The standard deviation of the forecast error for the forecasts given above.

4-2. Suppose that for an item the moving-average model is in use with $N = 4$ and where the prior four demand entries are $x_T = 16$, $x_{T-1} = 9$, $x_{T-2} = 14$, and $x_{T-3} = 11$. Find:
a. The forecast for each of the next two time periods.
b. The cumulative forecast for the next four time periods.
c. The standard deviation of the forecast errors for the forecasts above assuming that $\sigma = 5$.
d. If at the next time period, $x_{T+1} = 12$, then find updated estimates for parts a, b, and c.

4-3. A forecaster at $T = 3$ wishes to apply the moving-average model with $N = 5$ to an item with so far only three demand entries ($x_1 = 8$, $x_2 = 20$, $x_3 = 12$). Also, he/she estimates the standard deviation of the item to be $\hat{\sigma} = 6$.
a. Find the forecast for the next time period and the standard deviation of the forecast error using the early-stage method applied in Example 4.4.
b. If $x_4 = 9$, update the estimates above.
c. Update the estimates if $x_5 = 15$.
d. Update the estimates if $x_6 = 10$.

4-4. Suppose that a forecaster wishes to use the forecast model

$$\hat{x}_T(\tau) = \tfrac{1}{9}x_T + \tfrac{2}{9}x_{T-1} + \tfrac{3}{9}x_{T-2} + \tfrac{2}{9}x_{T-3} + \tfrac{1}{9}x_{T-4}$$

Find the weighted average age (A) of the demands used in the forecast model.

4-5. Suppose that a forecaster is currently using the moving-average model with $N = 6$ to forecast the cumulative demands for each of the next 12 time periods.
a. What percent decrease is observed in the standard deviation of the forecast error when N is increased to 12?
b. How large a value of N is needed to reduce the current standard deviation of the forecast error by 25%? By 35%?

4-6. Consider an item that is being forecast by the single smoothing model with $\alpha = 0.2$. Now if $S_{T-1} = 100$ and $x_T = 80$, find
a. The forecast for each of the next three time periods.
b. The cumulative forecast for the next three time periods.
c. The standard deviation of the errors for the forecasts given above when $\sigma = 20$.
d. Update the results in parts a, b, and c if at the next time period the demand is 90 pieces.

4-7. Suppose that a new item becomes available in the inventory, and single smoothing with $\alpha = 0.2$ is to be used. Give the forecasts that would result for each of the first four time periods when $x_1 = 10$, $x_2 = 20$, $x_3 = 15$, and $x_4 = 22$, and when

a. At the outset ($T = 1$), the initial smoothed entry is $S_0 = x_1$.

b. At the outset ($T = 1$), the mean of the demand is estimated as $\hat{\mu} = 15$.

4-8. Consider a new item that is to be forecasted by single smoothing with $\alpha = 0.2$. If the first eight demand entries are (8, 12, 11, 9, 6, 12, 20, and 14), show the one-period-ahead forecasts that would result for each of these time periods when the following early stage schemes are applied.

a. The initial smoothed value (S_0) is set equal to x_1.

b. The smoothing parameter is set equal to the maximum of (α or $1/t$).

c. At the outset, $\hat{\mu} = 10$ is used as an estimate of the mean demand, and the smoothing parameters are set by the following rules:

$$\alpha_t = \begin{cases} 0.5 & \text{for } t = 1 \\ 0.4 & \text{for } t = 2, 3 \\ 0.3 & \text{for } t = 4, 5, 6 \\ 0.2 & \text{for } t \geq 7 \end{cases}$$

d. The scheme used in Example 4.12 is applied with the first five demand entries and $\alpha = 0.20$ is in use.

4-9. Consider a forecast of an item where $\alpha = 0.3$ is in use and where at the outset, S_0 is set equal to the first demand entry (i.e., $S_0 = x_1$). Show what weights are actually given to each of the demand entries at $T = 1, 2, 3$, and 4. [Note that at $T = 2$, for example,

$$\begin{aligned} S_2 &= \alpha x_2 + (1 - \alpha)S_1 \\ &= \alpha x_2 + \alpha(1 - \alpha)x_1 + (1 - \alpha)^2 S_0 \end{aligned}$$

4-10. In applying single smoothing, how much more weight is given to the most current six demand entries ($t = T$ to $T - 5$) over the next-most-current six entries ($t = T - 6$ to $T - 11$), when:

a. $\alpha = 0.1$ b. $\alpha = 0.2$

c. $\alpha = 0.3$ d. $\alpha = 0.5$

4-11. Suppose that an item is being forecasted by the following model:

$$\begin{aligned} \hat{x}_t(\tau) = {} & 0.05x_T + 0.10x_{T-1} + 0.15x_{T-2} + 0.20x_{T-3} \\ & + 0.20x_{T-4} + 0.15x_{T-5} + 0.10x_{T-6} + 0.05x_{T-7} \end{aligned}$$

Find the value of α in single smoothing that gives the same average age (A) as the forecast model above. What is the closest corresponding value of N in the moving-average model?

4-12. Suppose that the moving-average model with $N = 12$ is in use to forecast the cumulative demand for each of the next six time periods. Find an approximate value of α that gives a standard deviation of the forecast error which corresponds to the moving-average model with $N = 12$.

4-13. Assume that an item is being forecasted by single smoothing with $\alpha = 0.2$ and of interest is the cumulative six-period-ahead forecast.

a. Show the percent change in the standard deviation of the forecast error when the smoothing parameter is increased to $\alpha = 0.3$.

b. What is the associated change when the smoothing parameter is lowered to $\alpha = 0.1$?

The following demand entries are used for Problems 4-14 to 4-16:

t	x_t	t	x_t
1	12	7	16
2	9	8	11
3	10	9	20
4	20	10	14
5	23	11	8
6	19	12	16

4-14. With the above data, find the one-period-ahead forecasts using the naive model.

4-15. With the above data, find the one-period-ahead forecasts using the moving-average model with $N = 4$. Start the first forecast at $t = 4$.

4-16. With the above data find the one-period-ahead forecasts using single smoothing with $\alpha = 0.2$. Start the system by assuming that the mean demand is $\hat{\mu} = 20$.

REFERENCES

[1] Holt, C. C., "Forecasting Seasonal and Trends by Exponentially Weighted Moving Averages." Carnegie Institute of Technology, Pittsburgh, Pa., 1957.

[2] Brown, R. G., *Statistical Forecasting for Inventory Control.* New York: McGraw-Hill Book Company, 1959.

REFERENCES FOR FURTHER STUDY

Brown, R. G., *Smoothing, Forecasting and Prediction of Discrete Time Series.* Englewood Cliffs, N.J.: Prentice-Hall, Inc., 1962.

Brown, R. G., and R. F. Meyer, "The Fundamental Theorem of Exponential Smoothing." *Operations Research*, Vol. 9, 1961.

5

TREND MODELS

Many items in an inventory follow demand patterns where the level is gradually rising or falling at a steady pace from time period to time period. This is characteristic of the trend demand pattern where the level at time t is of the form $\mu_t = a + bt$. In this respect, a represents the intercept or the level at $t = 0$, and b is the slope. The trend models that are used to forecast such items seek a straight-line fit through the demand entries of the past and project the line forward to forecast the demands for future time periods.

Four trend forecasting models are described in this chapter. These are the linear regression model, the double-moving-average model, the double smoothing model, and the single smoothing model with linear trend. The models are also commonly classified as linear or linear trend models.

In the development to follow, it is convenient to denote the level at the most current time period (T) by a_T, where

$$a_T = \mu_T = a + bT$$

With this, the level at the jth time period of the past becomes

$$\mu_{T-j} = a_T - bj \qquad j = 0, 1, 2, \ldots, T - 1$$

and the level for the τth future time period is

$$\mu_{T+\tau} = a_T + b\tau \qquad \tau = 1, 2, 3, \ldots$$

Defining the level in this way simplifies matters later, since the intercept (a_T) always corresponds to the most current time period, T.

The role of the forecasting models is to estimate a_T and b from the history of demands. These estimates are denoted as $\hat{a}_T$ for a_T and $\hat{b}$ for b. With the estimates available, the forecast for the τth future demand is

$$\hat{x}_T(\tau) = \hat{a}_T + \hat{b}\tau$$

The forecast for the sum of the following τ time periods becomes

$$\begin{aligned}\hat{X}_T(\tau) &= \hat{x}_T(1) + \hat{x}_T(2) + \ldots + \hat{x}_T(\tau) \\ &= \tau\hat{a}_T + \frac{\tau(\tau+1)}{2}\hat{b}\end{aligned}$$ [1]

5-1 LINEAR REGRESSION MODEL

Perhaps the oldest method of forecasting the future demands for items with a trend demand pattern is by way of the *linear regression model.* The N most recent demand entries ($x_{T-N+1}, \ldots, x_{T-1}, x_T$) are used with equal weight to seek estimates of a_T and b.

In this model the least-squares method is used to seek the estimate of a_T and b. With this procedure the estimates yield residual errors with the property that the sum of the square of the residual errors is minimum. The residual error at $t = T - j$ is

$$\begin{aligned}e_{T-j} &= \hat{x}_{T-j} - x_{T-j} \\ &= (\hat{a}_T - \hat{b}j) - x_{T-j}\end{aligned}$$

and the sum of squares of the residual errors

$$\begin{aligned}S(e) &= \sum_{j=0}^{N-1} (\hat{x}_{T-j} - x_{T-j})^2 \\ &= \sum_{j=0}^{N-1} (\hat{a}_T - j\hat{b} - x_{T-j})^2\end{aligned}$$

---

[1]Note that $(1 + 2 + 3 + \ldots + \tau) = \dfrac{\tau(\tau+1)}{2}$.

The estimates of a_T and b are now found. These become

$$\hat{b} = \frac{\dfrac{-(N-1)}{2}\sum x_{T-j} + \sum (jx_{T-j})}{\dfrac{N(N-1)^2}{4} - \dfrac{N(N-1)(2N-1)}{6}}$$

$$\hat{a}_T = \frac{\sum x_{T-j} + \hat{b}\,\dfrac{N(N-1)}{2}}{N} = \bar{x} + \hat{b}\,\frac{N-1}{2}$$

where the summations range from $j = 0$ to $j = N - 1$.[2] Using $\hat{a}_T$ and $\hat{b}$, the forecast for the τth future period is obtained from

$$\hat{x}_T(\tau) = \hat{a}_T + \hat{b}\tau$$

Example 5.1

Suppose that $T = 10$, $N = 10$, and forecasts for the next two time periods are needed. Assume that the first 10 demand entries are 60, 70, 85, 60, 88, 68, 106, 75, 86, and 124.

Here $x_1 = 60$ and $x_{10} = 124$. The summations required in seeking the forecasts are:

$$\sum_{j=0}^{9} x_{10-j} = 124 + 86 + 75 + \ldots + 60 = 822$$

$$\sum_{j=0}^{9} jx_{10-j} = 0 \times 124 + 1 \times 86 + 2 \times 75 + \ldots + 9 \times 60 = 3321$$

So now the estimates of b and a_{10} are

$$\hat{b} = \frac{\dfrac{-9}{2}(822) + (3321)}{\dfrac{10(81)}{4} - \dfrac{10(9)(19)}{6}} = 4.58$$

$$\hat{a}_{10} = 82.2 + 4.58(\tfrac{9}{2}) = 102.81$$

The forecast equation is

$$\hat{x}_{10}(\tau) = 102.81 + 4.58\tau$$

and for $\tau = 1$ and 2 the forecasts are

$$\hat{x}_{10}(1) = 102.81 + 4.58 = 107.39$$
$$\hat{x}_{10}(2) = 102.81 + 4.58(2) = 111.97$$

---

[2]Hence,

$$\sum x_{T-j} = x_T + x_{T-1} + \ldots + x_{T-N+1}$$
$$\sum jx_{T-j} = 0x_T + 1x_{T-1} + \ldots + (N-1)x_{T-N+1}$$

Also, the cumulative forecast for $\tau = 1$ and 2 is

$$\hat{X}_{10}(2) = 107.39 + 111.97 = 219.36$$

The example continues in Table 5-1, where a listing of the forecasts for all time periods until $t = 24$ is shown. A plot of the one-period-ahead forecasts is given in Figure 5-1.

Table 5-1. Forecasts Using the Linear Regression Model with $N = 10$

t	x_t	$\hat{a}_t$	$\hat{b}$	$\hat{x}_t(1)$	$\hat{x}_t(2)$	$\hat{X}_t(2)$
1	60					
2	70					
3	85					
4	60					
5	88					
6	68					
7	106					
8	75					
9	86					
10	124	102.8	4.58	107.4	112.0	219.4
11	122	112.1	5.27	117.4	122.7	240.1
12	87	107.9	3.97	111.9	115.9	227.8
13	89	106.5	3.57	110.1	113.7	223.8
14	120	110.7	3.15	113.8	116.9	230.7
15	134	121.9	4.62	126.5	131.2	257.7
16	121	122.1	3.50	125.7	129.2	254.9
17	93	117.5	2.75	120.2	122.9	243.1
18	113	114.2	1.17	115.3	116.5	231.8
19	125	115.1	0.52	115.7	116.2	231.9
20	136	125.4	2.53	127.9	130.4	258.3
21	142	136.6	4.59	141.3	145.9	287.2
22	117	131.2	2.72	133.9	136.9	270.8
23	132	129.7	1.42	131.1	132.5	263.6
24	141	135.2	2.17	137.3	139.5	276.8

Mathematical Basis. The sum of squared residual errors is

$$S(e) = \sum_{j=0}^{N-1} (\hat{a}_T - j\hat{b} - x_{T-j})^2$$

With the least-squares methods, $\hat{a}_T$ and $\hat{b}$ are obtained by taking

$$\frac{\partial S(e)}{\partial \hat{a}_T} = 2 \sum_{j=0}^{N-1} (\hat{a}_T - j\hat{b} - x_{T-j}) \overset{!}{=} 0$$

$$\frac{\partial S(e)}{\partial \hat{b}} = -2 \sum_{j=0}^{N-1} j(\hat{a}_T - j\hat{b} - x_{T-j}) = 0$$

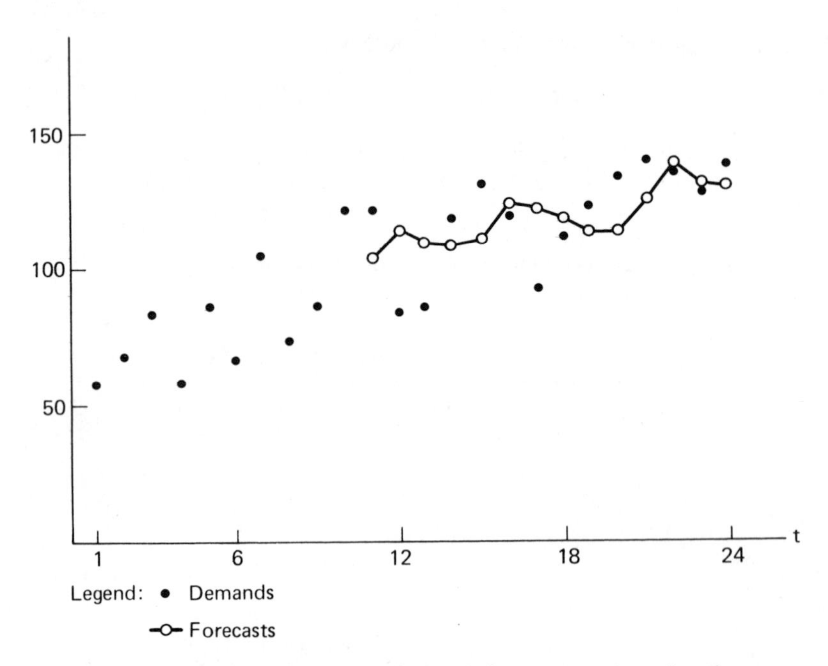

Figure 5-1. Plot of one-period-ahead forecasts using the linear regression model with $N = 10$.

The results yield

$$\sum x_{T-j} = \hat{a}_T N - \hat{b} \sum j$$
$$\sum j x_{T-j} = \hat{a}_T \sum j - \hat{b} \sum j^2$$

where the summations range from $j = 0$ to $N - 1$. Hence,

$$\hat{b} = \frac{\sum j \sum x_{T-j} - N \sum j x_{T-j}}{-(\sum j)^2 + N \sum (j^2)}$$
$$\hat{a}_T = \frac{\sum x_{T-j} + \hat{b} \sum j}{N}$$

where

$$\sum j = \frac{N(N-1)}{2} \quad \text{and} \quad \sum j^2 = \frac{N(N-1)(2N-1)}{6}$$

5-2 DOUBLE-MOVING-AVERAGE MODEL

The *double-moving-average model* is an extension of the moving-average model of Chapter 4 and is used with items that follow a trend demand pattern. Two moving averages (M_T and $M_T^{(2)}$) are measured at time period T and are used to generate the forecasts. The forecaster selects a parameter N, which gives the number of time periods upon which the moving averages are to be based.

The moving averages are calculated from

$$M_T = \frac{x_T + x_{T-1} + \ldots + x_{T-N+1}}{N}$$

$$M_T^{(2)} = \frac{M_T + M_{T-1} + \ldots + M_{T-N+1}}{N}$$

M_T is the average of the N most recent demand entries and $M_T^{(2)}$ is the average of the N most recent averages. To simplify the calculations, the following recursive relations may be used:

$$M_T = M_{T-1} + \frac{x_T - x_{T-N}}{N}$$

$$M_T^{(2)} = M_{T-1}^{(2)} + \frac{M_T - M_{T-N}}{N}$$

The estimates of b and a_T become

$$\hat{b} = \frac{2}{N-1}[M_T - M_T^{(2)}]$$

and

$$\hat{a}_T = 2M_T - M_T^{(2)}$$

These are now applied to yield the forecast relation for the τth future time period, i.e.,

$$\hat{x}_T(\tau) = \hat{a}_T + \hat{b}\tau$$

Example 5.2

Suppose that the double-moving-average model with $N = 6$ is to be used to generate forecasts for $\tau = 1$ and 2. Assume that the current time period is $T = 11$ and the first 11 demand entries are 60, 70, 85, 60, 88, 68, 106, 75, 86, 124, and 122, where $x_1 = 60$ and $x_{11} = 122$.

The moving averages for $t = 6$ to 11 are found as follows:

$$M_6 = \frac{60 + 70 + 85 + 60 + 88 + 68}{6} = 71.83$$

$$M_7 = 71.83 + \frac{106 - 60}{6} = 79.50$$

$$M_8 = 79.50 + \frac{75 - 70}{6} = 80.33$$

$$M_9 = 80.33 + \frac{86 - 85}{6} = 80.50$$

$$M_{10} = 80.50 + \frac{124 - 60}{6} = 91.17$$

$$M_{11} = 91.17 + \frac{122 - 88}{6} = 96.84$$

Now the double moving average at $T = 11$ is

$$M_{11}^{(2)} = \frac{71.83 + 79.50 + 80.33 + 80.50 + 91.17 + 96.84}{6} = 83.36$$

Hence, the estimates of b and a_{11} become

$$\begin{aligned}\hat{b} &= \frac{2}{N-1}[M_{11} - M_{11}^{(2)}] \\ &= \tfrac{2}{5}(96.84 - 83.36) = 5.39 \\ \hat{a}_{11} &= 2M_{11} - M_{11}^{(2)} \\ &= 2(96.84) - 83.36 = 110.32\end{aligned}$$

The forecasts sought are obtained from

$$\hat{x}_{11}(\tau) = 110.32 + 5.39\tau$$

So with $\tau = 1$ and 2, the relation above gives

$$\begin{aligned}\hat{x}_{11}(1) &= 110.32 + 5.39 = 115.71 \\ \hat{x}_{11}(2) &= 110.32 + 5.39(2) = 121.10\end{aligned}$$

Finally, the cumulative forecast with $\tau = 2$ is

$$\hat{X}_{11}(2) = 115.71 + 121.10 = 236.81$$

The example continues in Table 5-2 for the subsequent time periods. A plot of the one-period-ahead forecasts is shown in Figure 5-2.

Mathematical Basis. The averages M_T and $M_T^{(2)}$ are

$$M_T = \frac{1}{N}\sum_{j=0}^{N-1} x_{T-j}$$

$$M_T^{(2)} = \frac{1}{N}\sum_{j=0}^{N-1} M_{T-j}$$

The expected values of the averages above become

$$E(M_T) = \frac{1}{N}\sum_{j=0}^{N-1}(a_T - jb) = a_T - \left(\frac{N-1}{2}\right)b$$

$$E(M_T^{(2)}) = \frac{1}{N}\sum_{j=0}^{N-1}\left[a_T - jb - \left(\frac{N-1}{2}\right)b\right] = a_T - (N-1)b$$

Table 5-2. FORECASTS USING THE DOUBLE-MOVING-AVERAGE MODEL WITH $N = 6$

t	x_t	M_t	$M_t^{(2)}$	$\hat{a}_t$	$\hat{b}$	$\hat{x}_t(1)$	$\hat{x}_t(2)$	$\hat{X}_t(2)$
1	60							
2	70							
3	85							
4	60							
5	88							
6	66	71.83						
7	106	79.50						
8	75	80.33						
9	86	80.50						
10	124	91.17						
11	122	96.84	83.36	110.32	5.39	115.7	121.1	236.8
12	87	100.00	88.06	111.94	4.78	116.7	121.5	238.2
13	89	97.17	91.00	103.34	2.47	105.8	108.3	214.1
14	120	104.67	95.06	114.28	3.84	118.1	121.9	240.0
15	134	112.67	100.42	124.92	4.90	129.8	134.7	264.5
16	121	112.17	103.92	120.42	3.30	123.7	127.0	250.7
17	93	107.33	105.67	180.99	0.66	109.7	110.4	220.1
18	113	111.67	107.62	115.72	1.62	117.3	118.9	236.2
19	125	117.67	113.81	121.53	1.54	123.1	124.6	247.7
20	136	120.33	113.64	127.02	2.68	129.7	132.4	262.1
21	142	121.67	115.14	128.20	2.61	130.8	133.4	264.2
22	117	121.00	116.61	125.39	1.76	127.2	129.0	256.2
23	132	127.50	119.97	135.03	3.01	138.0	141.0	279.0
24	141	132.17	123.39	140.95	3.51	144.5	148.0	292.5

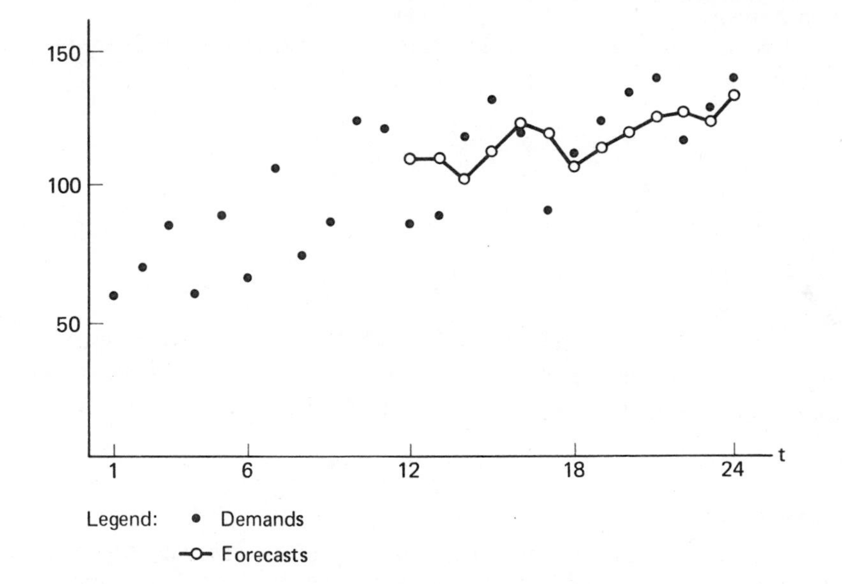

Figure 5-2. Plot of one-period-ahead forecasts using the double moving average model with $N = 6$.

Using the method of moments, M_T, $M_T^{(2)}$, $\hat{a}_T$, and $\hat{b}$ are substituted for $E(M_T)$, $E(M_T^{(2)})$, a_T, and b, respectively. This gives

$$M_T = \hat{a}_T - \left(\frac{N-1}{2}\right)\hat{b}$$

$$M_T^{(2)} = \hat{a}_T - (N-1)\hat{b}$$

Hence,

$$\hat{a}_T = 2M_T - M_T^{(2)}$$

$$\hat{b} = \frac{2}{N-1}[M_T - M_T^{(2)}]$$

5-3 DOUBLE SMOOTHING MODEL

Double smoothing [1] is an extension of single smoothing and is applied with items that follow a trend demand pattern. A smoothing parameter α is chosen and is used to assign weights to the demand entries of the past. As in single smoothing, a higher weight is assigned to the more current demand entries.

A distinct advantage of double smoothing over the linear regression and double-moving-average models is that far fewer data are required to maintain the system. At the current time period (T), the data needed by the forecaster are x_T, $\hat{a}_{T-1}$, and $\hat{b}_{T-1}$. Here, $\hat{a}_{T-1}$ and $\hat{b}_{T-1}$ are the estimates of a and b, respectively, as of $t = T - 1$.

With these data, the coefficients are updated by the following two relations:

$$\hat{a}_T = x_T + (1-\alpha)^2 e_T$$

$$\hat{b}_T = \hat{b}_{T-1} - \alpha^2 e_T$$

where $e_T = (\hat{x}_{T-1}(1) - x_T)$ is the one-period-ahead forecast error at time T and

$$\hat{x}_{T-1}(1) = \hat{a}_{T-1} + \hat{b}_{T-1}$$

Note that the current one-period-ahead forecast error is used as the principal basis from which the coefficients a and b are updated. As before, the current forecast for the τth future time period becomes

$$\hat{x}_T(\tau) = \hat{a}_T + \hat{b}_T\tau$$

Example 5.3

Suppose that double smoothing with $\alpha = 0.1$ is in use to forecast an item's demands. Assume that at time period T the prior estimates of the coefficients are $\hat{a}_{T-1} = 10$ and $\hat{b}_{T-1} = 1$. Now if the current demand is $x_T = 9$, find the forecasts for each of the next three time periods.

Note that the one-period-ahead forecast at $t = T - 1$ is $\hat{x}_{T-1}(1) = 10 + 1 = 11$. This gives $e_T = (11 - 9) = 2$, whereby the updated coefficient estimates are

$$\hat{a}_T = 9 + (0.9)^2(2) = 10.62$$

$$\hat{b}_T = 1 - (0.1)^2(2) = 0.98$$

Hence,

$$\hat{x}_T(\tau) = 10.62 + 0.98\tau$$

$$\hat{x}_T(1) = 10.62 + 0.98 = 11.60$$

$$\hat{x}_T(2) = 10.62 + 0.98(2) = 12.58$$

$$\hat{x}_T(3) = 10.62 + 0.98(3) = 13.56$$

Initial Forecast

As in single smoothing, a difficulty arises in getting started since at $T = 1$, values of $\hat{a}_0$ and $\hat{b}_0$ are needed but are not available to the forecaster. Three methods by which the system can be initialized are described below. These are:

1. Initialize using x_1.
2. Estimate the level and slope at $t = 0$.
3. Fit a straight line through the past demand entries.

In the first two methods, no demand entries prior to the current time period ($T = 1$) are used to initialize the system. The third method uses demand entries of the past in order to get started.

INITIALIZE USING x_1. If at $t = 1$, the only information known to the forecaster is the current demand entry x_1, this observation alone is used to yield the initial estimates. Here $\hat{a}_1$ is set equal to x_1, and $\hat{b}_1$ is set to zero. This gives $\hat{x}_1(\tau) = x_1$ as the initial forecast. Thereafter, as each new demand entry becomes available, new estimates of $\hat{a}_T$ and $\hat{b}_T$ are found in the standard manner.

This method of initializing should be used with caution since the initial estimate of the slope is zero and the initial estimate of the level is x_1. Both of these estimates may differ widely from their true values. At $T = 1$, the slope is probably a positive value since the demand pattern of a new item is hopefully on the rise. When a small value of the smoothing parameter (α) is used, the future forecasts are highly influenced by the initial estimates $\hat{a}_1$ and $\hat{b}_1$. A larger setting of α seems reasonable in the early stages since the coefficient estimates will rely more heavily on the current demands. However, the larger setting of α will cause wider fluctuations in forecasts from time period to time period.

Example 5.4

Suppose that $x_1 = 3$ and at the outset the coefficients are estimated by $\hat{a}_1 = x_1 = 3$ and $\hat{b}_1 = 0$. Table 5-3 shows how this example proceeds for six time periods when $\alpha = 0.1$. In Table 5-4 are corresponding results when $\alpha = 0.3$ for $t = 1, 2,$ and 3; $\alpha = 0.2$ for $t = 4$ and 5; and $\alpha = 0.1$ for $t = 6$. Notice in this example how the forecasts are following the demands more closely with the higher α settings (Table 5-4) than with the consistently smaller value of $\alpha = 0.1$. This is true even though the initial estimate of the slope is obviously too low.

Table 5-3. Example 5.4 Results Using $\alpha = 0.10$

t	x_t	$\hat{a}_t$	$\hat{b}_t$	$\hat{x}_t(1)$
1	3	3.00	0.00	3.00
2	8	3.95	0.05	4.00
3	12	5.52	0.13	5.65
4	20	8.39	0.27	8.66
5	18	10.42	0.37	10.79
6	25	13.49	0.51	14.00

Table 5-4. Example 5.4 Results Using $\alpha = 0.3$ for $t = 1, 2, 3$; $\alpha = 0.2$ for $t = 4, 5$; and $\alpha = 0.1$ for $t = 6$

t	x_t	α	$\hat{a}_t$	$\hat{b}_t$	$\hat{x}_t(1)$
1	3	0.3	3.00	0.00	3.00
2	8	0.3	5.55	0.45	6.00
3	12	0.3	9.06	0.99	10.05
4	20	0.2	3.63	1.39	15.02
5	18	0.2	16.09	1.51	17.60
6	25	0.1	19.01	1.58	20.59

Estimate the Level and Slope. A second method of initializing is possible when estimates of $\hat{a}_0$ and $\hat{b}_0$ are available to the forecaster. With these, the updated coefficient estimates at $t = 1$ are

$$\hat{a}_1 = x_1 + (1 - \alpha)^2 e_1$$
$$\hat{b}_1 = \hat{b}_0 - \alpha^2 e_1$$

where

$$e_1 = \hat{x}_0(1) - x_1 \quad \text{and} \quad \hat{x}_0(1) = \hat{a}_0 + \hat{b}_0$$

Example 5.5

Suppose that the initial estimates of the level and slope for an item are $\hat{a}_0 = 0$ and $\hat{b}_0 = 4$. Assuming that $x_1 = 3$ and double smoothing is used with $\alpha = 0.1$, find $\hat{x}_1(\tau)$.

Carrying out the procedure above yields $e_1 = 4 - 3 = 1$ and

$$\hat{a}_1 = 3 + 0.81(1) = 3.81$$
$$\hat{b}_1 = 4 - 0.01(1) = 3.99$$
$$\hat{x}_1(\tau) = 3.81 + 3.99\tau$$

The results continue in Table 5-5 for six time periods. Notice how much more accurate the one-period-ahead forecasts are in this example than are the corresponding results of Tables 5-3 and 5-4.

Table 5-5. EXAMPLE 5.5 RESULTS

t	x_t	$\hat{a}_t$	$\hat{b}_t$	$\hat{x}_t(1)$
1	3	3.81	3.99	7.80
2	8	7.83	3.99	11.82
3	12	11.86	3.99	15.85
4	20	16.64	4.04	20.68
5	18	20.17	4.01	24.18
6	25	24.33	4.02	28.35

FITTING A STRAIGHT LINE THROUGH PAST DEMANDS. Demand entries from the past are often available for an item when an initial double smoothing forecast is sought. This occurs, for example, when the item has been in the inventory for some time, although no forecasts have previously been generated. With these data, $x_1, x_2, \ldots, x_T$, say, the forecaster can generate the initial estimates $\hat{a}_T$ and $\hat{b}_T$ that are needed. Thereafter, double smoothing is carried forward in the ordinary manner.

Various methods are possible to arrive at the initial estimates of the coefficients. One way is to plot the demands of the past and hand-fit a straight line to obtain the estimates $\hat{a}_T$ and $\hat{b}_T$. Another is to fit the line by use of linear regression or the double-moving-average model. In any event, the estimates $\hat{a}_T$ and $\hat{b}_T$ are obtained and used to generate the forecasts at time T.

At the next time period ($t = T + 1$), the demand entry x_{T+1} becomes available and is used to find $\hat{a}_{T+1}$ and $\hat{b}_{T+1}$. Thereupon, the new forecasts are generated as shown earlier.

Example 5.6

Suppose that $T = 10$ and the first 10 demand entries are 10, 20, 35, 10, 38, 18, 56, 25, 36, and 74. Find the initial estimates $\hat{a}_{10}$ and $\hat{b}_{10}$ by linear regression and use these results to find $\hat{x}_{10}(1)$.

Applying the linear regression method of Section 5-1 to the 10 observations, the results yield

$$\hat{a}_{10} = 52.81$$
$$\hat{b}_{10} = 4.58$$

With these results, the one-period-ahead forecast is

$$\hat{x}_{10}(1) = 52.81 + 4.58 = 57.39$$

Continuing with this example, suppose that at the next time period ($T = 11$) the demand is $x_{11} = 65$ and that double smoothing with $\alpha = 0.1$ is to be used. With $e_{11} = -7.61$, then

$$\hat{a}_{11} = 65 + 0.81(-7.61) = 58.84$$
$$\hat{b}_{11} = 4.58 - 0.01(-7.61) = 4.66$$

and

$$\hat{x}_{11}(\tau) = 58.84 + 4.66\tau$$

Example 5.7

Consider the 24 demand entries shown in Table 5-1. Suppose that at the outset the forecaster estimates the coefficient as of $t = 0$ to be $\hat{a}_0 = 60$ and $\hat{b}_0 = 3$. Now using $\alpha = 0.1$, find the one-period-ahead forecasts for each of the 24 time periods.

The results for $t = 1$ to 24 are listed in Table 5-6 and the plot of the one-period-ahead forecasts is shown in Figure 5-3.

Table 5-6. One-Period-Ahead Forecasts Using Double Smoothing with $\alpha = 0.1$

t	x_t	$\hat{a}_t$	$\hat{b}_t$	$\hat{x}_t(1)$
1	60	62.43	2.97	65.4
2	70	66.27	3.02	69.3
3	85	72.28	3.17	75.5
4	60	72.52	3.02	75.5
5	88	77.90	3.14	81.0
6	68	78.57	3.01	81.6
7	106	86.21	3.26	89.5
8	75	86.72	3.11	89.8
9	86	89.10	3.07	92.2
10	124	98.23	3.39	101.6
11	122	105.49	3.60	109.1
12	87	104.90	3.38	108.3
13	89	104.60	3.18	107.8
14	120	110.11	3.30	113.4
15	134	117.32	3.51	120.8
16	121	120.87	3.51	124.4
17	93	118.41	3.20	121.6
18	113	119.98	3.11	123.1
19	125	123.45	3.13	126.6
20	136	128.36	3.22	131.6
21	142	133.57	3.33	136.9
22	117	133.12	3.13	136.3
23	132	135.44	3.09	138.5
24	141	139.00	3.11	142.1

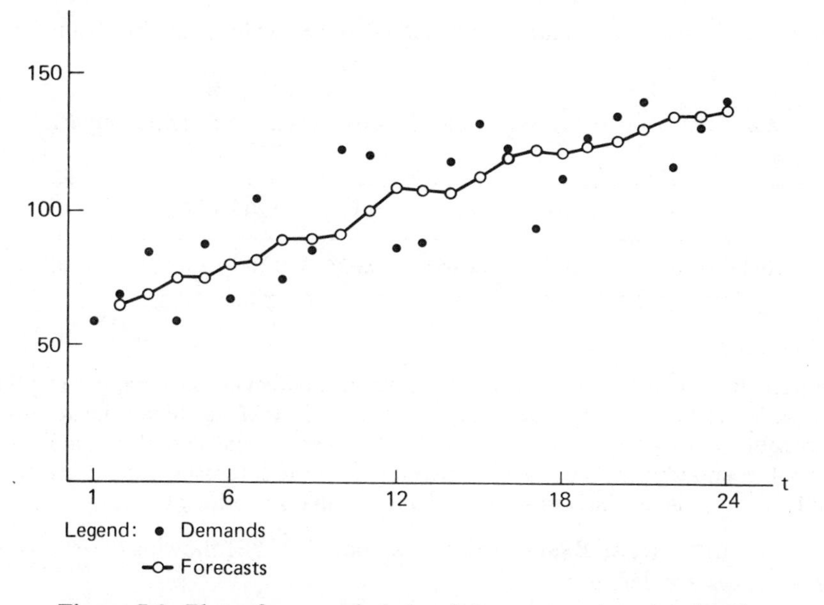

Figure 5-3. Plot of one-period-ahead forecasts using the double smoothing model with $\alpha = 0.1$.

The Smoothing Parameter

In this section a description is given of choosing the smoothing parameter. To begin, it can be shown that the level at time T for the single smoothing model is estimated by

$$\hat{a}_T = x_T + (1 - \alpha)[\hat{x}_{T-1}(1) - x_T]$$

The corresponding level using double smoothing is

$$\hat{a}_T = x_T + (1 - \alpha)^2[\hat{x}_{T-1}(1) - x_T]$$

Note that the preceding two relations for $\hat{a}_T$ are quite similar. These relations are used to choose a value for α in double smoothing which corresponds to an equivalent smoothing parameter (say α_0) in single smoothing. For this purpose the forecaster seeks an α where

$$(1 - \alpha)^2 = 1 - \alpha_0$$

The results give

$$\alpha = 1 - \sqrt{1 - \alpha_0}$$

Table 5-7 shows how α and α_0 are related for α_0 values ranging from 0.05 to 0.30.

Table 5-7. EQUIVALENT SMOOTHING PARAMETERS IN SINGLE AND DOUBLE SMOOTHING

Single smoothing	0.05	0.10	0.15	0.20	0.25	0.30
Double smoothing	0.025	0.051	0.078	0.106	0.134	0.163

Example 5.8

Suppose that the forecaster has some items where single smoothing is applicable but also wishes to apply double smoothing to other items. If he has been using $\alpha = 0.1$ successfully in single smoothing and would like an equivalent smoothing parameter in double smoothing, then, by way of Table 5-7, the forecaster should use $\alpha = 0.051$ for those items that are associated with double smoothing forecasts.

MATHEMATICAL BASIS. In double smoothing, the following two smoothing averages are defined:

(1)
$$S_T = \alpha x_T + (1 - \alpha)S_{T-1}$$
$$S_T^{(2)} = \alpha S_T + (1 - \alpha)S_{T-1}^{(2)}$$

These are rewritten as

(2)
$$S_T = \alpha \sum_{j=0}^{\infty} (1 - \alpha)^j x_{T-j}$$
$$S_T^{(2)} = \alpha \sum_{j=0}^{\infty} (1 - \alpha)^j S_{T-j}$$

The expected values of the averages are

(3)
$$E(S_T) = \alpha \sum_{j=0}^{\infty} (1 - \alpha)^j (a_T - jb)$$
$$= a_T - \frac{1 - \alpha}{\alpha} b$$
$$E(S_T^{(2)}) = \alpha \sum_{j=0}^{\infty} (1 - \alpha)^j \left(a_T - \frac{1 - \alpha}{\alpha} b - jb\right)$$
$$= a_T - \frac{2(1 - \alpha)}{\alpha} b$$

Using the method of moments, S_T, $S_T^{(2)}$, $\hat{a}_T$, and $\hat{b}$ are substituted for $E(S_T)$, $E(S_T^{(2)})$, a_T, and b to yield

(4)
$$S_T = \hat{a}_T - \frac{1-\alpha}{\alpha}\hat{b}$$
$$S_T^{(2)} = \hat{a}_T - \frac{2(1-\alpha)}{\alpha}\hat{b}$$

Hence,

(5)
$$\hat{a}_T = 2S_T - S_T^{(2)}$$
$$\hat{b} = \frac{\alpha}{1-\alpha}(S_T - S_T^{(2)})$$

Note also that

$$\begin{aligned}\hat{a}_T &= 2S_T - S_T^{(2)} \\ &= 2[\alpha x_T + (1-\alpha)S_{T-1}] - \alpha^2 x_T - \alpha(1-\alpha)S_{T-1} - (1-\alpha)S_{T-1}^{(2)}\end{aligned}$$

Now, in the above, substitute

$$S_{T-1} = \hat{a}_{T-1} - \frac{1-\alpha}{\alpha}\hat{b}_{T-1}$$
$$S_{T-1}^{(2)} = \hat{a}_{T-1} - 2\frac{1-\alpha}{\alpha}\hat{b}_{T-1}$$

where $\hat{b}_{T-1}$ is the estimate of b at $t = T-1$. This gives

$$\begin{aligned}\hat{a}_T &= x_T + (1-\alpha)^2(\hat{a}_{T-1} + \hat{b}_{T-1} - x_T) \\ &= x_T + (1-\alpha)^2 e_T\end{aligned}$$

In the same manner,

$$\begin{aligned}\hat{b}_T &= \frac{\alpha}{1-\alpha}(S_T - S_T^{(2)}) \\ &= \frac{\alpha}{1-\alpha}[\alpha x_T + (1-\alpha)S_{T-1} - \alpha^2 x_T - \alpha(1-\alpha)S_{T-1} - (1-\alpha)S_{T-1}^{(2)}] \\ &= \alpha^2 x_T + \alpha(1-\alpha)S_{T-1} - \alpha S_{T-1}^{(2)} \\ &= \hat{b}_{T-1} - \alpha^2(\hat{a}_{T-1} + \hat{b}_{T-1} - x_T) \\ &= \hat{b}_{T-1} - \alpha^2 e_T\end{aligned}$$

5-4 SINGLE SMOOTHING MODEL WITH LINEAR TREND

Another trend model that applies smoothing techniques is called *single smoothing with linear trend* [2]. As before, a smoothing parameter α $(0 < \alpha < 1)$ is chosen and is used to assign weights to the past demands. This model is similar to the double smoothing model since the smoothing is carried on with the level and slope directly.

In this model $\hat{a}_T$ and $\hat{b}_T$ denote the current estimates of the level and the slope, respectively. These are updated with the following relations:

$$\hat{a}_T = \alpha x_T + (1 - \alpha)(\hat{a}_{T-1} + \hat{b}_{T-1})$$

$$\hat{b}_T = \alpha(\hat{a}_T - \hat{a}_{T-1}) + (1 - \alpha)\hat{b}_{T-1}$$

The forecast for future time period $T + \tau$ is

$$\hat{x}_T(\tau) = \hat{a}_T + \hat{b}_T\tau$$

Example 5.9

Suppose that $\alpha = 0.1$, $\hat{a}_{T-1} = 100$, $\hat{b}_{T-1} = 4$, and $x_T = 95$. Find the forecasts for time periods $T + 1$ and $T + 2$.

The level and slope are updated by

$$\hat{a}_T = 0.1(95) + 0.9(104) = 103.10$$

$$\hat{b}_T = 0.1(103.1 - 100) + 0.9(4) = 3.91$$

Hence, the forecasts needed are

$$\hat{x}_T(1) = 103.1 + 3.91 = 107.01$$

$$\hat{x}_T(2) = 103.1 + 3.91(2) = 110.92$$

Initial Forecasts

This model has the same difficulties in seeking the initial forecasts as double smoothing, since at $t = 1$, there are no estimates of a_0 and b_0. As before, three methods are described to yield the initial forecasts and are essentially the same as are cited in double smoothing.

The first method begins by setting $\hat{a}_1 = x_1$ and $\hat{b}_1 = 0$. So, as before, at $t = 1$ the slope is zero and the level is the first demand entry x_1. Hence, $\hat{x}_1(\tau) = \hat{a}_1 + \hat{b}_1\tau$.

The second method requires the forecaster to arrive at estimates of a_0 and b_0. These estimates $(\hat{a}_0, \hat{b}_0)$ are used with x_1 to find $\hat{a}_1$ and $\hat{b}_1$ as follows:

$$\hat{a}_1 = \alpha x_1 + (1 - \alpha)(\hat{a}_0 + \hat{b}_0)$$

$$\hat{b}_1 = \alpha(\hat{a}_1 - \hat{a}_0) + (1 - \alpha)\hat{b}_0$$

So now

$$\hat{x}_1(\tau) = \hat{a}_1 + \hat{b}_1\tau$$

Finally, in the third method the current time period is T and the past known demands are $(x_1, \ldots, x_T)$. Linear regression may be used to find estimates of a_T and b_T by fitting a straight line through the past demands. The initial forecast is $\hat{x}_T(\tau) = \hat{a}_T + \hat{b}_T\tau$. At the subsequent time period $(t = T + 1)$, the coefficients are updated in the standard manner, i.e.,

$$\hat{a}_{T+1} = \alpha x_{T+1} + (1 - \alpha)(\hat{a}_T + \hat{b}_T)$$

$$\hat{b}_{T+1} = \alpha(\hat{a}_{T+1} - \hat{a}_T) + (1 - \alpha)\hat{b}_T$$

Example 5.10

Once again consider the demands listed in Table 5-1. Assume at the outset that the forecaster estimates $\hat{a}_0 = 60$ and $\hat{b}_0 = 3$ and will use $\alpha = 0.1$ in seeking one-period-ahead forecasts. The one-period-ahead forecasts for the 24 time periods that result are listed in Table 5-8 and are plotted in Figure 5-4.

Table 5-8. ONE-PERIOD-AHEAD FORECASTS USING THE SINGLE SMOOTHING MODEL WITH LINEAR TREND AND $\alpha = 0.1$

t	x_t	$\hat{a}_t$	$\hat{b}_t$	$\hat{x}_t(1)$
1	60	62.70	2.97	65.67
2	70	66.10	3.01	69.12
3	85	70.70	3.17	73.88
4	60	72.49	3.03	75.52
5	88	76.77	3.16	79.93
6	68	78.74	3.04	81.77
7	106	84.20	3.28	87.48
8	75	86.23	3.16	89.39
9	86	89.05	3.12	92.17
10	124	95.35	3.44	98.79
11	122	101.11	3.67	104.79
12	87	103.01	3.49	106.50
13	89	104.75	3.32	108.07
14	120	109.27	3.44	112.71
15	134	114.83	3.65	118.49
16	121	118.74	3.68	122.42
17	93	119.47	3.38	122.86
18	113	121.87	3.28	125.16
19	125	125.14	3.28	128.42
20	136	129.18	3.36	132.54
21	142	133.49	3.45	136.94
22	117	134.94	3.25	138.20
23	132	137.58	3.19	140.77
24	141	140.79	3.19	143.99

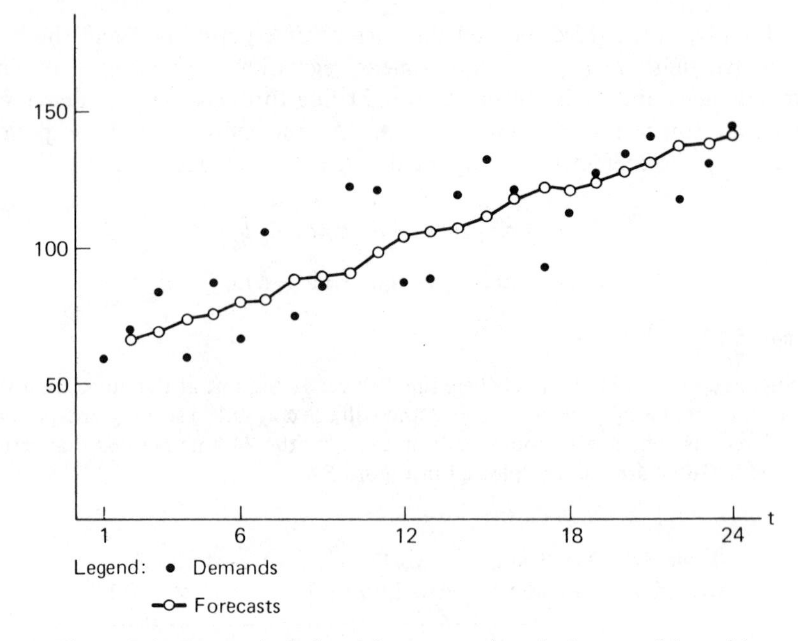

Figure 5-4. One-period-ahead forecasts using single smoothing with linear trend and $\alpha = 0.1$.

Mathematical Basis. The smoothed value for the level can be written as

$$\begin{aligned}\hat{a}_T &= \alpha x_T + (1-\alpha)(\hat{a}_{T-1} + \hat{b}_{T-1}) \\ &= \alpha \sum_{j=0}^{\infty} (1-\alpha)^j x_{T-j} + \sum_{j=1}^{\infty} (1-\alpha)^j b_{T-j}\end{aligned}$$

Taking expected values,

$$\begin{aligned}E(\hat{a}_T) &= \alpha \sum_{j=0}^{\infty} (1-\alpha)^j (a_T - jb) + b \sum_{j=1}^{\infty} (1-\alpha)^j \\ &= a_T\end{aligned}$$

The corresponding results for the slope are

$$\begin{aligned}\hat{b}_T &= \alpha(\hat{a}_T - \hat{a}_{T-1}) + (1-\alpha)\hat{b}_{T-1} \\ &= \alpha\hat{a}_T - \alpha^2 \sum_{j=0}^{\infty} (1-\alpha)^j \hat{a}_{T-j-1} \\ E(\hat{b}_T) &= \alpha a_T - \alpha^2 \sum_{j=0}^{\infty} (1-\alpha)^j [a_T - (j+1)b] \\ &= b\end{aligned}$$

Hence, the coefficients and the forecasts are unbiased.

PROBLEMS

5-1. Consider an item with the following five demands ($x_1 = 5$, $x_2 = 8$, $x_3 = 15$, $x_4 = 12$, and $x_5 = 14$). Find the forecast for each of the next three time periods using the linear regression model with $N = 5$. Give the cumulative forecast for this time horizon.

5-2. Using the results of Problem 5-1, find the fitted values of x_t for $t = 1, \ldots, 5$. Now calculate the residual errors and the sum of squared errors, $S(e)$.

5-3. Use the data of Table 5-1 to find $\hat{a}_{10}$ and $\hat{b}$ at $T = 10$ by using the linear regression model with $N = 8$. Find the forecasts for $t = 11$ and 12 and compare the results with those in the table.

5-4. Using the data of Problem 5-1, find the forecast for time periods 6, 7, and 8 when the double-moving-average model with $N = 3$ is in use.

5-5. Use the results of Problem 5-4 to show the corresponding updated forecasts at $T = 6$ when $x_6 = 18$.

5-6. Using the data of Table 5-2, assume that you are at time period $t = 11$. Now apply the double-moving-average model with $N = 5$ to find the forecasts for time periods $t = 12$ and 13.

5-7. Assume that you are using the double-moving-average model with $N = 3$. Show what weights are given to each of the past demand entries in finding $M_T^{(2)}$.

5-8. Suppose that an item is being forecast by the double smoothing model with $\alpha = 0.2$, and the most current demand entry is 10 pieces. Now if the estimates of a and b of the just-prior time period are $\hat{a}_{T-1} = 8$ and $\hat{b}_{T-1} = 1$, find the forecasts for the subsequent two time periods ($T + 1$ and $T + 2$).

5-9. A new item is introduced into the inventory and forecast by the double smoothing model with $\alpha = 0.3$. With no prior information known on the item, the initial forecast uses only the first demand entry of $x_1 = 10$ to initialize the system and find the one-period-ahead forecast. Show how the one-period-ahead forecasts are updated at $t = 2$, 3, and 4 when the corresponding demand entries are $x_2 = 12$, $x_3 = 15$, and $x_4 = 19$.

5-10. Using the data of Problem 5-9, find the one-period-ahead forecasts when at the outset, estimates of the level and slope are available for $t = 0$, and are $\hat{a}_0 = 5$ and $\hat{b}_0 = 4$, respectively. As before, double smoothing with $\alpha = 0.3$ is in use.

5-11. Suppose that an item is to be forecasted for the first time at $T = 9$. Assume that the first 9 entries of this item are the following: 1, 3, 5, 4, 6, 12, 10, 8, and 13.

a. Apply the double-moving-average model with $N = 5$ to find the initial estimates of a and b as of $t = 9$. Use this result to forecast the next two periods' demand.

b. Now use double smoothing with $\alpha = 0.1$ to update the estimates at $t = 10$ when the new demand entry is $x_{10} = 14$.

5-12. Find the smoothing parameter in single smoothing that corresponds to a smoothing parameter of $\alpha = 0.15$ in double smoothing. If $\alpha = 0.15$ in single smoothing, what is the corresponding smoothing parameter to use in double smoothing?

5-13. Double smoothing is often described in the following manner. Two smoothed averages (S_t and $S_t^{(2)}$) are updated at each time period as follows:

$$S_t = \alpha x_t + (1 - \alpha)S_{t-1}$$

$$S_t^{(2)} = \alpha S_t + (1 - \alpha)S_{t-1}^{(2)}$$

These yield the coefficient estimates

$$\hat{a}_t = 2S_t - S_t^{(2)}$$

$$\hat{b} = \frac{\alpha}{1 - \alpha}(S_t - S_t^{(2)})$$

whereby the forecast equation is

$$\hat{x}_t(\tau) = \hat{a}_t + \hat{b}\tau$$

Assume now that an item is being forecast by this method with $\alpha = 0.1$. Suppose that at the current time period T, $S_{T-1} = 48.8$, $S_{T-1}^{(2)} = 12.3$, and $x_T = 100$. Find $\hat{a}_T$, $\hat{b}$, and the forecast for the next two time periods.

5-14. Suppose that an item is being forecast by the single smoothing model with linear trend, using a smoothing parameter of $\alpha = 0.2$. Assume that at the current time period T, the most recent demand entry is $x_T = 60$, and that the estimates of the level and slope at $T - 1$ are $\hat{a}_{T-1} = 50$ and $\hat{b}_{T-1} = 5$. Find the current forecast for the one-period-ahead demand.

5-15. A new item is introduced into the inventory and is to be forecasted with the single smoothing model with linear trend, using $\alpha = 0.3$. With no prior information known for the item, the forecaster decides to initialize the forecasts using the first period's demand of $x_1 = 20$.

a. Find the forecasts at $t = 1$ for the next four time periods.

b. If at $t = 2$, the demand entry is $x_2 = 25$, find the updated forecasts for the next four time periods.

5-16. Consider an item with the following 10 prior demands: 1, 2, 2, 3, 5, 4, 5, 4, 6, and 7. Assume that no prior estimates of the level and slope are available for the item, but now forecasts for the next three time periods are needed:

a. Using the linear regression model with $N = 10$, find estimates of a_T and b and generate the forecasts needed as of $T = 10$.

b. Assume at the next time period ($T = 11$) that the demand is $x_{11} = 8$. Using the single smoothing with linear trend model and $\alpha = 0.2$, find the update coefficients and the corresponding forecasts for the next three time periods.

5-17. Using $\beta = 1 - \alpha$, show in the double smoothing model that S_T and $S_T^{(2)}$ as defined in Problem 5-13 become

$$S_T = \alpha \sum_{j=0}^{\infty} \beta^j x_{T-j}$$

$$S_T^{(2)} = \alpha^2 \sum_{j=0}^{\infty} (j+1)\beta^j x_{T-j}$$

5-18. Note in double smoothing that S_T and $S_T^{(2)}$ are found given $\hat{a}_T$ and $\hat{b}_T$ by the relations

$$S_T = \hat{a}_T - \frac{1-\alpha}{\alpha}\hat{b}_T$$

$$S_T^{(2)} = \hat{a}_T - 2\frac{1-\alpha}{\alpha}\hat{b}_T$$

Now if $\hat{a}_T = 100$, $\hat{b}_T = 2$, and $\alpha = 0.2$, find the corresponding smoothed averages (S_T and $S_T^{(2)}$). Do the same using $\alpha = 0.1$ and compare the results.

REFERENCES

[1] BROWN, R. G., *Smoothing, Forecasting and Prediction of Discrete Time Series.* Englewood Cliffs, N.J.: Prentice-Hall, Inc., 1962, pp. 128–132.

[2] MONTGOMERY, D. C., "An Application of Statistical Forecasting Techniques in an Inventory Control Policy." *Production and Inventory Management*, First Quarter, 1969, pp. 66–74.

REFERENCE FOR FURTHER STUDY

BROWN, R. G., *Statistical Forecasting for Inventory Control.* New York: McGraw-Hill Book Company, 1959.

6

QUADRATIC MODELS

The level for some items in the inventory may follow a demand pattern that is gradually shifting in a nonlinear manner over time. In these situations the quadratic demand pattern is often suitable for use in defining the level at time t, whereas $\mu_t = a + bt + ct^2$. For this demand pattern the component $(a + bt)$ is linear and the component (ct^2) is nonlinear. The total effect is near linear when the coefficient c is close to zero and is more nonlinear as c deviates from zero. Example plots of the quadratic demand pattern are given in Figure 3-3.

Two quadratic forecasting models are described in this chapter, the quadratic regression and triple smoothing models. The quadratic regression model is an extension of the linear regression model of Chapter 5, and triple smoothing is an extension of the double smoothing model of that chapter.

It is noted that the double-moving-average model of Chapter 5 can also be extended to a triple-moving-average model for use with quadratic demand patterns. However, because the model requires many data and has found limited use, it is not formulated in this chapter.

6-1 QUADRATIC REGRESSION MODEL

In using the quadratic regression model, the forecaster first specifies the number of demand entries (N) he/she wishes to use in the fitting process. Thereupon, the N most recent demand entries are collected and for notational ease in this model, T, the current time period, is set equal to N. Hence, the demand entries are $x_1, x_2, \ldots, x_T$, with x_T the most current entry.

It is also convenient in this model to transform the time axis from t to t' where over the past N time periods, t' ranges from $-\frac{N-1}{2}$ to $\frac{N-1}{2}$. In this way the corresponding values of t and t' are the following:

t	1	2	...	$T-1$	T
t'	$-\frac{N-1}{2}$	$-\frac{N-3}{2}$	...	$\frac{N-3}{2}$	$\frac{N-1}{2}$

For example, with $N = 3$,

t	1	2	3
t'	−1	0	1

and with $N = 4$,

t	1	2	3	4
t'	−1.5	−0.5	0.5	1.5

The principal reason for this transformation is to place zero at the center of the t' axis.

The estimates of the coefficients a, b, and c are obtained by the following relations, where the summations range from $t' = -\frac{N-1}{2}$ to $\frac{N-1}{2}$:

$$\hat{b} = \frac{\sum t'x_t}{\sum t'^2}$$

$$\hat{c} = \frac{\sum x_t \sum t'^2 - N \sum t'^2 x_t}{(\sum t'^2)^2 - N \sum t'^4}$$

$$\hat{a} = \frac{\sum x_t - \hat{c} \sum t'^2}{N}$$

Table 6-1 is a guide showing how the summations above are generated for the case when $T = N$.

With the estimates ($\hat{a}$, $\hat{b}$, and $\hat{c}$) now available, the forecasts for the τth future time period can be found. This is by the relation

$$\hat{x}_T(\tau) = \hat{a} + \hat{b}\left(\tau + \frac{N-1}{2}\right) + \hat{c}\left(\tau + \frac{N-1}{2}\right)^2$$

Table 6-1. SUMMATIONS NEEDED IN THE QUADRATIC REGRESSION MODEL

t	t'	x_t	$t'x_t$	t'^2	t'^2x_t	t'^4
1	$-\frac{N-1}{2}$	x_1	$-\frac{N-1}{2}x_1$	$\left(\frac{N-1}{2}\right)^2$	$\left(\frac{N-1}{2}\right)^2 x_1$	$\left(\frac{N-1}{2}\right)^4$
2	$-\frac{N-3}{2}$	x_2	$-\frac{N-3}{2}x_2$	$\left(\frac{N-3}{2}\right)^2$	$\left(\frac{N-3}{2}\right)^2 x_2$	$\left(\frac{N-3}{2}\right)^4$
.						
.						
.						
$T-1$	$\frac{N-3}{2}$	x_{T-1}	$\frac{N-3}{2}x_{T-1}$	$\left(\frac{N-3}{2}\right)^2$	$\left(\frac{N-3}{2}\right)^2 x_{T-1}$	$\left(\frac{N-3}{2}\right)^4$
T	$\frac{N-1}{2}$	x_T	$\frac{N-1}{2}x_T$	$\left(\frac{N-1}{2}\right)^2$	$\left(\frac{N-1}{2}\right)^2 x_T$	$\left(\frac{N-1}{2}\right)^4$
Sums		$\sum x_t$	$\sum t'x_t$	$\sum t'^2$	$\sum t'^2x_t$	$\sum t'^4$

In the above, the constant $\frac{N-1}{2}$ is added to τ, because in the fitting process the intercept is located at the exact center of the time axis. In this way the estimate of the level at the current time period T is

$$\hat{\mu}_T = \hat{a} + \hat{b}\left(\frac{N-1}{2}\right) + \hat{c}\left(\frac{N-1}{2}\right)^2$$

Example 6.1

Consider an item with the following demands as of $T = 11$: 10, 12, 12, 16, 15, 19, 18, 20, 22, 21, and 23. Use the quadratic regression model with $N = 11$ to seek forecasts for the next two time periods ($t = 12, 13$).

The following table gives all the summations required to find the coefficients $\hat{a}$, $\hat{b}$, and $\hat{c}$.

t	t'	x_t	$t'x_t$	t'^2	t'^2x_t	t'^4
1	−5	10	−50	25	250	625
2	−4	12	−48	16	192	256
3	−3	12	−36	9	108	81
4	−2	16	−32	4	64	16
5	−1	15	−15	1	15	1
6	0	19	0	0	0	0
7	1	18	18	1	18	1
8	2	20	40	4	80	16
9	3	22	66	9	198	81
10	4	21	84	16	336	256
11	5	23	115	25	575	625
Sums	0	188	142	110	1836	1958

The estimates of the coefficients are

$$\hat{b} = \frac{142}{110} = 1.291$$

$$\hat{c} = \frac{188 \times 110 - 11 \times 1836}{(110)^2 - 11 \times 1958} = -0.0513$$

$$\hat{a} = \frac{188 + 0.0513 \times 110}{11} = 17.604$$

Hence, the current level (at $T = 11$) is estimated by

$$\hat{\mu}_{11} = 17.604 + 1.291(5) - 0.0513(5)^2 = 22.777$$

and the forecast equation is

$$\hat{x}_{11}(\tau) = 17.604 + 1.291(\tau + 5) - 0.0513(\tau + 5)^2$$

Now for $\tau = 1$ and 2, the forecasts are

$$\hat{x}_{11}(1) = 17.604 + 1.291(6) - 0.0513(6^2) = 23.503$$

$$\hat{x}_{11}(2) = 17.604 + 1.291(7) - 0.0513(7^2) = 24.127$$

and the cumulative two-period-ahead forecast is

$$\hat{X}_{11}(2) = 23.503 + 24.127 = 47.630$$

A plot of the fit and the forecasts is shown in Figure 6-1.

Mathematical Basis. The sum of squared residual errors for the quadratic regression model is

$$S(e) = \sum_{t'} (x_{t'} - \hat{x}_{t'})^2$$

$$= \sum_{t'} (x_{t'} - \hat{a} - \hat{b}t' - \hat{c}t'^2)^2$$

Using the least-squares method, the estimates for a, b, and c are found by first obtaining

$$\frac{\partial S(e)}{\partial \hat{a}} = -2 \sum_{t'} (x_{t'} - \hat{a} - \hat{b}t' - \hat{c}t'^2)$$

$$\frac{\partial S(e)}{\partial \hat{b}} = -2 \sum_{t'} t'(x_{t'} - \hat{a} - \hat{b}t' - \hat{c}t'^2)$$

$$\frac{\partial S(e)}{\partial \hat{c}} = -2 \sum_{t'} t'^2(x_{t'} - \hat{a} - \hat{b}t' - \hat{c}t'^2)$$

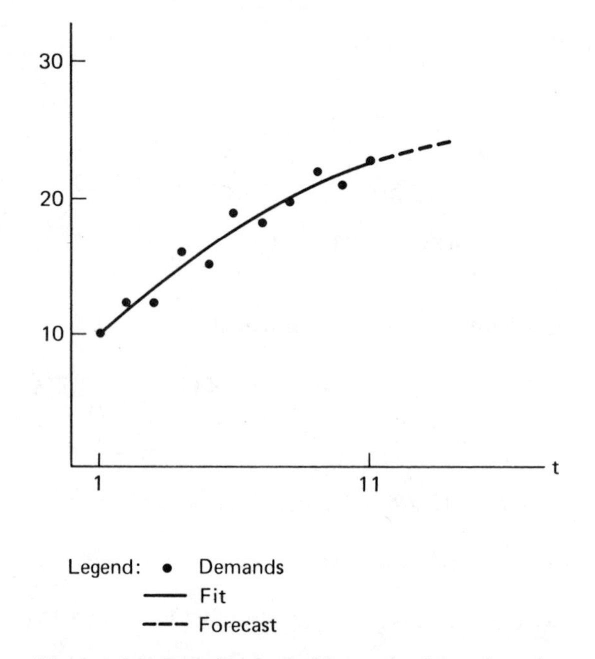

Figure 6-1. Results from Example 6.1 using the quadratic regression model.

Setting these partial derivatives to zero yields

$$\sum x_{t'} = N\hat{a} + \sum t'\hat{b} + \sum t'^2\hat{c}$$
$$\sum t'x_{t'} = \sum t'\hat{a} + \sum t'^2\hat{b} + \sum t'^3\hat{c}$$
$$\sum t'^2x_{t'} = \sum t'^2\hat{a} + \sum t'^3\hat{b} + \sum t'^4\hat{c}$$

where the summations range from $-\dfrac{N-1}{2}$ to $\dfrac{N-1}{2}$.

Since $\sum t' = \sum t'^3 = 0$, then

$$\sum x_{t'} = N\hat{a} + \sum t'^2\hat{c}$$
$$\sum t'x_{t'} = \sum t'^2\hat{b}$$
$$\sum t'^2x_{t'} = \sum t'^2\hat{a} + \sum t'^4\hat{c}$$

Solving for $\hat{a}$, $\hat{b}$, and $\hat{c}$ yields the results shown earlier.

6-2 TRIPLE SMOOTHING MODEL

Triple smoothing [1] is an extension to double smoothing and is used as a forecast model for items whose demand pattern is quadratic. As in the other smoothing models, a smoothing parameter (α) is selected by the forecaster and is used to assign weights to the demand entries of the past. In the smoothing manner, the more recent entries are given higher weights in seeking estimates of the coefficients a, b, and c.

In triple smoothing, the current time period (T) is always assumed as the origin of time. In this way the level at time $t = T + k$ is

$$\mu_{T+k} = a + bk + \tfrac{1}{2}ck^2$$

Note that the level at the current time period T is $\mu_T = a$, since $k = 0$. For the jth prior time period, the level becomes

$$\mu_{T-j} = a - jb + \tfrac{1}{2}cj^2$$

and for the τth future time period it is

$$\mu_{T+\tau} = a + \tau b + \tfrac{1}{2}c\tau^2$$

The coefficient c is divided by 2 to be consistent with the Taylor expansion series (Brown [2]).

The estimates of the coefficients (a, b, c) at time T are denoted as $\hat{a}_T$, $\hat{b}_T$, and $\hat{c}_T$, respectively, and the data required to find the current estimates are the following:

$$x_T = \text{current demand}$$

$$\hat{a}_{T-1}, \hat{b}_{T-1}, \hat{c}_{T-1} = \text{estimates from the prior time period}$$

Note that the one-period-ahead forecast at $T - 1$ is

$$\hat{x}_{T-1}(1) = \hat{a}_{T-1} + \hat{b}_{T-1} + \tfrac{1}{2}\hat{c}_{T-1}$$

and the current one-period-ahead forecast error is

$$e_T = \hat{x}_{T-1}(1) - x_T$$

Now the updated coefficient estimates can be found. These are

$$\hat{a}_T = x_T + (1 - \alpha)^3 e_T$$

$$\hat{b}_T = \hat{b}_{T-1} + \hat{c}_{T-1} - 1.5\alpha^2(2 - \alpha)e_T$$

$$\hat{c}_T = \hat{c}_{T-1} - \alpha^3 e_T$$

With $\hat{a}_T$, $\hat{b}_T$, and $\hat{c}_T$ now available, the forecast for the τth future time period can be generated. This is by

$$\hat{x}_T(\tau) = \hat{a}_T + \hat{b}_T\tau + \tfrac{1}{2}\hat{c}_T\tau^2$$

As before, the forecast for the sum of the next τ time periods is

$$\hat{X}_T(\tau) = \hat{x}_T(1) + \hat{x}_T(2) + \dots + \hat{x}_T(\tau)$$

Example 6.2

Find the updating equation for $\hat{a}_T$, $\hat{b}_T$, and $\hat{c}_T$ when $\alpha = 0.1$.

Since

$$(1 - \alpha)^3 = 0.9^3 = 0.7290$$

$$1.5\alpha^2(2 - \alpha) = 1.5(0.1)^2(1.9) = 0.0285$$

$$\alpha^3 = 0.1^3 = 0.0010$$

$$\hat{x}_{T-1}(1) = \hat{a}_{T-1} + \hat{b}_{T-1} + 0.5\hat{c}_{T-1}$$

$$e_T = \hat{x}_{T-1}(1) - x_T$$

then

$$\hat{a}_T = x_T + 0.7290e_T$$

$$\hat{b}_T = \hat{b}_{T-1} + \hat{c}_{T-1} - 0.0285e_T$$

$$\hat{c}_T = \hat{c}_{T-1} - 0.0010e_T$$

Example 6.3

Suppose that $\alpha = 0.1$ and at time T, $\hat{a}_{T-1} = 50$, $\hat{b}_{T-1} = 2$, $\hat{c}_{T-1} = 0.1$, and $x_T = 56$. Find the updated coefficient estimates and cumulative forecasts for the next three time periods.

Since

$$\hat{x}_{T-1}(1) = 50 + 2 + \tfrac{1}{2}(0.1) = 52.05$$

then

$$e_T = 52.05 - 56 = -3.95$$

$$\hat{a}_T = 56 + 0.7290(-3.95) = 53.1205$$

$$\hat{b}_T = 2 + 0.1 - 0.0285(-3.95) = 2.2126$$

$$\hat{c}_T = 0.1 - 0.0010(-3.95) = 0.1040$$

and

$$\hat{x}_T(\tau) = 53.1205 + 2.2126\tau + \tfrac{1}{2}(0.1040)\tau^2$$

Now with $\tau = 1, 2, 3$,

$$\hat{x}_T(1) = 53.1205 + 2.2126(1) + \tfrac{1}{2}(0.1040)(1) = 55.39$$

$$\hat{x}_T(2) = 53.1205 + 2.2126(2) + \tfrac{1}{2}(0.1040)(4) = 57.75$$

$$\hat{x}_T(3) = 53.1205 + 2.2126(3) + \tfrac{1}{2}(0.1040)(9) = 60.23$$

and

$$\hat{X}_T(3) = 55.39 + 57.75 + 60.23 = 173.37$$

Initial Forecasts

As in single and double smoothing, a difficulty arises at the outset because no prior estimates of the coefficients are available. In this section three methods of initializing the system are described. These are essentially the same as cited in double smoothing.

Initialize Using x_1 Only

Suppose at the outset that the forecaster has knowledge of x_1 (the first demand entry), but of nothing else. He may begin the forecasting process by setting $\hat{a}_1 = x_1$, $\hat{b}_1 = 0$, and $\hat{c}_1 = 0$. This yields forecasts of

$$\hat{x}_1(\tau) = x_1$$

for time period $T = 1$. Thereafter, the coefficients are updated as shown earlier.

Example 6.4

Suppose that the first demand entry is $x_1 = 10$. Use this entry to find the initial forecast at $T = 1$.

Since $x_1 = 10$, then $\hat{a}_1 = 10$, $\hat{b}_1 = 0$, $\hat{c}_1 = 0$, and $\hat{x}_1(\tau) = 10$.

Initialize with Estimates of the Coefficients at $t = 0$

A more refined manner of initializing the system is possible when estimates of the coefficients for $t = 0$ are available to the forecaster. The estimates are denoted by $\hat{a}_0$, $\hat{b}_0$, and $\hat{c}_0$. At $T = 1$, the entry x_1 becomes known and the current forecast error is found. This is

$$e_1 = \hat{x}_0(1) - x_1$$

where

$$\hat{x}_0(1) = \hat{a}_0 + \hat{b}_0 + \tfrac{1}{2}\hat{c}_0$$

So now the coefficient estimates at $T = 1$ become

$$\hat{a}_1 = x_1 + (1 - \alpha)^3 e_1$$
$$\hat{b}_1 = \hat{b}_0 + \hat{c}_0 - 1.5\alpha^2(2 - \alpha)e_1$$
$$\hat{c}_1 = \hat{c}_0 - \alpha^3 e_1$$

Example 6.5

Suppose that triple smoothing with $\alpha = 0.1$ will be used on a new item. Assume that the first demand entry is $x_1 = 10$ and the estimate of expected demands as of $t = 0$ is given by the relation

$$\hat{\mu}_\tau = 8 + 1\tau + 0.15\tau^2$$

Find the forecast equation for the item as of $T = 1$.

Note that

$$\hat{a}_0 = 8, \qquad \hat{b}_0 = 1, \qquad \hat{c}_0 = 2(0.15) = 0.30$$

$$\hat{x}_0(1) = 8 + 1 + \tfrac{1}{2}(0.30) = 9.15$$

$$e_1 = 9.15 - 10 = -0.85$$

Hence,

$$\hat{a}_1 = 10 + 0.7290(-0.85) = 9.3804$$

$$\hat{b}_1 = 1 + 0.30 - 0.0285(-0.85) = 1.3242$$

$$\hat{c}_1 = 0.30 - 0.0010(-0.85) = 0.3009$$

and

$$\hat{x}_1(\tau) = 9.3804 + 1.3242\tau + \tfrac{1}{2}(0.3009)\tau^2$$

Initialize by Fitting Past Demands

Another method of initializing is possible when some demand entries of the past are available. Suppose that these are $x_1, x_2, \ldots, x_T$ where T is the current time period. If at this time the forecaster wishes to use triple smoothing, estimates of the coefficients at time T are required to proceed.

These estimates may be obtained by fitting a quadratic equation using the past demands to arrive at estimates of a, b, and c as of time T. The estimates are denoted as $\hat{a}_T$, $\hat{b}_T$, and $\hat{c}_T$, respectively, and are used to forecast the future demands at time T.

Now at $t = T + 1$, a new demand entry is available and the updated coefficients are found using the triple smoothing procedure. The new forecasts are generated and the system proceeds in the normal manner from this point on.

Example 6.6

Consider an item where $T = 11$ and the forecaster decides to use triple smoothing with $\alpha = 0.10$. Suppose also that no prior estimates of the coefficients are available to the forecaster; however, the first 11 demand entries are known and the same entries used in Example 6.1. Find the forecasts as of $T = 11$.

Using quadratic regression, the 11 demand entries are applied to find estimates of the coefficients as of $T = 11$. The solution has been obtained in Example 6.1 with the results

$$\hat{a} = 17.604, \qquad \hat{b} = 1.291, \qquad \hat{c} = -0.0513$$

Note, however, that these estimates pertain to a time scale t' and are different from that needed in triple smoothing. In triple smoothing, the level is $\mu_t = a + bj + 0.5cj^2$ with $j = 0$ at $t = T$. In the quadratic regression model, $\mu_t = a + bj + cj^2$, where $j = 0$ at $t = \frac{N+1}{2}$. With $T = N = 11$, then $j = 5$ at $t = 11$.

Because of these inconsistencies, the coefficients given above (from quadratic regression) cannot be used as they are. A transformation is needed to convert $\hat{a}$, $\hat{b}$, and $\hat{c}$ as shown to $\hat{a}_T$, $\hat{b}_T$ and $\hat{c}_T$ where the latter are the coefficients required in triple smoothing. The transformation is presented below and pertains to a general case where N demand entries are used in quadratic regression. This is

$$\hat{a}_T = \hat{a} + \left(\frac{N-1}{2}\right)\hat{b} + \left(\frac{N-1}{2}\right)^2 \hat{c}$$

$$\hat{b}_T = \hat{b} + (N-1)\hat{c}$$

$$\hat{c}_T = 2\hat{c}$$

So in this example with $T = N = 11$, the coefficients in triple smoothing are:

$$\hat{a}_{11} = 17.604 + 5(1.291) + 25(-0.0513) = 22.777$$

$$\hat{b}_{11} = 1.291 + 10(-0.0513) = 0.778$$

$$\hat{c}_{11} = 2(-0.0513) = -0.103$$

The forecasts at $T = 11$ are now

$$\hat{x}_{11}(\tau) = 22.777 + 0.778\tau - \tfrac{1}{2}(0.103)\tau^2$$

Example 6.7

Continuing with Example 6.6, show the one-period-ahead forecasts for the demands listed in Table 6-2. Recall that $\alpha = 0.10$ is the smoothing parameter.

Table 6-2. RESULTS FROM EXAMPLE 6.7 USING TRIPLE SMOOTHING WITH $\alpha = 0.10$

t	x_t	$\hat{a}_t$	$\hat{b}_t$	$\hat{c}_t$	$\hat{x}_t(1)$
1	10				
2	12				
3	12				
4	16				
5	15				
6	19				
7	18				
8	20				
9	22				
10	21				
11	23	22.777	0.778	−0.103	23.504
12	24	23.638	0.689	−0.103	24.276
13	24	24.201	0.578	−0.103	24.728
14	25	24.797	0.483	−0.103	25.229
15	27	25.709	0.430	−0.101	26.088

The results are listed in Table 6-2 and a plot of the one-period-ahead forecasts is shown in Figure 6-2.

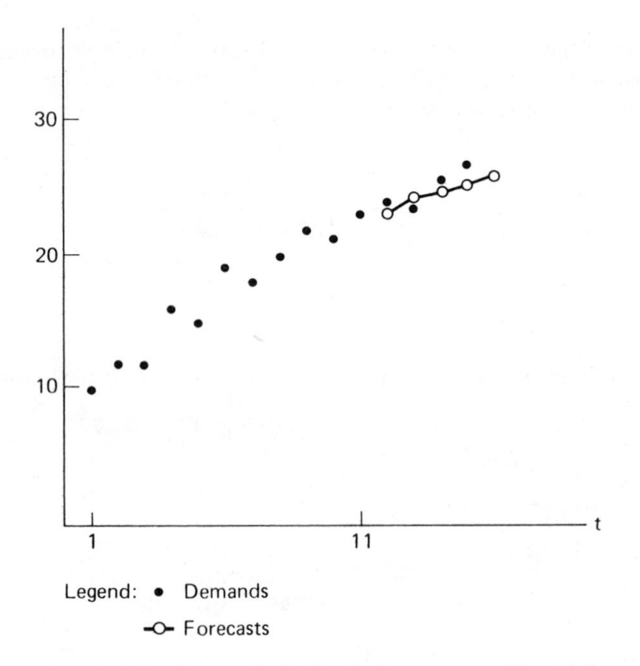

Figure 6-2. One-period-ahead forecasts with triple smoothing and $\alpha = 0.10$.

The Smoothing Parameter

In triple smoothing, the level at time T is related to the current demand (x_T) and the one-period-ahead forecast error $e_T = \hat{x}_{T-1}(1) - x_T$ by the relation

$$\hat{a}_T = x_T + (1 - \alpha)^3(e_T)$$

The corresponding result in single smoothing is

$$\hat{a}_T = x_T + (1 - \alpha)(e_T)$$

The two relations above are used to find a smoothing parameter in triple smoothing which corresponds to an equivalent setting in single smoothing. Letting α_0 represent the single smoothing parameter

$$(1 - \alpha)^3 = 1 - \alpha_0$$

or

$$\alpha = 1 - \sqrt[3]{1 - \alpha_0}$$

Some corresponding values are listed in Table 6-3.

Table 6-3. Equivalent Smoothing Parameters in Single and Triple Smoothing

Single smoothing	0.05	0.10	0.15	0.20	0.25	0.30
Triple smoothing	0.017	0.035	0.053	0.072	0.091	0.112

Example 6.8

Suppose that the forecaster is using $\alpha = 0.10$ in single smoothing successfully. Find the equivalent smoothing parameter in triple smoothing.

This is

$$\alpha = 1 - \sqrt[3]{1 - 0.10} = 0.035$$

Mathematical Basis. In triple smoothing, three smoothing averages are considered:

(1)
$$\begin{aligned} S_T &= \alpha x_T + (1-\alpha)S_{T-1} \\ S_T^{(2)} &= \alpha S_T + (1-\alpha)S_{T-1}^{(2)} \\ S_T^{(3)} &= \alpha S_T^{(2)} + (1-\alpha)S_{T-1}^{(3)} \end{aligned}$$

These may be written

(2)
$$\begin{aligned} S_T &= \alpha \sum_{j=0}^{\infty} (1-\alpha)^j x_{T-j} \\ S_T^{(2)} &= \alpha \sum_{j=0}^{\infty} (1-\alpha)^j S_{T-j} \\ S_T^{(3)} &= \alpha \sum_{j=0}^{\infty} (1-\alpha)^j S_{T-j}^{(2)} \end{aligned}$$

to find the following expected values:

(3)
$$\begin{aligned} E(S_T) &= \alpha \sum_{j=0}^{\infty} (1-\alpha)^j \left(a - bj + \frac{c}{2} j^2\right) \\ &= a - \left(\frac{1-\alpha}{\alpha}\right) b + \left[\frac{(1-\alpha)(2-\alpha)}{2\alpha^2}\right] c \\ E(S_T^{(2)}) &= \alpha \sum_{j=0}^{\infty} (1-\alpha)^j \left[a - b\left(j + \frac{1-\alpha}{\alpha}\right)\right. \\ &\quad \left. + \frac{c}{2}\left(j^2 + \frac{(2-\alpha)(1-\alpha)}{\alpha^2}\right) + 2j\left(\frac{1-\alpha}{\alpha}\right)\right] \\ &= a - \frac{2(1-\alpha)}{\alpha} b + \frac{2(1-\alpha)(3-2\alpha)}{2\alpha^2} c \end{aligned}$$

$$E(S_T^{(3)}) = \alpha \sum_{j=0}^{\infty} (1-\alpha)^j \left[a - b\left(j + \frac{2(1-\alpha)}{\alpha}\right) + \frac{c}{2}\left(j^2 + \frac{2(1-\alpha)(3-2\alpha)}{\alpha^2}\right) + 4j\left(\frac{1-\alpha}{\alpha}\right)\right]$$

$$= a - \frac{3(1-\alpha)}{\alpha} b + \frac{3(1-\alpha)(4-3\alpha)}{2\alpha^2} c$$

Using the method of moments to solve for the coefficients a, b, and c, the following substitutions are made: S_T, $S_T^{(2)}$, and $S_T^{(3)}$ for $E(S_T)$, $E(S_T^{(2)})$, and $E(S_T^{(3)})$ and $\hat{a}_T$, $\hat{b}_T$, and $\hat{c}_T$ for a, b, and c.
Hence,

$$(4) \qquad S_T = \hat{a}_T - \frac{1-\alpha}{\alpha}\hat{b}_T + \frac{(1-\alpha)(2-\alpha)}{2\alpha^2}\hat{c}_T$$

$$S_T^{(2)} = \hat{a}_T - \frac{2(1-\alpha)}{\alpha}\hat{b}_T + \frac{2(1-\alpha)(3-2\alpha)}{2\alpha^2}\hat{c}_T$$

$$S_T^{(3)} = \hat{a}_T - \frac{3(1-\alpha)}{\alpha}\hat{b}_T + \frac{3(1-\alpha)(4-3\alpha)}{2\alpha^2}\hat{c}_T$$

This gives three equations and three unknowns. Solving for $\hat{a}_T$, $\hat{b}_T$, and $\hat{c}_T$ yields

$$(5) \qquad \hat{a}_T = 3S_T - 3S_T^{(2)} + S_T^{(3)}$$

$$\hat{b}_T = \frac{\alpha}{2(1-\alpha)^2}(6-5\alpha)S_T - \frac{2\alpha}{2(1-\alpha)^2}(5-4\alpha)S_T^{(2)} + \frac{\alpha}{2(1-\alpha)^2}(4-3\alpha)S_T^{(3)}$$

$$\hat{c}_T = \frac{\alpha^2}{(1-\alpha)^2}S_T - \frac{2\alpha^2}{(1-\alpha)^2}S_T^{(2)} + \frac{\alpha^2}{(1-\alpha)^2}S_T^{(3)}$$

Note at this time that (1) can be written as

$$(6) \qquad S_T = \alpha x_T + (1-\alpha)S_{T-1}$$

$$S_T^{(2)} = \alpha^2 x_T + \alpha(1-\alpha)S_{T-1}^{(1)} + (1-\alpha)S_{T-1}^{(2)}$$

$$S_T^{(3)} = \alpha^3 x_T + \alpha^2(1-\alpha)S_{T-1}^{(1)} + \alpha(1-\alpha)S_{T-1}^{(2)} + (1-\alpha)S_{T-1}^{(3)}$$

Now substituting (6) into (5) gives

$$(7) \quad \hat{a}_T = [1-(1-\alpha)^3]x_T + (1-\alpha)(3-3\alpha+\alpha^2)S_{T-1} + (1-\alpha)(\alpha-3)S_{T-1}^{(2)} + (1-\alpha)S_{T-1}^{(3)}$$

$$\hat{b}_T = \frac{\alpha}{2(1-\alpha)}[3\alpha(2-\alpha)(1-\alpha)^2 x_T + 3(1-\alpha)^3(2-\alpha)S_{T-1} + (1-\alpha)(-10+12\alpha-3\alpha^2)S_{T-1}^{(2)} + (1-\alpha)(4-3\alpha)S_{T-1}^{(3)}]$$

$$\hat{c}_T = \frac{\alpha^2}{(1-\alpha)^2}[\alpha(1-\alpha)^2 x_T + (1-\alpha)^3 S_{T-1}$$
$$+ (1-\alpha)(\alpha-2)S_{T-1}^{(2)} - (1-\alpha)S_{T-1}^{(3)}]$$

But since

(8)
$$S_{T-1} = \hat{a}_{T-1} - \frac{1-\alpha}{\alpha}\hat{b}_{T-1} + \frac{(1-\alpha)(2-\alpha)}{2\alpha^2}\hat{c}_{T-1}$$
$$S_{T-1}^{(2)} = \hat{a}_{T-1} - \frac{2(1-\alpha)}{\alpha}\hat{b}_{T-1} + \frac{2(1-\alpha)(3-2\alpha)}{2\alpha^2}\hat{c}_{T-1}$$
$$S_{T-1}^{(3)} = \hat{a}_{T-1} - \frac{3(1-\alpha)}{\alpha}\hat{b}_{T-1} + \frac{3(1-\alpha)(4-3\alpha)}{2\alpha^2}\hat{c}_{T-1}$$

then $\hat{a}_T$, $\hat{b}_T$, and $\hat{c}_T$ are related to $\hat{a}_{T-1}$, $\hat{b}_{T-1}$, and $\hat{c}_{T-1}$. These are by substituting (8) into (7), i.e.,

(9)
$$\hat{a}_T = x_T + (1-\alpha)^3(e_T)$$
$$\hat{b}_T = \hat{b}_{T-1} + \hat{c}_{T-1} - 1.5\alpha^2(2-\alpha)(e_T)$$
$$\hat{c}_T = \hat{c}_{T-1} - \alpha^3(e_T)$$

where

$$e_T = \hat{x}_{T-1}(1) - x_T$$

PROBLEMS

6-1. Consider an item with the following nine demands for $t = 1, \ldots, 9$: 1, 2, 4, 4, 5, 7, 7, 8, and 8. Using the quadratic regression model with $N = 9$, calculate the coefficient estimates and find the forecast for time period $t = 10$.

6-2. Suppose an item has the following eight demands for $t = 1, \ldots, 8$: 3, 5, 6, 9, 10, 11, 12, and 13. Using the quadratic regression model with $N = 8$, find the forecast of the demand for $t = 9$.

6-3. Use the result of Problem 6-2 to generate the fitted demands: (i.e., find $\hat{x}_t$ for $t = 1, \ldots, 8$). Now calculate the residual errors and the sum of squared errors $S(e)$.

6-4. Suppose that at $T = 13$, the quadratic regression model with $N = 13$ is in use with the results

$$\hat{x}_T(\tau) = 100 + 5(\tau + 6) + 0.1(\tau + 6)^2$$

In this situation the coefficients estimates $\hat{a} = 100$, $\hat{b} = 5$, and $\hat{c} = 0.1$ are calculated with the intercept at $t = 7$. Find the coefficients that are needed when the forecast equation is transformed to

$$\hat{x}_T(\tau) = \hat{a}_T + \hat{b}_T\tau + 0.5\hat{c}_T\tau^2$$

where the intercept is at $T = 13$ (the most current time period).

6-5. Use the results of Problem 6-4 to compare the forecast of the demand at time period 14, by using the two forecast equations.

6-6. Suppose triple smoothing with $\alpha = 0.1$ is to be applied in finding the forecast for the next two time periods. Assume that at the current time period (T) $\hat{a}_{T-1} = 50$, $\hat{b}_{T-1} = 3$, $\hat{c}_{T-1} = -0.1$, and $x_T = 55$. Find the updated coefficient estimates and the forecasts needed.

6-7. A forecaster has been using double smoothing with $\alpha = 0.20$ successfully for items that have a trend demand pattern. Find the corresponding smoothing parameter that should be used for items with a quadratic demand pattern and where triple smoothing will be used.

6-8. Triple smoothing is frequently presented in the following manner. Three smoothing averages (S_t, $S_t^{(2)}$, $S_t^{(3)}$) are updated at each time period t in the following manner:

$$S_t = \alpha x_t + (1 - \alpha)S_{t-1}$$
$$S_t^{(2)} = \alpha S_t + (1 - \alpha)S_{t-1}^{(2)}$$
$$S_t^{(3)} = \alpha S_t^{(2)} + (1 - \alpha)S_{T-1}^{(3)}$$

The coefficients are estimated by

$$\hat{a}_t = 3S_t - 3S_t^{(2)} + S_t^{(3)}$$

$$\hat{b}_t = \frac{\alpha}{2(1-\alpha)^2}[(6 - 5\alpha)S_t - 2(5 - 4\alpha)S_t^{(2)} + (4 - 3\alpha)S_t^{(3)}]$$

$$\hat{c}_t = \frac{\alpha^2}{(1-\alpha)^2}(S_t - 2S_t^{(2)} + S_t^{(3)})$$

and the forecasts are found by

$$\hat{x}_t(\tau) = \hat{a}_t + \hat{b}_t\tau + 0.5\hat{c}_t\tau^2$$

Now suppose that triple smoothing with $\alpha = 0.1$ is to be used for an item. Assume that at the current time period T the following is known for the item: $S_{T-1} = 14.96$, $S_{T-1}^{(2)} = 7.34$, $S_{T-1}^{(3)} = 2.12$, and $x_T = 23$. Find the coefficient estimates and the cumulative forecasts of the next three time periods.

6-9. Using $\beta = 1 - \alpha$, show that

$$S_T = \alpha \sum_{j=0}^{\infty} \beta^j x_{T-j}$$

$$S_T^{(2)} = \alpha \sum_{j=0}^{\infty} \beta^j S_{T-j}$$

$$S_T^{(3)} = \alpha \sum_{j=0}^{\infty} \beta^j S_{T-j}^{(2)}$$

6-10. Using the results of Problem 6-9, show that

$$S_T^{(2)} = \alpha^2 \sum_{j=0}^{\infty} (j + 1)\beta^j x_{T-j}$$

$$S_T^{(3)} = \alpha^3 \sum_{j=0}^{\infty} \frac{(j + 1)(j + 2)}{2} \beta^j x_{T-j}$$

6-11. Given $\hat{a}_T$, $\hat{b}_T$, and $\hat{c}_T$, the corresponding values of S_T, $S_T^{(2)}$, and $S_T^{(3)}$ can be found by equation (4) of the Mathematical Basis in Section 6-2. Now if $\hat{a}_T = 120$, $\hat{b}_T = 3$, $\hat{c}_T = -0.8$, and $\alpha = 0.2$, find S_T, $S_T^{(2)}$, and $S_T^{(3)}$.

REFERENCES

[1] BROWN, R. G., *Smoothing, Forecasting and Prediction of Discrete Time Series.* Englewood Cliffs, N.J., Prentice-Hall, Inc., 1962, pp. 136–142.

[2] BROWN, R. G., *Smoothing, Forecasting and Prediction of Discrete Time Series.* Englewood Cliffs, N.J.: Prentice-Hall, Inc., 1962, p. 134.

7

REGRESSION, DISCOUNTING, AND ADAPTIVE SMOOTHING MODELS

This chapter describes forecasting models that are applicable for use with a wider variety of demand patterns than were shown earlier. The forecasting models are the regression, discounted regression, and adaptive smoothing models. The demand patterns are of any type where the expected demand (or level at time t) can be defined by the relation

$$\mu_t = a_1 f_1(t) + a_2 f_2(t) + \ldots + a_k f_k(t)$$

In the above, the expected demand includes k terms with k unknown coefficients $(a_1, a_2, \ldots, a_k)$ and k known functions of time $(f_1(t), \ldots, f_k(t))$. Each function $f_i(t)$, $i = 1, 2, \ldots, k$, is defined using a relationship with time t. Some common examples of the functions are $1, t, t^2, t^3, e^t, e^{ct}, \sin ct$, and $\cos ct$, where c is a fixed constant.

In this manner the horizontal demand pattern is defined as

$$\mu_t = a_1 = a_1 f_1(t)$$

where $f_1(t) = 1$. In the trend demand pattern

$$\mu_t = a_1 + a_2 t = a_1 f_1(t) + a_2 f_2(t)$$

with $f_1(t) = 1$ and $f_2(t) = t$. The quadratic demand pattern becomes

$$\mu_t = a_1 + a_2 t + a_3 t^2 = a_1 f_1(t) + a_2 f_2(t) + a_3 f_3(t)$$

where $f_1(t) = 1, f_2(t) = t$, and $f_3(t) = t^2$. Some other examples are

$$\begin{aligned}\mu_t &= a_1 + a_2 \sin ct + a_3 \cos ct \\ &= a_1 f_1(t) + a_2 f_2(t) + a_3 f_3(t)\end{aligned}$$

where $f_1(t) = 1, f_2(t) = \sin ct$, and $f_3(t) = \cos ct$, and

$$\begin{aligned}\mu_t &= a_1 t + a_2 e^{ct} \\ &= a_1 f_1(t) + a_2 f_2(t)\end{aligned}$$

where $f_1(t) = t$ and $f_2(t) = e^{ct}$.

The role of the forecasting model is to find a fit through the past demands $(x_1, x_2, \ldots, x_T)$ that corresponds to the demand pattern assumed by the forecaster. This entails finding estimates for $a_1, a_2, \ldots, a_k$ which are denoted as $\hat{a}_1, \hat{a}_2, \ldots, \hat{a}_k$, respectively.

In the regression model, equal weight is given to each of the past demands to seek the fit. A second model is also by way of regression except that a shift in time is used so that the current time period always acts as the origin. Again equal weight is assigned to each of the past demands.

A third model uses discounting and is called the *discounted regression model.* With this method, more weight is assigned to the more recent demand entries. Also, a shift in the time axis is used to make calculations more convenient.

The fourth model is *adaptive smoothing*, where discounting and a shift in time is applied. Adaptive smoothing models are the more popular of the four models shown, since they are easier to use. Only the most current demand (x_T) is needed in order to update the forecasts, and the calculations require but a few steps.

7-1 REGRESSION MODEL

In the regression model, the first T demand entries $(x_1, x_2, \ldots, x_T)$ are used in the fitting process. For an assumed demand pattern of the type

$$\mu_t = a_1 f_1(t) + a_2 f_2(t) + \ldots + a_k f_k(t)$$

the following data are required in carrying out the calculations:

x_t	$f_1(t)$	$f_2(t) \ldots f_k(t)$
x_1	$f_1(1)$	$f_2(1) \ldots f_k(1)$
x_2	$f_1(2)$	$f_2(2) \ldots f_k(2)$
.	.	. .
.	.	. .
.	.	. .
x_T	$f_1(T)$	$f_2(T) \ldots f_k(T)$

Using the data above, the k equations listed below are generated:

$$g_1(T) = \hat{a}_1 F_{11}(T) + \hat{a}_2 F_{12}(T) + \ldots + \hat{a}_k F_{1k}(T)$$

$$g_2(T) = \hat{a}_1 F_{21}(T) + \hat{a}_2 F_{22}(T) + \ldots + \hat{a}_k F_{2k}(T)$$

$$\vdots$$

$$g_k(T) = \hat{a}_1 F_{k1}(T) + \hat{a}_2 F_{k2}(T) + \ldots + \hat{a}_k F_{kk}(T)$$

where

$$g_i(T) = x_1 f_i(1) + x_2 f_i(2) + \ldots + x_T f_i(T)$$

$$= \sum_{t=1}^{T} x_t f_i(t) \qquad i = 1, 2, \ldots, k$$

$$F_{in}(T) = f_i(1) f_n(1) + f_i(2) f_n(2) + \ldots + f_i(T) f_n(T)$$

$$= \sum_{t=1}^{T} f_i(t) f_n(t) \qquad i = 1, 2, \ldots, k \text{ and } n = 1, 2, \ldots, k$$

The relations above give k equations and k unknowns and are now used to seek $\hat{a}_1, \hat{a}_2, \ldots, \hat{a}_k$. In this endeavor, the relations are conveniently listed in their equivalent matrix form, i.e.,

$$\begin{bmatrix} g_1(T) \\ g_2(T) \\ \cdot \\ \cdot \\ \cdot \\ g_k(T) \end{bmatrix} = \begin{bmatrix} F_{11}(T) & F_{12}(T) \ldots F_{1k}(T) \\ F_{21}(T) & F_{22}(T) \ldots F_{2k}(T) \\ \cdot & \\ \cdot & \\ \cdot & \\ F_{k1}(T) & F_{k2}(T) \ldots F_{kk}(T) \end{bmatrix} \begin{bmatrix} \hat{a}_1 \\ \hat{a}_2 \\ \cdot \\ \cdot \\ \cdot \\ \hat{a}_k \end{bmatrix}$$

The estimates of the coefficients are obtained by the matrix relation

$$\begin{bmatrix} \hat{a}_1 \\ \hat{a}_2 \\ \cdot \\ \cdot \\ \cdot \\ \hat{a}_k \end{bmatrix} = \begin{bmatrix} F_{11}(T) & F_{12}(T) \ldots F_{1k}(T) \\ F_{21}(T) & F_{22}(T) \ldots F_{2k}(T) \\ \cdot & \\ \cdot & \\ \cdot & \\ F_{k1}(T) & F_{k2}(T) \ldots F_{kk}(T) \end{bmatrix}^{-1} \begin{bmatrix} g_1(T) \\ g_2(T) \\ \cdot \\ \cdot \\ \cdot \\ g_k(T) \end{bmatrix}$$

where $[\ \ ]^{-1}$ represents the inverse of the matrix contained within the brackets.

With $\hat{a}_1, \hat{a}_2, \ldots, \hat{a}_k$ now available, the forecasts for the τth future time period is

$$\hat{x}_T(\tau) = \hat{a}_1 f_1(T+\tau) + \hat{a}_2 f_2(T+\tau) + \ldots + \hat{a}_k f_k(T+\tau)$$

As each new demand entry becomes available, the summations $g_i(T)$ and $F_{in}(T)$ are updated and new estimates of the coefficients are found. These are applied as before to generate updated forecasts.

Example 7.1

Consider an item with the following history of demands: 21, 25, 28, 29, 31, 40, 41, 43, 51, and 53. Here, $x_1 = 21$, $x_{10} = 53$, and $T = 10$. Find the forecasts for $\tau = 1$, 2, 3, and 4, when the expected demands are assumed to follow the form

$$\mu_t = a_1 + a_2 e^{0.1t}$$

and $e = 2.7183$.

In this situation $k = 2$, $f_1(t) = 1$, and $f_2(t) = e^{0.1t}$. The data available become

t	x_t	$f_1(t)$	$f_2(t)$
1	21	1	$e^{0.1} = 1.1052$
2	25	1	$e^{0.2} = 1.2214$
3	28	1	$e^{0.3} = 1.3499$
4	29	1	$e^{0.4} = 1.4918$
5	31	1	$e^{0.5} = 1.6487$
6	40	1	$e^{0.6} = 1.8221$
7	41	1	$e^{0.7} = 2.0138$
8	43	1	$e^{0.8} = 2.2255$
9	51	1	$e^{0.9} = 2.4596$
10	53	1	$e^{1.0} = 2.7183$

The summations required are shown in the following table:

t	$x_t f_1(t)$	$x_t f_2(t)$	$f_1(t) f_1(t)$	$f_1(t) f_2(t)$	$f_2(t) f_2(t)$
1	21	$21 \cdot e^{0.1}$	$1 \cdot 1$	$1e^{0.1}$	$e^{0.1}e^{0.1}$
2	25	$25 \cdot e^{0.2}$	$1 \cdot 1$	$1e^{0.2}$	$e^{0.2}e^{0.2}$
3	28	$28 \cdot e^{0.3}$	$1 \cdot 1$	$1e^{0.3}$	$e^{0.3}e^{0.3}$
4	29	$29 \cdot e^{0.4}$	$1 \cdot 1$	$1e^{0.4}$	$e^{0.4}e^{0.4}$
5	31	$31 \cdot e^{0.5}$	$1 \cdot 1$	$1e^{0.5}$	$e^{0.5}e^{0.5}$
6	40	$40 \cdot e^{0.6}$	$1 \cdot 1$	$1e^{0.6}$	$e^{0.6}e^{0.6}$
7	41	$41 \cdot e^{0.7}$	$1 \cdot 1$	$1e^{0.7}$	$e^{0.7}e^{0.7}$
8	43	$43 \cdot e^{0.8}$	$1 \cdot 1$	$1e^{0.8}$	$e^{0.8}e^{0.8}$
9	51	$51 \cdot e^{0.9}$	$1 \cdot 1$	$1e^{0.9}$	$e^{0.9}e^{0.9}$
10	53	$53 \cdot e^{1.0}$	$1 \cdot 1$	$1e^{1.0}$	$e^{1.0}e^{1.0}$
Sums	362	706.5691	10	18.0563	35.2463
	$g_1(10)$	$g_2(10)$	$F_{11}(10)$	$F_{12}(10) = F_{21}(10)$	$F_{22}(10)$

Hence, the two equations that result are

$$362 = 10\hat{a}_1 + 18.0563\hat{a}_2$$

$$706.5691 = 18.0563\hat{a}_1 + 35.2463\hat{a}_2$$

In matrix form the equations become

$$\begin{bmatrix} 362 \\ 706.5691 \end{bmatrix} = \begin{bmatrix} 10 & 18.0563 \\ 18.0563 & 35.2463 \end{bmatrix} \begin{bmatrix} \hat{a}_1 \\ \hat{a}_2 \end{bmatrix}$$

and the solution[1] is

$$\begin{bmatrix} \hat{a}_1 \\ \hat{a}_2 \end{bmatrix} = \begin{bmatrix} 10 & 18.0563 \\ 18.0563 & 35.2463 \end{bmatrix}^{-1} \begin{bmatrix} 362 \\ 706.5691 \end{bmatrix}$$

$$= \begin{bmatrix} 1.3334 & -0.6831 \\ -0.6831 & 0.3783 \end{bmatrix} \begin{bmatrix} 362 \\ 706.5691 \end{bmatrix}$$

$$= \begin{bmatrix} 0.0334 \\ 20.0129 \end{bmatrix}$$

So now the forecast equation is

$$\hat{x}_{10}(\tau) = 0.0334 + 20.129e^{0.1(10+\tau)}$$

[1]It is noted that the solution can easily be obtained without use of matrix methods.

With $\tau = 1, 2, 3$, and 4, the forecasts[2] are:

$$\hat{x}_{10}(1) = 0.0334 + 20.0129e^{1.1} = 60.1554$$
$$\hat{x}_{10}(2) = 0.0334 + 20.0129e^{1.2} = 66.4785$$
$$\hat{x}_{10}(3) = 0.0334 + 20.0129e^{1.3} = 73.4667$$
$$\hat{x}_{10}(4) = 0.0334 + 20.0129e^{1.4} = 81.1897$$

A plot of the fit and the forecasts are shown in Figure 7-1.

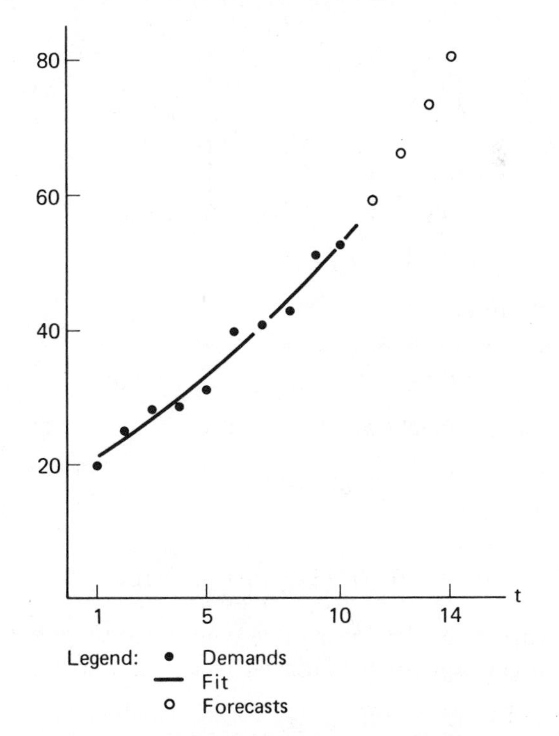

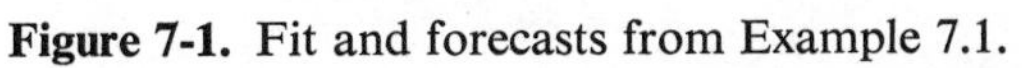

Figure 7-1. Fit and forecasts from Example 7.1.

MATHEMATICAL BASIS. The sum of squares of the residual errors is

$$S(e) = \sum_{t=1}^{T} (\hat{x}_t - x_t)^2$$

where

$$\hat{x}_t = \sum_{i=1}^{k} \hat{a}_i f_i(t)$$

---

[2]Note that $e^{1.1} = 3.004$, $e^{1.2} = 3.320$, $e^{1.3} = 3.669$, and $e^{1.4} = 4.055$.

Using the least-squares method, the estimates of the coefficients are obtained by taking

$$\frac{\partial S(e)}{\partial \hat{a}_i} = -2 \sum_{t=1}^{T} [f_i(t)(\hat{a}_1 f_1(t) + \ldots + \hat{a}_k f_k(t) - x_t)]^2 = 0$$

for $i = 1, 2, \ldots, k$. These yield the matrix relation

$$g(T) = F(T)\hat{a}$$

where

$$g(T) = \begin{bmatrix} g_1(T) \\ \cdot \\ \cdot \\ \cdot \\ g_k(T) \end{bmatrix}, \qquad F(T) = \begin{bmatrix} F_{11}(T) \ldots F_{1k}(T) \\ \cdot \\ \cdot \\ \cdot \\ F_{k1}(T) \ldots F_{kk}(T) \end{bmatrix}$$

$$\hat{a} = \begin{bmatrix} \hat{a}_1 \\ \cdot \\ \cdot \\ \cdot \\ \hat{a}_k \end{bmatrix}$$

Hence, the estimates sought are found from

$$\hat{a} = F(T)^{-1} g(T)$$

7-2 REGRESSION WITH TIME-SHIFT MODEL

As shown in earlier chapters, it is often convenient to shift the time axis so that the origin corresponds to the most recent time period. This is accomplished by letting

$$t = T - j$$

So for the time periods of the past, t and $-j$ are related by

t	1	2	3	$\ldots$	$T-1$	T
$-j$	$-(T-1)$	$-(T-2)$	$-(T-3)$		-1	0

For the future time periods, τ is used in place of $-j$, and t and τ are related by

t	$T+1$	$T+2$	$T+3 \ldots$
τ	1	2	3

The regression model of Section 7-1 is used as before, with these new indices. In this manner, the expected demand for the jth prior time period is

$$\mu_{T-j} = a_1 f_1(-j) + a_2 f_2(-j) + \ldots + a_k f_k(-j)$$

and for the τth future time period, it is

$$\mu_{T+\tau} = a_1 f_1(\tau) + a_2 f_2(\tau) + \ldots + a_k f_k(\tau)$$

The data required to carry out the calculations are as follows:

x_t	$f_1(-j)$	....	$f_k(-j)$
x_1	$f_1(-(T-1))$		$f_k(-(T-1))$
x_2	$f_1(-(T-2))$		$f_k(-(T-2))$
.			
.			
.			
x_{T-1}	$f_1(-1)$		$f_k(-1)$
x_T	$f_1(0)$		$f_k(0)$

These data are now used to generate the following k equations.

$$g_1(T) = \hat{a}_1 F_{11}(T) + \hat{a}_2 F_{12}(T) + \ldots + \hat{a}_k F_{1k}(T)$$
$$\vdots$$
$$g_k(T) = \hat{a}_1 F_{k1}(T) + \hat{a}_2 F_{k2}(T) + \ldots + \hat{a}_k F_{kk}(T)$$

where

$$g_i(T) = x_1 f_i(-(T-1)) + x_2 f_i(-(T-2)) + \ldots + x_T f_i(0)$$

for $i = 1, 2, \ldots, k$, and

$$F_{in}(T) = f_i(-(T-1)) f_n(-(T-1)) + \ldots + f_i(0) f_n(0)$$

for $i = 1, 2, \ldots, k$ and $n = 1, 2, \ldots, k$. Note that $F_{in}(T) = F_{ni}(T)$.

Proceeding as before, the relations above are placed in matrix form to give

$$\begin{bmatrix} g_1(T) \\ \cdot \\ \cdot \\ \cdot \\ g_k(T) \end{bmatrix} = \begin{bmatrix} F_{11}(T) & \ldots & F_{1k}(T) \\ \cdot & & \cdot \\ \cdot & & \cdot \\ \cdot & & \cdot \\ F_{k1}(T) & \ldots & F_{kk}(T) \end{bmatrix} \begin{bmatrix} \hat{a}_1 \\ \cdot \\ \cdot \\ \cdot \\ \hat{a}_k \end{bmatrix}$$

whereby the estimates of the coefficients are found by

$$\begin{bmatrix} \hat{a}_1 \\ \cdot \\ \cdot \\ \cdot \\ \hat{a}_k \end{bmatrix} = \begin{bmatrix} F_{11}(T) \dots F_{1k}(T) \\ \cdot \quad \cdot \\ \cdot \quad \cdot \\ \cdot \quad \cdot \\ F_{k1}(T) \dots F_{kk}(T) \end{bmatrix}^{-1} \begin{bmatrix} g_1(T) \\ \cdot \\ \cdot \\ \cdot \\ g_k(T) \end{bmatrix}$$

So now the forecasts for the τth future time period are

$$\hat{x}_T(\tau) = \hat{a}_1 f_1(\tau) + \hat{a}_2 f_2(\tau) + \dots + \hat{a}_k f_k(\tau)$$

Example 7.2

Consider once more the 10 demand entries from Example 7.1. Assume, as before, that

$$\mu_t = a_1 + a_2 e^{0.1t}$$

Now using regression with a time shift, find the forecasts for $t = 11, 12, 13$, and 14. The data available to the forecaster are

t	$-j$	x_t	$f_1(-j)$	$f_2(-j)$
1	−9	21	1	$e^{-0.9} = 0.4066$
2	−8	25	1	$e^{-0.8} = 0.4493$
3	−7	28	1	$e^{-0.7} = 0.4966$
4	−6	29	1	$e^{-0.6} = 0.5488$
5	−5	31	1	$e^{-0.5} = 0.6065$
6	−4	40	1	$e^{-0.4} = 0.6703$
7	−3	41	1	$e^{-0.3} = 0.7408$
8	−2	43	1	$e^{-0.2} = 0.8187$
9	−1	51	1	$e^{-0.1} = 0.9048$
10	0	53	1	$e^{0} = 1.0000$

Now the summations required become

$$g_1(T) = 21 + \dots + 53 = 362$$

$$g_2(T) = 21e^{-0.9} + \dots + 53e^0 = 259.9263$$

$$F_{11}(T) = 1 + \dots + 1 = 10$$

$$F_{12}(T) = 1 \cdot e^{-0.9} + \dots + 1 \cdot e^0 = 6.6424$$

$$F_{21}(T) = F_{12}(T) = 6.6424$$

$$F_{22}(T) = (e^{-0.9})^2 + \dots + (e^0)^2 = 4.7698$$

These yield the following two equations:

$$362 = 10\hat{a}_1 + 6.6424\hat{a}_2$$
$$259.9263 = 6.6424\hat{a}_1 + 4.7698\hat{a}_2$$

Solving for $\hat{a}_1$ and $\hat{a}_2$ as before gives

$$\hat{a}_1 = 0.0281$$
$$\hat{a}_2 = 54.4475$$

Hence, the forecast equation is

$$\hat{x}_{10}(\tau) = 0.0281 + 54.4475e^{0.1\tau}$$

With $\tau = 1, 2, 3$, and 4, then

$$\hat{x}_{10}(1) = 0.0281 + 54.4475e^{0.1} = 60.2019$$
$$\hat{x}_{10}(2) = 0.0281 + 54.4475e^{0.2} = 66.5304$$
$$\hat{x}_{10}(3) = 0.0281 + 54.4475e^{0.3} = 73.5244$$
$$\hat{x}_{10}(4) = 0.0281 + 54.4475e^{0.4} = 81.2542$$

7-3 DISCOUNTED REGRESSION MODELS

As in the smoothing models, many forecasters wish to generate forecasts that give a higher weight to the more recent demand entries. A common way of doing this is to assign relatively decreasing weights to the older demands. A discounting rate β $(0 < \beta < 1)$ is selected whereby the weight assigned for time period $t - 1$ is β times as large as that given at time t. For the most current time period (T) the weight is 1, at $T - 1$ it is β, at $T - 2$ it is β^2, and so on. The weights are applied to the square of the residual errors. Letting $e_{T-j} = (\hat{x}_{T-j} - x_{T-j})$ represent the residual error at $t = T - j$, the discounted sum of squared residual errors is

$$S(e) = e_T^2 + \beta e_{T-1}^2 + \beta^2 e_{T-2}^2 + \ldots + \beta^{T-1} e_1^2$$
$$= \sum_{j=0}^{T-1} \beta^j e_{T-j}^2$$

This method of assigning weights is called *discounting*.

Discounting can be used whether or not a shift in the time axis is carried on. For convenience later, the discounting regression method shown here is for a situation when a time shift is applied.

As in Section 7-2, the expected demand for the jth prior time period is

$$\mu_{T-j} = a_1 f_1(-j) + a_2 f_2(-j) + \ldots + a_k f_k(-j)$$

and for the τth future time period, it is

$$\mu_{T+\tau} = a_1 f_1(\tau) + a_2 f_2(\tau) + \ldots + a_k f_k(\tau)$$

The data needed to carry out the calculations (as of time T) are the following:

x_t	β^j	$f_1(-j)$	....	$f_k(-j)$
x_1	β^{T-1}	$f_1(-(T-1))$		$f_k(-(T-1))$
x_2	β^{T-2}	$f_1(-(T-2))$		$f_k(-(T-2))$
.	.	.		.
.	.	.		.
.	.	.		.
x_{T-1}	β	$f_1(-1)$		$f_k(-1)$
x_T	1	$f_1(0)$		$f_k(0)$

In order to find the estimates $(\hat{a}_1, \hat{a}_2, \ldots, \hat{a}_k)$, the following k equations are generated from the data listed above:

$$\begin{gathered} g_1(T) = F_{11}(T)\hat{a}_1 + \ldots + F_{1k}(T)\hat{a}_k \\ \vdots \\ g_k(T) = F_{k1}(T)\hat{a}_1 + \ldots + F_{kk}(T)\hat{a}_k \end{gathered}$$

where

$$g_i(T) = \sum_{j=0}^{T-1} \beta^j x_{T-j} f_i(-j) \qquad i = 1, 2, \ldots, k$$

$$F_{in}(T) = \sum_{j=0}^{T-1} \beta^j f_i(-j) f_n(-j) \qquad \begin{aligned} i &= 1, 2, \ldots, k \\ n &= 1, 2, \ldots, k \end{aligned}$$

As in the previous models of this chapter, $F_{in}(T) = F_{ni}(T)$. In matrix form, the estimates sought become

$$\begin{bmatrix} \hat{a}_1 \\ \cdot \\ \cdot \\ \cdot \\ \hat{a}_k \end{bmatrix} = \begin{bmatrix} F_{11}(T) \ldots F_{1k}(T) \\ \cdot \qquad \cdot \\ \cdot \qquad \cdot \\ \cdot \qquad \cdot \\ F_{k1}(T) \ldots F_{kk}(T) \end{bmatrix}^{-1} \begin{bmatrix} g_1(T) \\ \cdot \\ \cdot \\ \cdot \\ g_k(T) \end{bmatrix}$$

Now the forecast for the τth future time period is found. This is

$$\hat{x}_T(\tau) = \hat{a}_1 f_1(\tau) + \ldots + \hat{a}_k f_k(\tau)$$

Example 7.3

Using the same demands as in Example 7.1, and assuming that $\mu_t = a_1 + a_2 e^{0.1t}$, find the forecasts for $t = 11, 12, 13$, and 14 using the discounted regression model with $\beta = 0.9$.

Since $f_1(-j) = 1$ and $f_2(-j) = e^{-0.1j}$, the data available are the following:

t	j	x_t	β^j	$f_1(-j)$	$f_2(-j)$
1	9	21	$0.9^9 = 0.3874$	1	$e^{-0.9} = 0.4066$
2	8	25	$0.9^8 = 0.4305$	1	$e^{-0.8} = 0.4493$
3	7	28	$0.9^7 = 0.4783$	1	$e^{-0.7} = 0.4966$
4	6	29	$0.9^6 = 0.5314$	1	$e^{-0.6} = 0.5488$
5	5	31	$0.9^5 = 0.5905$	1	$e^{-0.5} = 0.6065$
6	4	40	$0.9^4 = 0.6561$	1	$e^{-0.4} = 0.6703$
7	3	41	$0.9^3 = 0.7290$	1	$e^{-0.3} = 0.7408$
8	2	43	$0.9^2 = 0.8100$	1	$e^{-0.2} = 0.8187$
9	1	51	$0.9^1 = 0.9000$	1	$e^{-0.1} = 0.9048$
10	0	53	$0.9^0 = 1.0000$	1	$e^0 = 1.0000$

Now the summations required are generated. These are

$$g_1(T) = 0.9^9 \cdot 21 \cdot 1 + \ldots + 0.9^0 \cdot 53 \cdot 1 = 255.8694$$

$$g_2(T) = 0.9^9 \cdot 21 \cdot e^{-0.9} + \ldots + 0.9^0 \cdot 53 e^0 = 197.1316$$

$$F_{11}(T) = 0.9^9 \cdot 1 \cdot 1 + \ldots + 0.9^0 \cdot 1 \cdot 1 = 6.5132$$

$$F_{12}(T) = 0.9^9 \cdot 1 \cdot e^{-0.9} + \ldots + 0.9^0 \cdot 1 \cdot e^0 = 4.6955$$

$$F_{22}(T) = 0.9^9 e^{-0.9} e^{-0.9} + \ldots + 0.9^0 e^0 e^0 = 3.6207$$

Because $F_{21}(T) = F_{12}(T) = 4.6955$, then

$$255.8694 = 6.5132\hat{a}_1 + 4.6955\hat{a}_2$$

$$197.1316 = 4.6955\hat{a}_1 + 3.6207\hat{a}_2$$

Solving for $\hat{a}_1$ and $\hat{a}_2$ yields

$$\hat{a}_1 = 0.5425$$

$$\hat{a}_2 = 53.7423$$

and the forecasts become

$$\hat{x}_{10}(\tau) = 0.5425 + 53.7423 e^{0.1\tau}$$

Now with $\tau = 1, 2, 3,$ and 4,

$$\hat{x}_{10}(1) = 0.5425 + 53.7423e^{0.1} = 60$$

$$\hat{x}_{10}(2) = 0.5425 + 53.7423e^{0.2} = 66$$

$$\hat{x}_{10}(3) = 0.5425 + 53.7423e^{0.3} = 73$$

$$\hat{x}_{10}(4) = 0.5425 + 53.7423e^{0.4} = 81$$

MATHEMATICAL BASIS. Since the sum of discounted residual squared errors is

$$S(e) = \sum_{j=0}^{T-1} \beta^j(\hat{x}_{T-j} - x_{T-j})^2$$

and

$$\hat{x}_{T-j} = \hat{a}_1 f_1(-j) + \ldots + \hat{a}_k f_k(-j)$$

then, using the least-squares method,

$$\frac{\partial S(e)}{\partial \hat{a}_i} = -2\sum_{j=0}^{T-1} \beta^j f_i(-j)[\hat{a}_1 f_1(-j) + \ldots + \hat{a}_k f_k(-j) - x_{T-j}]$$

for $i = 1, 2, \ldots, k$. Setting these partial derivatives equal to zero yields

$$g(T) = F(T)\hat{a}$$

where

$$g(T) = \begin{bmatrix} g_1(T) \\ \cdot \\ \cdot \\ \cdot \\ g_k(T) \end{bmatrix}$$

$$F(T) = \begin{bmatrix} F_{11}(T) \ldots F_{1k}(T) \\ \cdot \qquad\qquad \cdot \\ \cdot \qquad\qquad \cdot \\ \cdot \qquad\qquad \cdot \\ F_{k1}(T) \ldots F_{kk}(T) \end{bmatrix}$$

and

$$\hat{a} = \begin{bmatrix} \hat{a}_1 \\ \cdot \\ \cdot \\ \cdot \\ \hat{a}_k \end{bmatrix}$$

The elements $g_i(T)$ $(i = 1, 2, \ldots, k)$ and $F_{in}(T)$ $(i = 1, 2, \ldots, k$ and $n = 1, 2, \ldots, k)$ are as shown earlier. The estimates are found by

$$\hat{a} = F(T)^{-1}g(T)$$

7-4 ADAPTIVE SMOOTHING MODELS

Although the forecasting models that have been presented so far in this chapter allow much flexibility in both the choice of demand pattern and in assigning weights to the past demands, they lack the ease needed to process a large number of items at frequent time intervals. This is because of the large amount of data and calculations needed to generate the forecasts. The adaptive smoothing model [1] does overcome these difficulties and is presented in this section.

In adaptive smoothing, the time periods are shifted as shown earlier and discounting is used. The expected demand at $t = T - j$ is given by

$$\mu_{T-j} = a_1 f_1(-j) + \ldots + a_k f_k(-j)$$

The estimates at time T of $(a_1, a_2, \ldots, a_k)$ are denoted by $(\hat{a}_1(T), \hat{a}_2(T), \ldots, \hat{a}_k(T))$, respectively. For convenience, these are listed in vector form, where

$$\hat{a}(T) = \begin{bmatrix} \hat{a}_1(T) \\ \cdot \\ \cdot \\ \cdot \\ \hat{a}_k(T) \end{bmatrix}$$

The estimates are obtained by the relation

$$\hat{a}(T) = L'\hat{a}(T-1) + h(x_T - \hat{x}_{T-1}(1))$$

where $\hat{a}(T-1)$ are the corresponding estimates at $t = T - 1$, x_T is the current demand, and $\hat{x}_{T-1}(1)$ is the one-period-ahead forecast at $t = T - 1$. L is a matrix

$$L = \begin{bmatrix} L_{11} & \ldots & L_{1k} \\ \cdot & & \cdot \\ \cdot & & \cdot \\ \cdot & & \cdot \\ L_{k1} & \ldots & L_{kk} \end{bmatrix}$$

and h is a vector

$$h = \begin{bmatrix} h_1 \\ \cdot \\ \cdot \\ \cdot \\ h_k \end{bmatrix}$$

whose characteristics are described shortly. Finally, L' is the transpose of the matrix L.

L and h are found once and for all for a particular demand pattern and a specific discounting rate (β). Since they remain the same for all time periods, the only data required to update the coefficient estimates are $\hat{a}(T-1)$, x_T, and $\hat{x}_{T-1}(1)$. In this respect, the adaptive smoothing model is far more convenient to use than the models shown earlier in this chapter.

Having the estimates ($\hat{a}_1, \hat{a}_2, \ldots, \hat{a}_k$), available at time T, the forecast for the τth future time period is

$$\hat{x}_T(\tau) = \hat{a}_1(T)f_1(\tau) + \ldots + \hat{a}_k(T)f_k(\tau)$$

Note that the forecast with $\tau = 1$ is used to update the coefficient estimates at the subsequent time period.

Although adaptive smoothing is quite flexible in use, it is not suitable in all situations. There are two conditions that must be satisfied in order to apply the model. These conditions are described below and concern the matrices L and $F(T)$.

First Condition [Concerning L]

The first condition is associated with the expected demand of the items as of time t. Without the time shift this is

$$\mu_t = a_1 f_1(t) + \ldots + a_k f_k(t)$$

The expected demand can also be written in vector form by

$$\mu_t = a' f(t)$$

where

$$a' = [a_1, \ldots, a_k]$$

and

$$f(t) = \begin{bmatrix} f_1(t) \\ \cdot \\ \cdot \\ \cdot \\ f_k(t) \end{bmatrix}$$

The first condition is satisfied for those expected demands where a matrix L exists whereby

$$f(t) = Lf(t-1)$$

or

$$\begin{bmatrix} f_1(t) \\ \cdot \\ \cdot \\ \cdot \\ f_k(t) \end{bmatrix} = \begin{bmatrix} L_{11} & \dots & L_{1k} \\ \cdot & & \cdot \\ \cdot & & \cdot \\ \cdot & & \cdot \\ L_{k1} & \dots & L_{kk} \end{bmatrix} \begin{bmatrix} f_1(t-1) \\ \cdot \\ \cdot \\ \cdot \\ f_k(t-1) \end{bmatrix}$$

Example 7.4

The matrix L and the relation $f(t) = Lf(t-1)$ is as follows for various versions of the expected demand (μ_t):

μ_t	$f(t) = L \; f(t-1)$
1. $\mu_t = a_1$	$[1] = [1][1]$
2. $\mu_t = a_1 + a_2 t$	$\begin{bmatrix} 1 \\ t \end{bmatrix} = \begin{bmatrix} 1 & 0 \\ 1 & 1 \end{bmatrix} \begin{bmatrix} 1 \\ t-1 \end{bmatrix}$
3. $\mu_t = a_1 + a_2 t + a_3 t^2$	$\begin{bmatrix} 1 \\ t \\ t^2 \end{bmatrix} = \begin{bmatrix} 1 & 0 & 0 \\ 1 & 1 & 0 \\ 1 & 2 & 1 \end{bmatrix} \begin{bmatrix} 1 \\ t-1 \\ (t-1)^2 \end{bmatrix}$
4. $\mu_t = a_1 e^{ct}$	$[e^{ct}] = [e^c][e^{c(t-1)}]$
5. $\mu_t = a_1 \sin ct + a_2 \cos ct$	$\begin{bmatrix} \sin ct \\ \cos ct \end{bmatrix} = \begin{bmatrix} \cos c & \sin c \\ -\sin c & \cos c \end{bmatrix} \begin{bmatrix} \sin c(t-1) \\ \cos c(t-1) \end{bmatrix}$

Second Condition [Concerning $F(T)$]

The second condition concerns the matrix

$$F(T) = \begin{bmatrix} F_{11}(T) & \dots & F_{1k}(T) \\ \cdot & & \cdot \\ \cdot & & \cdot \\ \cdot & & \cdot \\ F_{k1}(T) & \dots & F_{kk}(T) \end{bmatrix}$$

This condition is satisfied when T is large enough so that $F(T) = F(T-1)$. In essence, each element in $F(T)$ is the same for all time periods from $T-1$ and beyond. When this condition is satisfied, F is used in place of $F(T)$ and the elements of F are denoted by F_{in}, i.e.,

$$F = \begin{bmatrix} F_{11} & \dots & F_{1k} \\ \cdot & & \cdot \\ \cdot & & \cdot \\ \cdot & & \cdot \\ F_{k1} & \dots & F_{kk} \end{bmatrix}$$

The element F_{in} is found by

$$F_{in} = \sum_{j=0}^{\infty} \beta^j f_i(-j) f_n(-j)$$

Although T generally is not large enough to satisfy the condition above, F can safely be used in many situations nevertheless. This occurs when the prior estimates of the coefficients $(\hat{a}_1(T-1), \dots, \hat{a}_k(T-1))$ are deemed fairly precise. Under these circumstances the estimates have reached an equilibrium-type status and the matrix F need not be updated any longer. Even at the outset, should these estimates be obtained in a reliable manner, the matrix F can be used.

SOLVING FOR *h*. Assuming that the two conditions of adaptive smoothing are satisfied, the vector h can now be generated. This is obtained by using F and the vector $f(t)$ with $t = 0$ in the following relation:

$$h = \begin{bmatrix} h_1 \\ \cdot \\ \cdot \\ \cdot \\ h_k \end{bmatrix} = \begin{bmatrix} F_{11} \dots F_{1k} \\ \cdot \quad\quad \cdot \\ \cdot \quad\quad \cdot \\ \cdot \quad\quad \cdot \\ F_{k1} \dots F_{kk} \end{bmatrix}^{-1} \begin{bmatrix} f_1(0) \\ \cdot \\ \cdot \\ \cdot \\ f_k(0) \end{bmatrix}$$

Example 7.5

Find h when $\mu_t = a_1$ and $\beta = 0.9$.

Here $f(t) = [f_1(t)] = [1]$ and $f_1(0) = 1$. Also,

$$F = [F_{11}] = \left[\sum_{j=0}^{\infty} \beta^j\right] = \left[\frac{1}{1-\beta}\right]$$

and

$$h = [F_{11}]^{-1}[f_1(0)] = \left[\frac{1}{1-\beta}\right]^{-1}[1] = [1-\beta][1]$$

$$= [1-\beta] = [0.1]$$

Example 7.6

Find h when $\mu_t = a_1 + a_2 t$ and $\beta = 0.9$.

Since the vector $f(t)$ is

$$f(t) = \begin{bmatrix} 1 \\ t \end{bmatrix}$$

then

$$f(0) = \begin{bmatrix} 1 \\ 0 \end{bmatrix}$$

Now

$$F = \begin{bmatrix} F_{11} & F_{12} \\ F_{21} & F_{22} \end{bmatrix} = \begin{bmatrix} \sum_{j=0}^{\infty} \beta^j & -\sum_{j=0}^{\infty} j\beta^j \\ -\sum_{j=0}^{\infty} j\beta^j & \sum_{j=0}^{\infty} j^2\beta^j \end{bmatrix}$$

$$= \begin{bmatrix} \frac{1}{1-\beta} & \frac{-\beta}{(1-\beta)^2} \\ \frac{-\beta}{(1-\beta)^2} & \frac{\beta(1+\beta)}{(1-\beta)^3} \end{bmatrix} = \begin{bmatrix} 10 & -90 \\ -90 & 1710 \end{bmatrix}$$

and

$$h = \begin{bmatrix} h_1 \\ h_2 \end{bmatrix} = \begin{bmatrix} 10 & -90 \\ -90 & 1710 \end{bmatrix}^{-1} \begin{bmatrix} 1 \\ 0 \end{bmatrix}$$

$$= \begin{bmatrix} 0.19 & 0.01 \\ 0.01 & 0.00111 \end{bmatrix} \begin{bmatrix} 1 \\ 0 \end{bmatrix}$$

$$= \begin{bmatrix} 0.19 \\ 0.01 \end{bmatrix}$$

Example 7.7

Show how to find $a(T)$ when $\mu_t = a_1$ and $\beta = 0.9$.

Since

$$\hat{a}(T) = L'\hat{a}(T-1) + h(x_T - \hat{x}_{T-1}(1))$$

and

$$L' = [1]' = [1]$$

$$\hat{a}(T-1) = [\hat{a}_1(T-1)]$$

$$h = [0.1]$$

$$\hat{x}_{T-1}(1) = \hat{a}_1(T-1)$$

then

$$\hat{a}(T) = [1][\hat{a}_1(T-1)] + [0.1][x_T - \hat{a}_1(T-1)]$$

$$= \hat{a}_1(T-1) + 0.1(x_T - \hat{a}_1(T-1))$$

Example 7.8

Assume in Example 7.7 that $\hat{a}_1(T-1) = 10$ and $x_T = 12$. Find $\hat{a}(T)$ and $\hat{x}_T(\tau)$.

$$\hat{a}(T) = 10 + 0.1(12 - 10) = 10.2$$

and

$$\hat{x}_T(\tau) = 10.2$$

Example 7.9

Show how to find $\hat{a}(T)$ when

$$\mu_t = a_1 + a_2 t \quad \text{and} \quad \beta = 0.9$$

Using the results of Example 7.6,

$$h = \begin{bmatrix} 0.19 \\ 0.01 \end{bmatrix}$$

Also, since

$$L = \begin{bmatrix} 1 & 0 \\ 1 & 1 \end{bmatrix}$$

then

$$L' = \begin{bmatrix} 1 & 1 \\ 0 & 1 \end{bmatrix}$$

and

$$\hat{a}(T) = \begin{bmatrix} \hat{a}_1(T) \\ \hat{a}_2(T) \end{bmatrix} = \begin{bmatrix} 1 & 1 \\ 0 & 1 \end{bmatrix} \begin{bmatrix} \hat{a}_1(T-1) \\ \hat{a}_2(T-1) \end{bmatrix} + \begin{bmatrix} 0.19 \\ 0.01 \end{bmatrix} (x_T - \hat{x}_{T-1}(1))$$

Carrying out the matrix calculations gives

$$\hat{a}_1(T) = \hat{a}_1(T-1) + \hat{a}_2(T-1) + 0.19(x_T - \hat{x}_{T-1}(1))$$
$$\hat{a}_2(T) = \hat{a}_2(T-1) + 0.01(x_T - \hat{x}_{T-1}(1))$$

Example 7.10

Suppose in Example 7.9 that $\hat{a}_1(T-1) = 10$, $\hat{a}_2(T-1) = 1$, and $x_T = 12$. Find $\hat{x}_T(\tau)$.

First note that the one-period-ahead forecast at $T - 1$ is

$$\hat{x}_{T-1}(1) = \hat{a}_1(T-1) + \hat{a}_2(T-1) \cdot 1$$
$$= 10 + 1(1) = 11$$

Hence,

$$\hat{a}_1(T) = 10 + 1 + 0.19(12 - 11) = 11.19$$
$$\hat{a}_2(T) = 1 + 0.01(12 - 11) = 1.01$$

and

$$\hat{x}_T(\tau) = 11.19 + 1.01\tau$$

Example 7.11

Show how to find $\hat{x}_T(\tau)$ when $\mu_t = a_1 e^{0.1t}$ and $\beta = 0.9$.

Note that $f(-j) = e^{-0.1j}$, since

$$\mu_{T-j} = a_1 e^{-0.1j}$$

Now

$$F = \sum_{j=0}^{\infty} \beta^j f(-j) f(-j) = \sum_{j=0}^{\infty} \beta^j e^{-0.2j}$$

$$= \frac{1}{1 - \beta/e^{0.2}} = \frac{1}{1 - 0.9/1.2214} = 3.8$$

$$F^{-1} = \frac{1}{3.8} = 0.2632$$

$$h = F^{-1} f_1(0) = 0.2632 e^0 = 0.2632$$

$$L = e^{0.1}$$

$$L' = e^{0.1} = 1.1052$$

$$\hat{a}(T) = L'\hat{a}(T-1) + h(x_T - \hat{x}_{T-1}(1))$$

$$= 1.1052\hat{a}(T-1) + 0.2632(x_T - \hat{x}_{T-1}(1))$$

and

$$\hat{x}_T(\tau) = \hat{a}(T)e^{0.1\tau}$$

Example 7.12

Using the results of Example 7.11 suppose that $x_T = 100$ and $\hat{a}(T-1) = 95$. Find $\hat{x}_T(\tau)$ for $\tau = 1$ and 2.

Since $\hat{a}(T-1) = 95$, then

$$\hat{x}_{T-1}(1) = 95e^{0.1} = 95(1.1052) = 104.99$$

Hence,

$$\hat{a}(T) = 1.1052(95) + 0.2632(100 - 104.99)$$

$$= 103.68$$

$$\hat{x}_T(1) = 103.68e^{0.1} = 103.68(1.1052) = 114.58$$

$$\hat{x}_T(2) = 103.68e^{0.2} = 103.68(1.2214) = 126.63$$

The Discounting Rate

Brown [2] shows that the discounting rate should be selected to conform with the number of terms (k) in the expected demand equation. Let β_1 be the discounting rate which seems appropriate when the expected demand contains one term (i.e., $k = 1$). Now let β_k represent the discounting rate for a demand pattern with k terms. The relationship between β_k and β_1 is

$$\beta_k = \beta_1^{1/k}$$

Some representative values of β_k and β_1 are listed in Table 7-1.

Table 7-1. VALUES OF β_k FOR $\beta_1 = 0.95, 0.90, 0.85$, AND 0.80

β_1	β_2	β_3	β_4	β_5	β_6
0.95	0.9747	0.9830	0.9873	0.9898	0.9915
0.90	0.9487	0.9655	0.9740	0.9791	0.9826
0.85	0.9220	0.9473	0.9602	0.9680	0.9733
0.80	0.8944	0.9283	0.9457	0.9564	0.9635

MATHEMATICAL BASIS. Using the demands $x_1, x_2, \ldots, x_T$ and the discounting rate β, the sum of discounted squares of residual errors is

$$S(e) = \sum_{j=0}^{T-1} \beta^j (x_{T-j} - \hat{x}_{T-j})^2$$

where

$$\hat{x}_{T-j} = \sum_{i=1}^{k} \hat{a}_i(T) f(-j)$$

With the least-squares method, the following partial derivatives are taken and set to zero.

$$\frac{\partial S(e)}{\partial \hat{a}_i(T)} = -2 \sum_{j=0}^{T-1} \beta^j f_i(-j)[x_{T-j} - \hat{a}_1 f_1(-j) - \ldots - \hat{a}_k f_k(-j)]$$

for $i = 1, 2, \ldots, k$. These results are used to generate the matrix relation

$$g(T) = F(T)\hat{a}(T)$$

where $g(T)$, $F(T)$, and $\hat{a}(T)$ are as shown earlier.

Now estimates for $\hat{a}(T)$ are

$$\hat{a}(T) = F^{-1}(T)g(T)$$

When $F = F(T)$, then

$$g(T) = F\hat{a}(T)$$

and

$$\hat{a}(T) = F^{-1}g(T)$$

The forecast for the τth future time period is

$$\hat{x}_T(\tau) = f(\tau)\hat{a}(\tau)'$$

It is noted that $g(T)$ is recursively related to $g(T-1)$ when $f(T-1) = L^{-1}f(T)$. Observe that

$$g(T-1) = x_{T-1}f(0) + \beta x_{T-2}f(-1) + \ldots + \beta^{T-2}x_1 f(-(T-2))$$

and

$$\begin{aligned} g(T) &= x_T f(0) + \beta x_{T-1}f(-1) + \ldots + \beta^{T-1}x_1 f(-(T-1)) \\ &= x_T f(0) + \beta L^{-1}[x_{T-1}f(0) + \ldots + \beta^{T-2}x_1 f(-(T-1))] \\ &= x_T f(0) + \beta L^{-1}g(T-1) \end{aligned}$$

Now, the estimates $\hat{a}(T)$ are recursively related to $\hat{a}(T-1)$. This is shown in the following manner. Since

$$g(T-1) = F\hat{a}(T-1)$$

then

$$\begin{aligned} g(T) &= x_T f(0) + \beta L^{-1}g(T-1) \\ &= x_T f(0) + \beta L^{-1}F\hat{a}(T-1) \end{aligned}$$

Hence,

$$\begin{aligned} \hat{a}(T) &= F^{-1}g(T) \\ &= F^{-1}[x_T f(0) + \beta L^{-1}F\hat{a}(T-1)] \\ &= F^{-1}f(0)x_T + \beta F^{-1}L^{-1}F\hat{a}(T-1) \\ &= hx_T + H\hat{a}(T-1) \end{aligned}$$

where

$$\begin{aligned} h &= F^{-1}f(0) \\ H &= \beta F^{-1}L^{-1}F \end{aligned}$$

The relation above for $\hat{a}(T)$ can be further simplified. Note that

$$\begin{aligned} L^{-1}F &= (L^{-1}F)(L'^{-1}L') \\ &= L^{-1}\sum_{j=0}^{\infty} \beta^j f(-j)f(-j)'L'^{-1}L' \\ &= \sum_{j=0}^{\infty} \beta^j (L^{-1}f(-j))(L^{-1}f(-j))'L' \\ &= \sum_{j=0}^{\infty} \beta^j f(-(j+1))f(-(j+1))'L' \\ &= \frac{1}{\beta}\left[\sum_{j=1}^{\infty} \beta^j f(-j)f(-j)' + f(0)f(0)' - f(0)f(0)'\right]L' \\ &= \frac{1}{\beta}[F - f(0)f(0)']L' \end{aligned}$$

Now H becomes

$$\begin{aligned}
H &= \beta F^{-1}L^{-1}F \\
&= \beta F^{-1}\frac{1}{\beta}[F - f(0)f(0)']L' \\
&= [F^{-1}F - F^{-1}f(0)f(0)']L' \\
&= L' - F^{-1}f(0)f(0)'L' \\
&= L' - hf(0)'L' \\
&= L' - hf(1)'
\end{aligned}$$

Finally,

$$\begin{aligned}
\hat{a}(T) &= hx_T + H\hat{a}(T-1) \\
&= hx_T + [L' - hf(1)]\hat{a}(T-1) \\
&= h[x_T - f(1)'\hat{a}(T-1)] + L'\hat{a}(T-1) \\
&= h[x_T - \hat{x}_{T-1}(1)] + L'\hat{a}(T-1)
\end{aligned}$$

PROBLEMS

7-1. Consider the following history of demands for an item where currently $T = 8$: 14, 17, 20, 25, 30, 34, 44, and 51. Assuming the expected demands follow a pattern of the following type:

$$\mu_t = a_1 + a_2 e^{0.2t}$$

where $e = 2.7183$, find the estimates of a_1 and a_2 using the regression model. Use the results to forecast the demands for one period ahead. Note that

$e^{0.2} = 1.221$	$e^{1.2} = 3.320$
$e^{0.4} = 1.492$	$e^{1.4} = 4.055$
$e^{0.6} = 1.822$	$e^{1.6} = 4.953$
$e^{0.8} = 2.225$	$e^{1.8} = 6.050$
$e^{1.0} = 2.718$	$e^{2.0} = 7.389$

7-2. Suppose that the history of demands for an item with $T = 5$ are the following: 5, 8, 9, 12, and 13, and the assumed pattern of expected demands is

$$\mu_t = a_1 + a_2 \ln t$$

where $\ln t$ = natural log of t. Using the regression model, find estimates of a_1 and a_2 and the one-period-ahead forecast. Note that:

$$\ln 1 = 0.000, \qquad \ln 2 = 0.693$$
$$\ln 3 = 1.099, \qquad \ln 4 = 1.386$$
$$\ln 5 = 1.609, \qquad \ln 6 = 1.792$$

7-3. An item has the following six demands: 3, 7, 10, 12, 17, and 18. Use the regression model to find the cumulative forecast for the next three time periods when the expected demand is the following:

$$\mu_t = bt$$

where b is a constant. Note that the intercept is assumed as zero.

7-4. Resolve Problem 7-1 using the regression model with a time shift. Note that $j = 8 - t$ and

$$\mu_{T-j} = a_1 + a_2 e^{-0.2j} \qquad j = 0, 1, \ldots, 7$$
$$\mu_{T+\tau} = a_1 + a_2 e^{0.2\tau} \qquad \tau = 1$$

Note that:

$$e^0 = 1.00 \qquad e^{-1.0} = 0.368$$
$$e^{-0.2} = 0.819 \qquad e^{-1.2} = 0.301$$
$$e^{-0.4} = 0.670 \qquad e^{-1.4} = 0.247$$
$$e^{-0.6} = 0.549 \qquad e^{-1.6} = 0.202$$
$$e^{-0.8} = 0.449 \qquad e^{0.2} = 1.221$$

7-5. Consider the expected demand model

$$\mu_{T-j} = a_1 - a_2 j \qquad j = 0, 1, 2, \ldots, T-1$$
$$\mu_{T+\tau} = a_1 + a_2 \tau \qquad \tau = 0, 1, 2, \ldots$$

Using the model with the demands: 4, 9, 12, 15, and 20, find the three-period-ahead forecast of demands. Note that $T = 5$.

7-6. Resolve Problem 7-2 using a discount parameter of $\beta = 0.8$. Note there is no time shift.

7-7. Resolve Problem 7-3 using a discount parameter of $\beta = 0.6$. Note there is no time shift.

7-8. Resolve Problem 7-1 using a discount parameter of $\beta = 0.9$. Note there is no time shift.

7-9. Resolve Problem 7-5 using a discount parameter of $\beta = 0.9$.

7-10. Resolve Problem 7-4 using a discount parameter of $\beta = 0.8$.

7-11. In accordance with the adaptive smoothing model, find L, $f(t)$, and $f(0)$ for the following expected demand relations:

a. $\mu_t = a + bt + \frac{1}{2}ct(t-1)$
b. $\mu_t = ae^{4t} + bte^{4t}$
c. $\mu_t = a + bt + ce^{0.5t}$
d. $\mu_t = a + bt + \frac{1}{2}ct(t-1) + de^{.03t}$

7-12. Suppose that $\mu_t = a_1$ and $\beta = 0.8$. Find $f(t)$, $f(0)$, F, h, L, F^{-1}, and L'.

7-13. Use the results of Problem 7-12 to find the following:
a. The updating relation to find $\hat{a}_1(T)$ given x_T and $\hat{a}_1(T-1)$.
b. The forecast equation at time T for the τth future demand.
c. If $x_T = 12$ and $\hat{a}_1(T-1) = 10$, find the forecast for the next time period.

7-14. Consider $\mu_t = a_1 + a_2 t$ with $\beta = 0.8$. Find $f(t)$, $f(0)$, F, h, and L.

7-15. In Problem 7-14,

$$F^{-1} = \begin{bmatrix} 0.36 & 0.04 \\ 0.04 & 0.01 \end{bmatrix}$$

Using the results of Problem 7-14, find the following:
a. The updating relation to find $\hat{a}_1(T)$ and $\hat{a}_2(T)$ when $\hat{a}_1(T-1)$, $\hat{a}_2(T-1)$, and x_T are known.
b. If $x_T = 100$, $\hat{a}_1(T-1) = 90$, and $\hat{a}_2(T-1) = 15$, find the updated forecast for the cumulative demand of the next three time periods.

7-16. Suppose that $\mu_t = a_1 e^{0.2t}$ and $\beta = 0.8$. Now find F, F^{-1}, h, L, and L'.

7-17. Use the results of Problem 7-16 to find the following:
a. The updating relation for $\hat{a}_1(T)$.
b. The forecast equation for the τth future time period.
c. The one-period-ahead forecast when $\hat{a}_1(T-1) = 100$ and $x_T = 120$.

REFERENCES

[1] Brown, R. G., *Smoothing, Forecasting and Prediction of Discrete Time Series.* Englewood Cliffs, N.J.: Prentice-Hall, Inc., 1962, Chap. 11.

[2] Brown, R. G., *Smoothing, Forecasting and Prediction of Discrete Time Series.* Englewood Cliffs, N.J.: Prentice-Hall, Inc., 1962, Chap. 12.

REFERENCE FOR FURTHER STUDY

Meyer, R. F., "An Adaptive Method for Routine Short Term Forecasting." International Federation of Operational Research Societies, Oslo, July 1963.

8

TRIGONOMETRIC MODELS

The adaptive smoothing models of Chapter 7 are flexible since they are applicable with a wide variety of demand patterns. An important set of such models are called *trigonometric models* and are used with demand patterns of the seasonal type. The models are called "trigonometric" because the time functions $[f_i(t), i = 1, 2, \ldots, k]$ includes a series of sine and cosine terms.

In this chapter various types of trigonometric models are described. The first applies when the level is horizontal in the long run but follows a simple sine-wave pattern for the span of a seasonal cycle. Here three terms are required to define the expected demand and as such the model is called the three-term model. The second model includes a trend effect and is called the four-term model—this is because four terms are needed to define the expected demand. In the third model a trend factor and two sine waves are used to define the expected demand. Since six terms are needed in this pattern, the model is called a six-term model. Finally, a summary of certain higher-order models are given. These are the eight-, ten-, and twelve-term models.

In this chapter k will represent the number of terms included in the model and β will designate the discounting rate. Brown [1] shows that the discounting rate for a k-term model should be related to the discounting rate that the forecaster desires for a one-term model (β_1). This relationship is

$$\beta = \beta_1^{1/k}$$

Hence, if $\beta_1 = 0.90$, then for a three-term model the discounting rate is $\beta = 0.9^{1/3} = 0.9655$. In the same way, the discounting rates for four- and six-term models become $\beta = 0.9^{1/4} = 0.9740$, and $\beta = 0.9^{1/6} = 0.9826$, respectively.

8-1 THREE-TERM MODEL

Consider a seasonal item whose demand pattern flows up and down in a sine-wave manner and where the average level is the same from year to year. In this sense the demand pattern is seasonal within a cycle and is horizontal over the long run. When the seasonal cycle covers one full year and the demand entries are observed at monthly time intervals, the cycle length is $M = 12$ months.

The expected demand at time t is defined by the following relation:

$$\mu_t = a_1 + a_2 \sin \omega t + a_3 \cos \omega t$$

where

$$\begin{aligned} a_1 &= \text{level over the long run} \\ a_2 &= \text{amplitude of } \sin \omega t \\ a_3 &= \text{amplitude of } \cos \omega t \\ \omega &= 30 \text{ degrees} \end{aligned}$$

Note that ω is determined by the number of time periods that are observed in one cycle length. In this situation the cycle covers one full year and a sine wave completes one full cycle in 360 degrees. Since 1 month is $\frac{1}{12}$ of a year, then $\omega = (\frac{1}{12})360° = 30°$ (30 degrees). If the demand entries were observed weekly, then $M = 52$ and $\omega = (1/52)360° = 6.92°$.

Also observe in the expected demand relation above that $f_1(t) = 1$, $f_2(t) = \sin \omega t$ and $f_3(t) = \cos \omega t$. Here, three terms are required to define the demand pattern, and the corresponding trigonometric model is called a three-term model.

Since in adaptive smoothing, the time axis is always shifted so that the current time period acts as the origin of time, the expected demand relation above is modified. For the jth time period of the past, it is

$$\mu_{T-j} = a_1 + a_2 \sin(-\omega j) + a_3 \cos(-\omega j)$$

and for the τth future time period

$$\mu_{T+\tau} = a_1 + a_2 \sin \omega\tau + a_3 \cos \omega\tau$$

At the current time period (T), the estimates of the coefficients are found and labeled as $\hat{a}_1(T)$, $\hat{a}_2(T)$, and $\hat{a}_3(T)$ for a_1, a_2, and a_3, respectively. When $M = 12$, these estimates are updated in the following manner:

$$\hat{a}_1(T) = h_1 e_T + \hat{a}_1(T-1)$$
$$\hat{a}_2(T) = h_2 e_T + 0.86603\hat{a}_2(T-1) - 0.5\hat{a}_3(T-1)$$
$$\hat{a}_3(T) = h_3 e_T + 0.5\hat{a}_2(T-1) + 0.86603\hat{a}_3(T-1)$$

Here

$\hat{a}_1(T-1)$, $\hat{a}_2(T-1)$, and $\hat{a}_3(T-1)$ are the corresponding estimates at $t = T-1$,

$e_T = (x_T - \hat{x}_{T-1}(1))$ is the most current one-period-ahead forecast error, and h_1, h_2, and h_3 are smoothing coefficients that depend on the discounting rate (β). For example, when $\beta = 0.7^{1/3}$, $0.8^{1/3}$, or $0.9^{1/3}$, then the values of h_1, h_2, and h_3 are as follows:

	$\beta = 0.7^{1/3}$	$\beta = 0.8^{1/3}$	$\beta = 0.9^{1/3}$
h_1	0.10479	0.06792	0.03347
h_2	0.06481	0.02707	0.00640
h_3	0.19521	0.13208	0.06653

The forecast for the τth future time period is

$$\hat{x}_T(\tau) = \hat{a}_1(T) + \hat{a}_2(T)\sin\omega\tau + \hat{a}_3(T)\cos\omega\tau$$

where the values of $\sin\omega\tau$ and $\cos\omega\tau$ are listed in Table 8-1 for $\tau = 1$ to 12. These values repeat themselves when τ is larger than 12. For example, at

Table 8-1. VALUES OF SIN $\omega\tau$ AND COS $\omega\tau$

τ	$\omega\tau$	$\sin\omega\tau$	$\cos\omega\tau$
1	30	0.500	0.866
2	60	0.866	0.500
3	90	1.000	0.000
4	120	0.866	−0.500
5	150	0.500	−0.866
6	180	0.000	−1.000
7	210	−0.500	−0.866
8	240	−0.866	−0.500
9	270	−1.000	0.000
10	300	−0.866	0.500
11	330	−0.500	0.866
12	360	−0.000	1.000

$\tau = 13$, $\sin 13\omega = \sin 1\omega$ and $\cos 13\omega = \cos 1\omega$. For $\tau = 14$, $\sin 14\omega = \sin 2\omega$ and $\cos 14\omega = \cos 2\omega$. Now, as before, the forecast for the sum of the following τ time periods is

$$\hat{X}_T(\tau) = \hat{x}_T(1) + \hat{x}_T(2) + \ldots + \hat{x}_T(\tau)$$

Example 8.1

Consider an item where $M = 12$ and the three-term model is in use with $\beta = 0.9^{1/3} = 0.9655$. Suppose at time T that the current demand is $x_T = 143$ and the prior estimates of the coefficients are $\hat{a}_1(T-1) = 100$, $\hat{a}_2(T-1) = 40$, and $\hat{a}_3(T-1) = 10$. Find the forecasts for the sum of the following three time periods.

First observe that the prior one-period-ahead forecast is

$$\hat{x}_{T-1}(1) = 100 + 40(0.500) + 10(0.866) = 128.66$$

This yields the forecast error

$$e_T = 143 - 128.66 = 14.34$$

Now, the updated coefficient estimates are found using the appropriate values for h_1, h_2, and h_3. These give

$$\begin{aligned}
\hat{a}_1(T) &= 0.03347(14.34) + (100) &&= 100.48 \\
\hat{a}_2(T) &= 0.00640(14.34) + 0.86603(40) - 0.5(10) &&= 29.74 \\
\hat{a}_3(T) &= 0.06653(14.34) + 0.5(40) + 0.86603(10) &&= 29.61
\end{aligned}$$

Hence, the forecast equation becomes

$$\hat{x}_T(\tau) = 100.48 + 29.74 \sin \omega\tau + 29.61 \cos \omega\tau$$

Using $\tau = 1, 2$, and 3,

$$\begin{aligned}
\hat{x}_T(1) &= 100.48 + 29.74(0.500) + 29.61(0.866) = 141.0 \\
\hat{x}_T(2) &= 100.48 + 29.74(0.866) + 29.61(0.500) = 141.0 \\
\hat{x}_T(3) &= 100.48 + 29.74(1.000) + 29.61(0.000) = 130.2
\end{aligned}$$

Finally, the forecast of demands for the following three time periods is

$$\hat{X}_T(3) = 141.0 + 141.0 + 130.2 = 412.2$$

Example 8.2

Continuing with Example 8.1, find the coefficient estimates and one-period-ahead forecasts for the demands listed in Table 8-2.

The results are listed in Table 8-2 and a plot of the one-period-ahead forecasts is shown in Figure 8-1.

Table 8-2. DEMANDS, COEFFICIENT ESTIMATES, AND ONE-PERIOD-AHEAD FORECASTS USING THE THREE-TERM MODEL

t	x_t	$\hat{a}_1(t)$	$\hat{a}_2(t)$	$\hat{a}_3(t)$	$\hat{x}_t(1)$
1	143	100.48	29.74	29.61	141.0
2	129	100.08	10.86	39.71	139.9
3	160	100.75	−10.32	41.16	131.2
4	134	100.84	−29.50	30.67	112.7
5	76	99.62	−41.12	9.37	87.2
6	130	101.05	−40.02	− 9.59	72.7
7	62	100.69	−29.93	−29.03	60.6
8	66	100.87	−11.37	−39.74	60.8
9	20	99.51	9.77	−42.81	67.3
10	66	99.46	29.86	−32.28	86.4
11	139	101.22	42.33	− 9.53	114.1
12	110	101.08	41.40	12.64	132.7
13	133	101.09	29.53	31.66	143.3
14	119	100.28	9.59	40.57	140.2
15	165	101.11	−11.82	41.58	131.2
16	104	100.20	−31.20	28.29	109.1
17	141	101.27	−40.96	11.02	90.3
18	90	101.26	−40.98	−10.96	71.3
19	57	100.78	−30.10	−30.93	58.9
20	71	101.18	−10.52	−41.04	60.4
21	10	99.50	11.08	−44.15	66.8
22	66	99.47	31.67	−32.75	86.9
23	99	99.87	43.87	−11.72	111.7
24	105	99.65	43.82	11.34	131.4

Initial Forecasts

As in the smoothing models of earlier chapters, a difficulty arises in initializing the system. This is because estimates of the coefficients from the prior time period are needed at the outset and may not be available to the forecaster.

In the situation where no prior demand entries are known, the forecaster must estimate the coefficients as best as possible. Using $T = 1$ for this initial time period, the estimates are $\hat{a}_1(0)$, $\hat{a}_2(0)$, and $\hat{a}_3(0)$. With these he can find $\hat{x}_0(1)$. Now applying x_1, the updated coefficient estimates are obtained as shown earlier.

A second situation occurs when there is some prior demand entries from which estimates of the coefficients can be generated. These estimates may be obtained using the method of least squares.

In this endeavor, consider the prior demand entries as of time T,

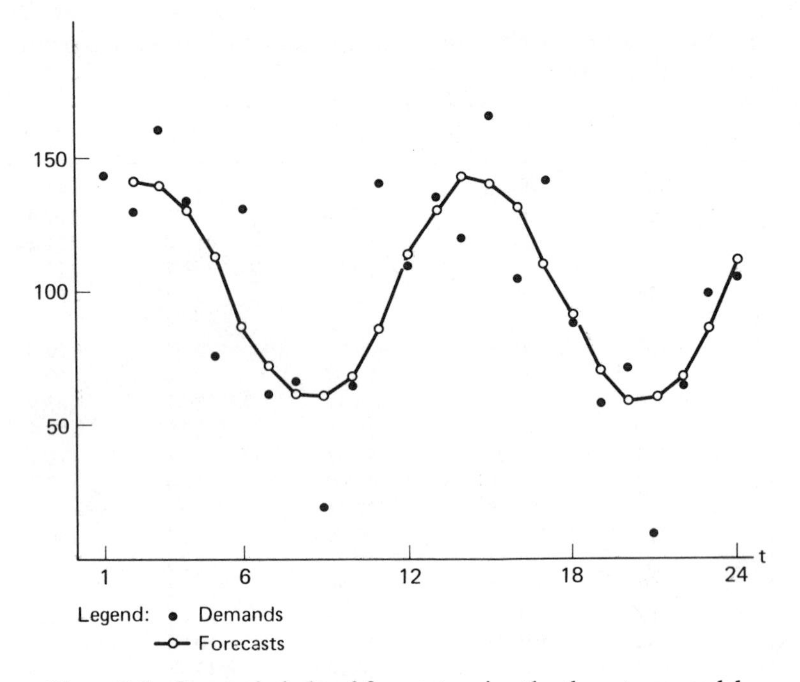

Figure 8-1. One-period-ahead forecasts using the three-term model.

$(x_1, x_2, \ldots, x_T)$. The least-squares equation becomes

$$\sum x_{T-j} = T\hat{a}_1 + \hat{a}_2 \sum \sin(-\omega j) + \hat{a}_3 \sum \cos(-\omega j)$$

$$\sum \sin(-\omega j) x_{T-j} = \hat{a}_1 \sum \sin(-\omega j) + \hat{a}_2 \sum \sin^2(-\omega j) + \hat{a}_3 \sum \sin(-\omega j)\cos(-\omega j)$$

$$\sum \cos(-\omega j) x_{T-j} = \hat{a}_1 \sum \cos(-\omega j) + \hat{a}_2 \sum \sin(-\omega j)\cos(-\omega j) + \hat{a}_3 \sum \cos^2(-\omega j)$$

where the summations range from $j = 0$ to $T - 1$. The solutions for $\hat{a}_1$, $\hat{a}_2$, and $\hat{a}_3$ are readily obtained and used as the estimates needed to get started. These become $\hat{a}_1(T) = \hat{a}_1$, $\hat{a}_2(T) = \hat{a}_2$, and $\hat{a}_3(T) = \hat{a}_3$.

Example 8.3

Assuming that $T = 12$, use the first 12 demand entries of Table 8-2 to estimate the coefficients needed.

The summations required are shown in Table 8-3 and yield the following three equations:

$$1235 = 12\hat{a}_1 + 0\hat{a}_2 + 0\hat{a}_3$$

$$262.43 = 0\hat{a}_1 + 6\hat{a}_2 + 0\hat{a}_3$$

$$102.20 = 0\hat{a}_1 + 0\hat{a}_2 + 6\hat{a}_3$$

Table 8-3. WORKSHEET TO INITIALIZE THE THREE-TERM MODEL (EXAMPLE 8.3)

$-j$	$T-j$	x_{T-j}	$\sin(-\omega j)$	$\cos(-\omega j)$	$x_{T-j}\cdot \sin(-\omega j)$	$x_{T-j}\cdot \cos(-\omega j)$	$\sin^2(-\omega j)$	$\cos^2(-\omega j)$	$\sin(-\omega j)\cdot \cos(-\omega j)$
−11	1	143	0.500	0.866	71.50	123.84	0.250	0.750	0.433
−10	2	129	0.866	0.500	111.71	64.50	0.750	0.250	0.433
−9	3	160	1.000	0.000	160.00	0.00	1.000	0.000	0.000
−8	4	134	0.866	−0.500	116.04	−67.00	0.750	0.250	−0.433
−7	5	76	0.500	−0.866	38.00	−65.82	0.250	0.750	−0.433
−6	6	130	0.000	−1.000	0.00	−130.00	0.000	1.000	0.000
−5	7	62	−0.500	−0.866	−31.00	−53.69	0.250	0.750	0.433
−4	8	66	−0.866	−0.500	−57.16	−33.00	0.750	0.250	0.433
−3	9	20	−1.000	0.000	−20.00	0.00	1.000	0.000	0.000
−2	10	66	−0.866	0.500	−57.16	33.00	0.750	0.250	−0.433
−1	11	139	−0.500	0.866	−69.50	120.37	0.250	0.750	−0.433
0	12	110	0.000	1.000	0.00	110.00	0.000	1.000	0.000
Sums		1235	0.000	0.000	262.43	102.20	6.000	6.000	0.000

Hence, the estimates at $T = 12$ are

$$\hat{a}_1(12) = \hat{a}_1 = \frac{1235}{12} = 102.92$$

$$\hat{a}_2(12) = \hat{a}_2 = \frac{262.43}{6} = 43.74$$

$$\hat{a}_3(12) = \hat{a}_3 = \frac{102.20}{6} = 17.03$$

Mathematical Basis. In this three-term model,

$$f(t) = \begin{bmatrix} 1 \\ \sin \omega t \\ \cos \omega t \end{bmatrix}$$

$$L = \begin{bmatrix} 1 & 0 & 0 \\ 0 & \cos \omega & \sin \omega \\ 0 & -\sin \omega & \cos \omega \end{bmatrix}$$

and

$$F = \begin{bmatrix} F_{11} & F_{12} & F_{13} \\ F_{21} & F_{22} & F_{23} \\ F_{31} & F_{32} & F_{33} \end{bmatrix}$$

where

$$F_{11} = \sum_{j=0}^{\infty} \beta^j$$

$$F_{12} = F_{21} = \sum_{j=0}^{\infty} \beta^j \sin(-\omega j)$$

$$F_{13} = F_{31} = \sum_{j=0}^{\infty} \beta^j \cos(-\omega j)$$

$$F_{22} = \sum_{j=0}^{\infty} \beta^j \sin^2(-\omega j)$$

$$F_{23} = F_{32} = \sum_{j=0}^{\infty} \beta^j \sin(-\omega j) \cos(-\omega j)$$

$$F_{33} = \sum_{j=0}^{\infty} \beta^j \cos^2(-\omega j)$$

These elements of F are easily generated using the relations listed in Appendix A. With F available,

$$h = F^{-1} f(0)$$

where

$$f(0) = \begin{bmatrix} 1 \\ 0 \\ 1 \end{bmatrix}$$

Hence, for $e_T = x_T - \hat{x}_{T-1}(1)$, then

$$\begin{bmatrix} \hat{a}_1(T) \\ \hat{a}_2(T) \\ \hat{a}_3(T) \end{bmatrix} = \begin{bmatrix} h_1 \\ h_2 \\ h_3 \end{bmatrix} e_T + \begin{bmatrix} 1 & 0 & 0 \\ 0 & \cos\omega & -\sin\omega \\ 0 & \sin\omega & \cos\omega \end{bmatrix} \begin{bmatrix} \hat{a}_1(T-1) \\ \hat{a}_2(T-1) \\ \hat{a}_3(T-1) \end{bmatrix}$$

8-2 FOUR-TERM MODEL

The four-term model is like the three-term model with an added term to account for the trend. The model accommodates demand patterns with an expected demand at time t defined by

$$\mu_t = a_1 + a_2\, t + a_3 \sin \omega t + a_4 \cos \omega t$$

As before, the time scale is shifted so that for the jth prior time period, the level is

$$\mu_{T-j} = a_1 - a_2 j + a_3 \sin(-\omega j) + a_4 \cos(-\omega j)$$

and for the τth future time period, it is

$$\mu_{T+\tau} = a_1 + a_2 \tau + a_3 \sin \omega\tau + a_4 \cos \omega\tau$$

The estimates of these coefficients at time T are designated by $\hat{a}_1(T)$, $\hat{a}_2(T)$, $\hat{a}_3(T)$, and $\hat{a}_4(T)$. Assuming that $M = 12$, these are obtained with the following relations:

$$\begin{aligned} \hat{a}_1(T) &= h_1 e_T + \hat{a}_1(T-1) + \hat{a}_2(T-1) \\ \hat{a}_2(T) &= h_2 e_T + \hat{a}_2(T-1) \\ \hat{a}_3(T) &= h_3 e_T + 0.86603\hat{a}_3(T-1) - 0.5\hat{a}_4(T-1) \\ \hat{a}_4(T) &= h_4 e_T + 0.5\hat{a}_3(T-1) + 0.86603\hat{a}_4(T-1) \end{aligned}$$

In the above, $\hat{a}_1(T-1)$, $\hat{a}_2(T-1)$, $\hat{a}_3(T-1)$, and $\hat{a}_4(T-1)$ are the corresponding estimates from $t = T - 1$, $e_T = x_T - \hat{x}_{T-1}(1)$ is the current one-period-ahead forecast error, and h_1, h_2, h_3, and h_4 are smoothing coefficients which depend on the discounting rate (β) that is chosen. The values of h_i ($i = 1, 2, 3, 4$) for common values of β are as follows:

	$\beta = 0.7^{1/4}$	$\beta = 0.8^{1/4}$	$\beta = 0.9^{1/4}$
h_1	0.15828	0.10216	0.05024
h_2	0.00685	0.00282	0.00066
h_3	0.06102	0.02557	0.00605
h_4	0.14172	0.09784	0.04976

Now the forecast for the τth future time period is

$$\hat{x}_T(\tau) = \hat{a}_1(T) + \hat{a}_2(T)\tau + \hat{a}_3(T) \sin \omega\tau + \hat{a}_4(T) \cos \omega\tau$$

The values for $\sin \omega\tau$ and $\cos \omega\tau$ are the same as those listed in Table 8.1 for the case when $M = 12$.

Example 8.4

Consider an item that is being forecast by a four-term model with $\beta = 0.9^{1/4}$. Suppose that at time T, $\hat{a}_1(T-1) = 40$, $\hat{a}_2(T-1) = 5$, $\hat{a}_3(T-1) = 40$, $\hat{a}_4(T-1) = 10$, and $x_T = 88$. Find the forecasts for the sum of the next three time periods.

The one-period-ahead forecast at $t = T - 1$ is

$$\hat{x}_{T-1}(1) = 40 + 5(1) + 40(0.500) + 10(0.866) = 73.66$$

and the current forecast error is

$$e_T = 88 - 73.66 = 14.34$$

So now the updated coefficient estimates are

$$\hat{a}_1(T) = 0.05024(14.34) + 40 + 5 = 45.72$$

$$\hat{a}_2(T) = 0.00066(14.34) + 5 = 5.01$$

$$\hat{a}_3(T) = 0.00605(14.34) + 0.86603(40) - 0.5(10) = 29.73$$

$$\hat{a}_4(T) = 0.04976(14.34) + 0.5(40) + 0.86603(10) = 29.37$$

and the forecast equation is

$$x_T(\tau) = 45.72 + 5.01\tau + 29.73 \sin \omega\tau + 29.37 \cos \omega\tau$$

With $\tau = 1, 2,$ and 3, the forecasts become

$$\hat{x}_T(1) = 45.72 + 5.01(1) + 29.73(0.500) + 29.37(0.866) = 91.0$$

$$\hat{x}_T(2) = 45.72 + 5.01(2) + 29.73(0.866) + 29.37(0.500) = 96.2$$

$$\hat{x}_T(3) = 45.72 + 5.01(3) + 29.73(1.00) + 29.37(0.000) = 90.5$$

Hence, the forecast for the sum of the following three time periods is

$$\hat{X}_T(3) = 91.0 + 96.2 + 90.5 = 277.7$$

Example 8.5

Continue with Example 8.4 for the demands listed in Table 8-4. Find the estimates of the coefficients and the one-period-ahead forecasts.

The results are listed in Table 8-4 and a plot of the one-period-ahead forecasts is shown in Figure 8-2.

Table 8-4. Demands, Coefficient Estimates, and One-Period-Ahead Forecasts Using the Four-Term Model

t	x_t	$a_1(t)$	$\hat{a}_2(t)$	$\hat{a}_3(t)$	$\hat{a}_4(t)$	$\hat{x}_t(1)$
1	88	45.72	5.01	29.73	29.37	91.0
2	79	50.13	5.00	10.98	39.70	95.0
3	115	56.13	5.01	−10.22	40.87	91.4
4	94	61.28	5.02	−29.27	30.41	78.0
5	41	64.43	4.99	−40.78	9.86	57.6
6	100	71.56	5.02	−39.99	−9.74	48.2
7	37	76.02	5.01	−29.83	−28.98	41.0
8	46	81.28	5.02	−11.31	−39.76	46.2
9	5	84.23	4.99	9.84	−42.14	57.6
10	56	89.13	4.99	29.58	−31.66	81.5
11	134	89.76	5.02	41.76	−10.01	114.0
12	110	101.58	5.02	41.15	12.01	137.6
13	138	106.62	5.02	29.63	31.00	153.3
14	129	110.42	5.00	10.01	40.45	155.5
15	180	116.66	5.02	−11.40	41.26	151.7
16	124	120.28	5.00	−30.67	28.65	134.8
17	160	126.86	5.02	−40.70	11.03	121.1
18	120	131.82	5.02	−40.76	−10.85	107.1
19	92	136.09	5.01	−29.97	−30.53	99.7
20	111	141.67	5.02	−10.62	−40.86	106.0
21	55	144.13	4.99	10.93	−43.23	117.1
22	116	149.05	4.98	31.07	−32.03	141.8
23	154	154.65	4.99	43.00	−11.60	171.1
24	165	159.34	4.99	43.00	11.15	195.5

Initial Forecasts

As in the three-term model, estimates of the coefficients are needed at the outset in order to get started. When no prior demand entries are available to the forecaster, the initial time period is $t = 0$. In this situation the forecaster uses his/her best judgment to arrive at estimates of $\hat{a}_1(0)$, $\hat{a}_2(0)$, $\hat{a}_3(0)$, and $\hat{a}_4(0)$. At $t = 1$ the demand entry x_1 becomes available and is used along

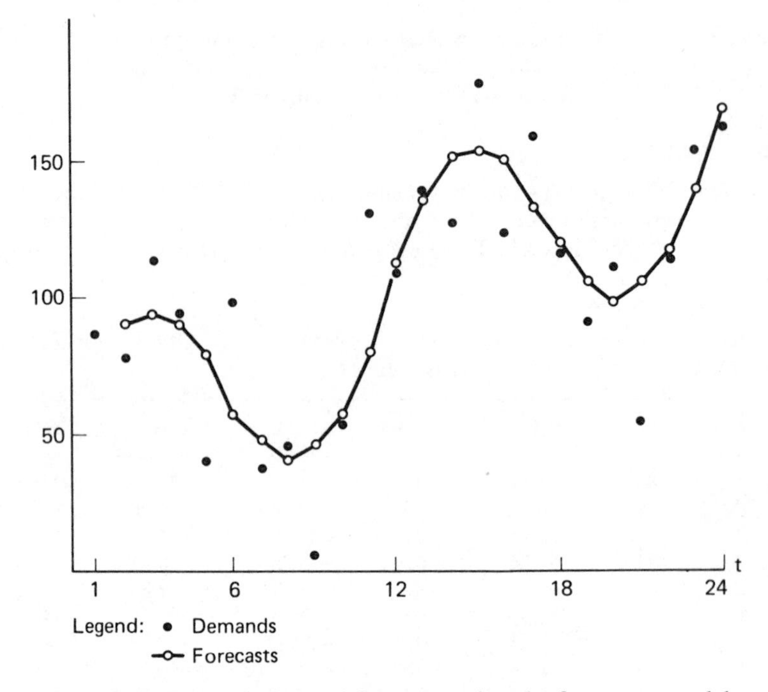

Figure 8-2. One-period-ahead forecasts using the four-term model.

with the prior estimates of coefficients to derive $\hat{a}_1(1)$, $\hat{a}_2(1)$, $\hat{a}_3(1)$, and $\hat{a}_4(1)$ from the relations given earlier. With these estimates the forecasts $\hat{x}_1(\tau)$ are generated and the system is on its way.

Example 8.6

Suppose that a new item will be forecasted using the four-term model with $\beta = 0.9^{1/4}$. No prior demand entries are available; however, the forecaster (using his/her best judgment) estimates the expected demand relation as

$$\hat{\mu}_\tau = 10 + 2\tau + 5 \sin \omega\tau + 0 \cos \omega\tau$$

Hence, the coefficient estimates are $\hat{a}_1(0) = 10$, $\hat{a}_2(0) = 2$, $\hat{a}_3(0) = 5$, and $\hat{a}_4(0) = 0$. Using the first demand entry $x_1 = 13$, find the updated estimates of the coefficients and the forecast equation.

Since

$$\hat{x}_0(1) = 10 + 2 + 5(0.500) + 0(0.866) = 14.5$$

then $e_1 = 13 - 14.5 = -1.5$ and

$$\hat{a}_1(1) = 0.05024(-1.5) + 10 + 2 = 11.925$$

$$\hat{a}_2(1) = 0.00066(-1.5) + 2 = 1.999$$

$$\hat{a}_3(1) = 0.00605(-1.5) + 0.86603(5) - 0.5(0) = 4.321$$
$$\hat{a}_4(1) = 0.04976(-1.5) + 0.5(5) + 0.86603(0) = 2.425$$

These yield

$$\hat{x}_1(\tau) = 11.925 + 1.999\tau + 4.321 \sin \omega\tau + 2.425 \cos \omega\tau$$

Another situation occurs when some past demands are known to the forecaster; however, prior estimates of the coefficients are not available. As before, the entries $(x_1, x_2, \ldots, x_T)$ as of the current time are used to estimate the unknown coefficients. Applying the least-squares method, the estimates are found by use of the following four equations, where the summations range from j equal 0 to $(T - 1)$:

$$\sum x_{T-j} = \hat{a}_1 T + \hat{a}_2 \sum(-j) + \hat{a}_3 \sum \sin(-\omega j) + \hat{a}_4 \sum \cos(-\omega j)$$
$$\sum -j x_{T-j} = \hat{a}_1 \sum(-j) + \hat{a}_2 \sum j^2 + \hat{a}_3 \sum -j \sin(-\omega j) + \hat{a}_4 \sum -j \cos(-\omega j)$$
$$\sum \sin(-\omega j) x_{T-j} = \hat{a}_1 \sum \sin(-\omega j) + \hat{a}_2 \sum -j \sin(-\omega j) + \hat{a}_3 \sum \sin(-\omega j)^2 + \hat{a}_4 \sum \sin(-\omega j)\cos(-\omega j)$$
$$\sum \cos(-\omega j) x_{T-j} = \hat{a}_1 \sum \cos(-\omega j) + \hat{a}_2 \sum -j \cos(-\omega j) + \hat{a}_3 \sum \sin(-\omega j)\cos(-\omega j) + \hat{a}_4 \sum \cos(-\omega j)^2$$

Biegle [2] cautions that this technique is applicable only if the cyclic pattern starts with its maximum or minimum at $t = 1$. This is not a consequence of the trigonometric terms of the forecasting function but of the term linear in t (i.e., $\hat{a}_2 t$). Of course, this effect is lessened as more cycles are used.

The estimates at time T are $\hat{a}_1(T) = \hat{a}_1$, $\hat{a}_2(T) = \hat{a}_2$, $\hat{a}_3(T) = \hat{a}_3$, and $\hat{a}_4(T) = \hat{a}_4$. These are applied to generate the forecasts at this time period. Afterward, as each new demand entry becomes available, the coefficients are updated in the normal manner.

Example 8.7

Suppose that $T = 24$ and the prior 24 demand entries are the following:

$x_1 - x_6$	$x_7 - x_{12}$	$x_{13} - x_{18}$	$x_{19} - x_{24}$
95	45	107	71
105	26	135	45
86	19	116	46
84	18	113	36
67	52	97	60
40	87	71	107

Find the estimates of the coefficients as of $T = 24$.

Using the method of least squares, the summations required are generated and yield the following four equations:

$$
\begin{aligned}
1728 &= 24\hat{a}_1 - 276\hat{a}_2 + 0\hat{a}_3 + 0\hat{a}_4 \\
-19{,}947 &= -276\hat{a}_1 + 4324\hat{a}_2 - 44.78\hat{a}_3 + 12\hat{a}_4 \\
-476.19 &= 0\hat{a}_1 - 44.78\hat{a}_2 + 12\hat{a}_3 + 0\hat{a}_4 \\
125.44 &= 0\hat{a}_1 + 12\hat{a}_2 + 0\hat{a}_3 + 12\hat{a}_4
\end{aligned}
$$

Solving by matrix methods yields

$$
\begin{aligned}
\hat{a}_1(24) &= \hat{a}_1 = 71.40 \\
\hat{a}_2(24) &= \hat{a}_2 = 1.78 \\
\hat{a}_3(24) &= \hat{a}_3 = 46.32 \\
\hat{a}_4(24) &= \hat{a}_4 = 8.68
\end{aligned}
$$

whereby the forecasts as of $T = 24$ are

$$\hat{x}_{24}(\tau) = 71.40 + 1.78\tau + 46.32 \sin \omega\tau + 8.68 \cos \omega\tau$$

Mathematical Basis. In the four-term model,

$$f(\tau) = \begin{bmatrix} 1 \\ \tau \\ \sin \omega\tau \\ \cos \omega\tau \end{bmatrix}$$

$$L = \begin{bmatrix} 1 & 0 & 0 & 0 \\ 1 & 1 & 0 & 0 \\ 0 & 0 & \cos \omega & \sin \omega \\ 0 & 0 & -\sin \omega & \cos \omega \end{bmatrix}$$

$$F = \begin{bmatrix} F_{11} & F_{12} & F_{13} & F_{14} \\ F_{21} & F_{22} & F_{23} & F_{24} \\ F_{31} & F_{32} & F_{33} & F_{34} \\ F_{41} & F_{42} & F_{43} & F_{44} \end{bmatrix}$$

and the elements of F are

$$
\begin{aligned}
F_{11} &= \sum \beta^j \\
F_{12} &= F_{21} = \sum -j\beta^j \\
F_{13} &= F_{31} = \sum \beta^j \sin(-\omega j) \\
F_{14} &= F_{41} = \sum \beta^j \cos(-\omega j)
\end{aligned}
$$

$$F_{22} = \sum j^2\beta^j$$

$$F_{23} = F_{32} = \sum -j\beta^j \sin(-\omega j)$$

$$F_{24} = F_{42} = \sum -j\beta^j \cos(-\omega j)$$

$$F_{33} = \sum \beta^j \sin(-\omega j)^2$$

$$F_{34} = F_{43} = \sum \beta^j \sin(-\omega j)\cos(-\omega j)$$

$$F_{44} = \sum \beta^j \cos(-\omega j)^2$$

where j ranges from 0 to ∞.

The elements of F above are generated using Appendix A. Now

$$h = F^{-1}f(0)$$

where

$$f(0) = \begin{bmatrix} 1 \\ 0 \\ 0 \\ 1 \end{bmatrix}$$

So when $e_T = (x_T - \hat{x}_{T-1}(1))$, then

$$\begin{bmatrix} \hat{a}_1(T) \\ \hat{a}_2(T) \\ \hat{a}_3(T) \\ \hat{a}_4(T) \end{bmatrix} = \begin{bmatrix} h_1 \\ h_2 \\ h_3 \\ h_4 \end{bmatrix} e_T + \begin{bmatrix} 1 & 1 & 0 & 0 \\ 0 & 1 & 0 & 0 \\ 0 & 0 & \cos\omega & -\sin\omega \\ 0 & 0 & \sin\omega & \cos\omega \end{bmatrix} \begin{bmatrix} \hat{a}_1(T-1) \\ \hat{a}_2(T-1) \\ \hat{a}_3(T-1) \\ \hat{a}_4(T-1) \end{bmatrix}$$

8-3 SIX-TERM MODEL

The six-term model is an extension of the four-term model to allow for more flexibility in the seasonal pattern. Two new terms are included and are $\sin 2\omega t$ and $\cos 2\omega t$. Note that where $\sin \omega t$ and $\cos \omega t$ complete one full cycle over a season of M time periods, their counterparts, $\sin 2\omega t$ and $\cos 2\omega t$, complete two full cycles. Recall that when $M = 12$, ω represents 360 degrees/12 = 30 degrees = 30°. Hence, $\sin 2\omega t = \sin 60°t$ and $\cos 2\omega t = \cos 60°t$.

The expected demand for the jth prior time period becomes

$$\mu_{T-j} = a_1 - a_2 j + a_3 \sin(-\omega j) + a_4 \cos(-\omega j) + a_5 \sin(-2\omega j) + a_6 \cos(-2\omega j)$$

and for the τth future period, it is

$$\mu_{T+\tau} = a_1 + a_2\tau + a_3 \sin \omega\tau + a_4 \cos \omega\tau + a_5 \sin 2\omega\tau + a_6 \cos 2\omega\tau$$

An example of the expected demand is depicted in Figure 8-3.

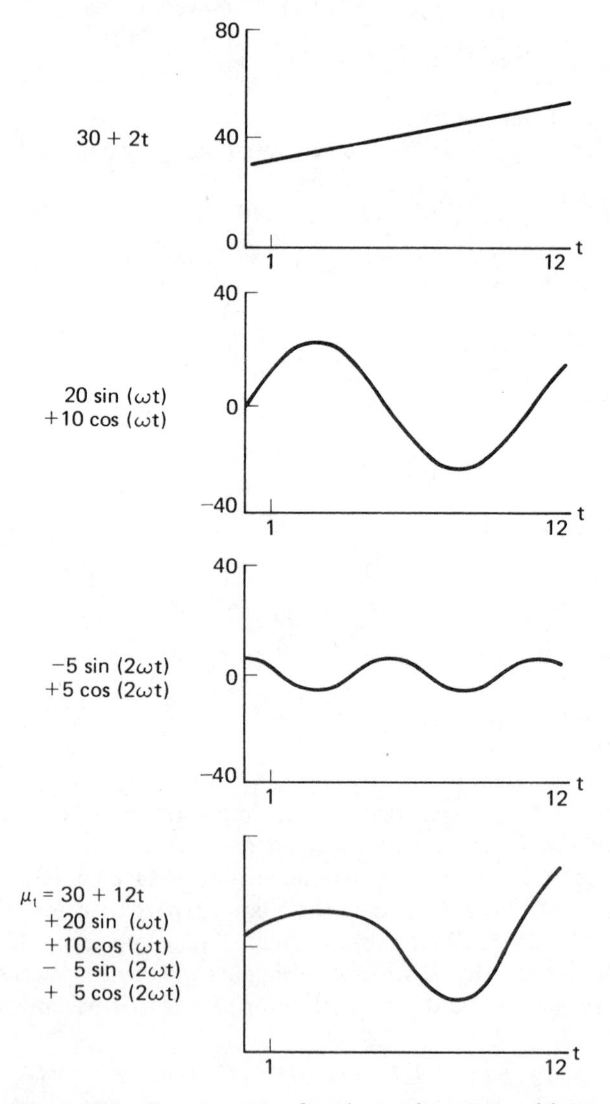

Figure 8-3. Components of a demand pattern with a six-term model.

As in the prior models, the coefficients are updated recursively. In this model the updating equations for $M = 12$ are

$$a_1(T) = h_1 e_T + a_1(T-1) + a_2(T-1)$$
$$a_2(T) = h_2 e_T + a_2(T-1)$$
$$a_3(T) = h_3 e_T + 0.866a_3(T-1) - 0.5a_4(T-1)$$
$$a_4(T) = h_4 e_T + 0.5a_3(T-1) + 0.866a_4(T-1)$$
$$a_5(T) = h_5 e_T + 0.5a_5(T-1) - 0.866a_6(T-1)$$
$$a_6(T) = h_6 e_T + 0.866a_5(T-1) + 0.5a_6(T-1)$$

where the h elements for $\beta = 0.8^{1/6}$ and $\beta = 0.9^{1/6}$ are as follows:

	$\beta = 0.8^{1/6} = 0.9635$	$\beta = 0.9^{1/6} = 0.9826$
h_1	0.06742	0.03342
h_2	0.00125	0.00029
h_3	0.00804	0.00189
h_4	0.06663	0.03333
h_5	0.01085	0.00257
h_6	0.06595	0.03325

Using the coefficients above, the forecasts at time T are generated from

$$\hat{x}_T(\tau) = \hat{a}_1(T) + \hat{a}_2(T)\tau + \hat{a}_3(T)\sin\omega\tau + \hat{a}_4(\tau)\cos\omega\tau$$
$$+ \hat{a}_5(T)\sin 2\omega\tau + \hat{a}_6(T)\cos 2\omega\tau$$

Also, the forecast for the sum of the following τ time periods is

$$\hat{X}_T(\tau) = \hat{x}_T(1) + \hat{x}_T(2) + \ldots + \hat{x}_T(\tau)$$

Example 8.8

Suppose the six-term model with $\beta = 0.9^{1/6} = 0.9826$ and $M = 12$ is in use. Assume the following is known at time T: $\hat{a}_1(T-1) = 100$, $\hat{a}_2(T-1) = 2$, $\hat{a}_3(T-1) = 20$, $\hat{a}_4(T-1) = 30$, $\hat{a}_5(T-1) = 5$, $\hat{a}_6(T-1) = -20$, and $x_T = 140$. Find the forecasts for the sum of the next two time periods.

Note that at $T-1$ the one-period-ahead forecast is

$$\hat{x}_{T-1}(1) = 100 + 2 + 20(0.500) + 30(0.866) + 5(0.866) - 20(0.5) = 132.31$$

whereby the current forecast error becomes

$$e_T = 140 - 132.31 = 7.69$$

Hence, the updated coefficients are

$$\hat{a}_1(T) = 0.03342(7.69) + 100 + 2 = 102.257$$
$$\hat{a}_2(T) = 0.00029(7.69) + 2 = 2.002$$
$$\hat{a}_3(T) = 0.00189(7.69) + 0.866(20) - 0.5(30) = 2.335$$
$$\hat{a}_4(T) = 0.03333(7.69) + 0.5(20) + 0.866(30) = 36.236$$
$$\hat{a}_5(T) = 0.00257(7.69) + 0.5(5) - 0.866(-20) = 19.840$$
$$\hat{a}_6(T) = 0.03325(7.69) + 0.866(5) + 0.5(-20) = -5.414$$

and the forecast equation is

$$\hat{x}_T(\tau) = 102.257 + 2.002\tau + 2.335 \sin \omega\tau + 36.236 \cos \omega\tau + 19.840 \sin 2\omega\tau - 5.414 \cos 2\omega\tau$$

where $\omega = 30$ degrees. With $\tau = 1$ and 2, then $\hat{x}_T(1) = 151.28$ and $\hat{x}_T(2) = 146.29$. The forecast for the sum of the following two time periods is $\hat{X}_T(2) = 151.28 + 146.29 = 297.57$.

Mathematical Basis. In the six-term model,

$$f(\tau) = \begin{bmatrix} 1 \\ \tau \\ \sin \omega\tau \\ \cos \omega\tau \\ \sin 2\omega\tau \\ \cos 2\omega\tau \end{bmatrix}$$

$$L = \begin{bmatrix} 1 & 0 & 0 & 0 & 0 & 0 \\ 1 & 1 & 0 & 0 & 0 & 0 \\ 0 & 0 & \cos \omega & \sin \omega & 0 & 0 \\ 0 & 0 & -\sin \omega & \cos \omega & 0 & 0 \\ 0 & 0 & 0 & 0 & \cos 2\omega & \sin 2\omega \\ 0 & 0 & 0 & 0 & -\sin 2\omega & \cos 2\omega \end{bmatrix}$$

and

$$F = \left[\begin{array}{cccc|cc} & & & & F_{15} & F_{16} \\ & & A & & F_{25} & F_{26} \\ & & & & F_{35} & F_{36} \\ & & & & F_{45} & F_{46} \\ \hline F_{51} & F_{52} & F_{53} & F_{54} & F_{55} & F_{56} \\ F_{61} & F_{62} & F_{63} & F_{64} & F_{65} & F_{66} \end{array}\right]$$

In the matrix F, A contains the same elements as in the four-term model. The remaining elements are

$$F_{15} = F_{51} = \sum \beta^j \sin(-2\omega j)$$
$$F_{16} = F_{61} = \sum \beta^j \cos(-2\omega j)$$
$$F_{25} = F_{52} = \sum j\beta^j \sin(-2\omega j)$$
$$F_{26} = F_{62} = -\sum j\beta^j \cos(-2\omega j)$$
$$F_{35} = F_{53} = \sum \beta^j \sin(-\omega j) \sin(-2\omega j)$$
$$F_{36} = F_{63} = \sum \beta^j \sin(-\omega j) \cos(-2\omega j)$$
$$F_{45} = F_{54} = \sum \beta^j \cos(-\omega j) \sin(-2\omega j)$$
$$F_{46} = F_{64} = \sum \beta^j \sin(-\omega j) \cos(-2\omega j)$$
$$F_{55} = \sum \beta^j \sin(-2\omega j)^2$$
$$F_{56} = F_{65} = \sum \beta^j \sin(-2\omega j) \cos(-2\omega j)$$
$$F_{66} = \sum \beta^j \cos(-2\omega j)^2$$

For the summations above, j ranges from 0 to ∞. Using Appendix A, the summations above can be expressed in closed form.

With F available, the vector h is

$$h = F^{-1}f(0)$$

and, as before, the coefficients are updated by

$$\hat{a}(T) = he_T + L'\hat{a}(T-1)$$

8-4 HIGHER-TERM MODELS

The trigonometric models can be extended to higher-term models in a straightforward manner. Letting the expected demand for the six-term model be $\mu_\tau(6)$, the expected demand for the eight-term model becomes

$$\mu_\tau(8) = \mu_\tau(6) + a_7 \sin 3\omega\tau + a_8 \cos 3\omega\tau$$

Now the 10-term model gives

$$\mu_\tau(10) = \mu_\tau(8) + a_9 \sin 4\omega\tau + a_{10} \cos 4\omega\tau$$

and the 12-term model yields

$$\mu_\tau(12) = \mu_\tau(10) + a_{11} \sin 5\omega\tau + a_{12} \cos 5\omega\tau$$

Table 8-5 lists the h vectors that correspond to the three models above for $\beta = 0.8^{1/k}$ and $\beta = 0.9^{1/k}$ with $k = 8$, 10, and 12. As before, k designates the number of terms included in each model.

Table 8-5. h VECTORS FOR 8-, 10-, AND 12-TERM MODELS USING $\beta = 0.8^{1/k}$ AND $\beta = 0.9^{1/k}$

	8 Terms		10 Terms		12 Terms	
β	$0.8^{1/8}$	$0.9^{1/8}$	$0.8^{1/10}$	$0.9^{1/10}$	$0.8^{1/12}$	$0.9^{1/12}$
h_1	0.05031	0.02503	0.04014	0.02002	0.03340	0.01667
h_2	0.00070	0.00016	0.00045	0.00011	0.00031	0.00007
h_3	0.00371	0.00087	0.00204	0.00048	0.00124	0.00029
h_4	0.05000	0.02500	0.04000	0.02000	0.03333	0.01667
h_5	0.00371	0.00087	0.00160	0.00038	0.00072	0.00017
h_6	0.05000	0.02500	0.04000	0.02000	0.03333	0.01667
h_7	0.00578	0.00137	0.00192	0.00045	0.00062	0.00015
h_8	0.05000	0.02500	0.04000	0.02000	0.03333	0.01667
h_9	—	—	0.00314	0.00074	0.00072	0.00017
h_{10}	—	—	0.04000	0.02000	0.03332	0.01667
h_{11}	—	—	—	—	0.00124	0.00029
h_{12}	—	—	—	—	0.03329	0.01666

PROBLEMS

8-1. In trigonometric models where the season length is 1 year, what values of ω should be used when:

a. The time periods are quarters (i.e., of 3-month duration)?
b. The time periods are of 4-week duration?
c. The time periods are of 2-week duration?

8-2. Consider the three-term model with $\beta = 0.7^{1/3}$ and where $M = 12$ is the number of time periods over a cycle. Suppose that at time $T = 10$, $x_{10} = 55$, $\hat{a}_1(9) = 50$, $\hat{a}_2(9) = 20$, and $\hat{a}_3(9) = -4$. Find:

a. The updated coefficient estimates as of $T = 10$.
b. The forecasts for the sum of the following four time periods.
c. What is the forecast for the 15th future time period?
d. If at $t = 11$, the demand is $x_{11} = 50$, find the updated forecast for the one-period-ahead demand.

8-3. Consider the following 12 demand entries for an item: 49, 62, 50, 52, 31, 20, 11, 8, 10, 18, 29, and 40. Use these demands to find the initial coefficient estimates for a three-term model with $M = 12$. The estimates should be obtained using the least-squares method.

8-4. Suppose that the four-term model with $\beta = 0.7^{1/4}$ is in use for an item. Assume that $M = 12$ and at $T = 8$, the following is known: $x_8 = 50$, $\hat{a}_1(7) = 40$, $\hat{a}_2(7) = 3$, $\hat{a}_3(7) = 16$, and $\hat{a}_4(7) = -5$.

a. Find the updated forecasts for the sum of the next two time periods.

b. If at $T = 9$, the demand becomes $x_9 = 58$, find the updated forecast for the sum of the next two time periods.

8-5. A new item to the inventory is to be forecast by the four-term model with $M = 12$ and $\beta = 0.8^{1/4}$. Suppose at the outset that the forecaster estimates the following expected demand equation for the item:

$$\mu_t = 50 + 5\tau + 0 \sin \omega\tau + 10 \cos \omega\tau$$

a. Find the forecast of demands for time periods 1, 2, and 3, assuming at the outset that $T = 0$.

b. If the first demand becomes $x_1 = 70$, find the updated coefficient estimates and the forecast for each of the following three time periods.

8-6. Suppose that the six-term model with $M = 12$ and $\beta = 0.8^{1/6}$ is in use for an item. Find the one-, two- three-, and four-period-ahead forecasts when the following is known as of time $T = 10$:

a. $x_9(\tau) = 60 + 3\tau + 50 \sin 30°\tau + 10 \cos 30°\tau$
$\qquad - 20 \sin 60°\tau + 5 \cos 60°\tau$

b. $x_{10} = 85$

8-7. List the updating equations for the coefficients of the eight-term model with $M = 12$ and $\beta = 0.9^{1/8}$.

REFERENCES

[1] Brown, R. G., *Smoothing, Forecasting and Prediction of Discrete Time Series.* Englewood Cliffs, N.J.: Prentice-Hall, Inc., 1962, Chap. 12.

[2] Biegle, J. E., *Production Control.* Englewood Cliffs, N.J.: Prentice-Hall, Inc., 1971, pp. 25–27.

9

SEASONAL MODELS

This chapter describes three forecasting models that are applicable to time series that have seasonal demand patterns. These include both the horizontal and trend seasonal demand patterns. In the models cited, smoothing of the past demand entries is used and in this way higher weights are assigned to the more current entries. This feature allows the forecasts to react quicker to more current shifts in the level or seasonal influences of the demands.

The first model described is the horizontal seasonal model which corresponds with items whose expected demand at time t is $\mu_t = \mu\rho_t$. In this respect μ is the average level over all time periods and ρ_t is the seasonal ratio at time t. The seasonal ratio for period t is found from the relation between μ_t and μ. The seasonal ratios always are greater or equal to zero and have an average value of 1 for all time periods in the cycle. In this way the average demand over a seasonal cycle is μ, whereby low and high expected demands over a cycle balance out.

The second model is the multiplicative trend seasonal model. This model is applicable when the expected demand for an item is $\mu_t = (a + bt)\rho_t$. Here a is the intercept or level at $t = 0$, b is the slope, and ρ_t is the seasonal ratio at time t. In this pattern, $(a + bt)$ defines the general trend of the expected demand. This trend is influenced by the seasonal ratio ρ_t to yield the specific expected demand at time t. Again, when $\rho_t > 1$, then μ_t is larger than the trend, and when $\rho_t < 1$, then μ_t is below the trend. With this pattern the seasonal influence is larger when the trend is at a higher level than when it is at a lower level.

The third model is the additive trend seasonal model. This model is used with items whose demand pattern is defined by $\mu_t = a + bt + \delta_t$. Again, $(a + bt)$ gives the trend; however, now the seasonal influence is seen to be additive and represented by δ_t. Here δ_t is called the seasonal increment, and over a full cycle the sum of all increments is zero. When $\delta_t > 0$, then $\mu_t > a + bt$, and when $\delta_t < 0$, then $\mu_t < a + bt$. In this pattern the seasonal influence is the same from cycle to cycle regardless of the magnitude of the level.

For convenience in this chapter, the time periods are assumed to be of monthly time intervals, and the length of the cycle (M) is assumed to be of 1 year's duration (i.e., $M = 12$ months). The models apply equally well to any other circumstances, however.

9-1 HORIZONTAL SEASONAL MODEL

Consider an item with a horizontal seasonal demand pattern (Figure 3-4) where within a year the demands flow in a seasonal manner, and between years the average monthly demands are the same. This section describes how such items are forecasted with use of the horizontal seasonal forecast model.

The model applies when the expected demand at month t is

$$\mu_t = \mu\rho_t$$

Here μ represents the average demand per month and ρ_t is the seasonal ratio at time t. The seasonal ratios are always greater or equal to zero and over a year their average value is 1. When $\rho_t = 1$, the expected demand in month $t(\mu_t)$ is the same as the average monthly demand over the year (μ). When $\rho_t < 1$, then μ_t is less than μ, and when $\rho_t > 1$, then μ_t is greater than μ.

Two phases are necessary in order to implement the model. The first is concerned with initializing the system and the second is with updating the forecasts. In the initializing phase the past demand entries $(x_1, x_2, \ldots, x_T)$ are used to find estimates of μ and ρ_t. The estimate of μ as of time T is $\hat{a}_T$, and the estimates of the seasonal ratios $\rho_{T+\tau}$ are $r_{T+\tau}$ for $\tau = 1, 2, 3, \ldots$. With these estimates available, the initializing phase is complete. From this time period on, the estimates above are updated as each new demand entry becomes available. The forecast for the τth future time period is

$$\hat{x}_T(\tau) = \hat{a}_T r_{T+\tau}$$

A description for each of the two phases above is given below.

Initialize

In order to get started, the forecaster must have available at least two full years of history in demand entries and the number of entries should be in increments of $M = 12$. Letting J represent the number of years of history available in prior demands, then

$$J = \frac{T}{12}$$

The demand history is conveniently grouped into J yearly sets as follows:

$$x_1, x_2, \ldots, x_{12}$$
$$x_{13}, x_{14}, \ldots, x_{24}$$
$$\vdots$$
$$x_{T-11}, X_{T-10}, \ldots, x_T$$

Here $(x_1, x_2, \ldots, x_{12})$ correspond to the first year of demand history, $(x_{13}, x_{14}, \ldots, x_{24})$ are with the second, and so on. Also, $(x_1, x_{13}, x_{25}, \ldots, x_{T-11})$ correspond to a particular calendar month, $(x_2, x_{14}, x_{26}, \ldots, x_{T-10})$ are with another, and so forth.

With the data above, 13 estimates are generated. These are the following:

$\hat{a}_T$ = estimate (as of time T) of the average monthly demand

$r_{T+\tau}$ $(\tau = 1, 2, \ldots, 12)$ = estimates of the seasonal ratios for months $T+1, T+2, \ldots, T+12$

Before proceeding with the initialization phase, the forecaster selects two smoothing parameters α and γ, where $(0 \leq \alpha \leq 1)$ and $(0 \leq \gamma \leq 1)$. The parameter α is used to find the estimate $\hat{a}_T$, and γ is applied in seeking $r_{T+\tau}$.

The following five steps are carried on in this initialization phase.

1. Find the average monthly demand for each of the J years. These are

$$\bar{x}_{(1)} = \tfrac{1}{12}(x_1 + \ldots + x_{12})$$
$$\bar{x}_{(2)} = \tfrac{1}{12}(x_{13} + \ldots + x_{24})$$
$$\vdots$$
$$\bar{x}_{(J)} = \tfrac{1}{12}(x_{T-11} + \ldots + x_T)$$

2. Calculate the seasonal ratio for month t by the relations

$$\tilde{r}_t = \begin{cases} \dfrac{x_t}{\bar{x}_{(1)}} & \text{for } t = 1, \ldots, 12 \\ \dfrac{x_t}{\bar{x}_{(2)}} & \text{for } t = 13, \ldots, 24 \\ \cdot & \\ \cdot & \\ \cdot & \\ \dfrac{x_t}{\bar{x}_{(J)}} & \text{for } t = T - 11, \ldots, T \end{cases}$$

3. Find the average seasonal ratio for each of the 12 calendar months. Here

$$\hat{r}_1 = \frac{1}{J}(\tilde{r}_1 + \tilde{r}_{13} + \ldots + \tilde{r}_{T-11})$$

$$\hat{r}_2 = \frac{1}{J}(\tilde{r}_2 + \tilde{r}_{14} + \ldots + \tilde{r}_{T-10})$$

$$\cdot$$

$$\cdot$$

$$\hat{r}_{12} = \frac{1}{J}(\tilde{r}_{12} + \tilde{r}_{24} + \ldots + \tilde{r}_T)$$

4. Let $\hat{a}_0 = \bar{x}_{(1)}$. Now, starting with $t = 1$ and continuing until $t = T$, apply the following recursive relations:

$$\hat{a}_t = \alpha\left(\frac{x_t}{\hat{r}_t}\right) + (1 - \alpha)\hat{a}_{t-1}$$

$$\hat{r}_{t+12} = \gamma\left(\frac{x_t}{\hat{a}_t}\right) + (1 - \gamma)\hat{r}_t$$

Note that at time t, the ratio $(x_t/\hat{r}_t)$ represents the current seasonally adjusted demand. This is smoothed with the corresponding prior average $(\hat{a}_{t-1})$ to yield an updated average $(\hat{a}_t)$.

Also note that $(x_t/\hat{a}_t)$ gives the seasonal ratio entry for month t. This is smoothed with $\hat{r}_t$ to generate the new seasonal ratio $\hat{r}_{t+12}$. The jump of 12 time periods is necessary because there are 12 months in the year. For example, when t corresponds to a January, then $t + 12$ is associated with the following January.

5. The most current 12 seasonal ratios

$$\hat{r}_{T+1}, \ldots, \hat{r}_{T+12}$$

are normalized so that their average is 1. This is performed by first finding the average

$$\bar{r} = \tfrac{1}{12}(\hat{r}_{T+1} + \ldots + \hat{r}_{T+12})$$

and then adjusting the ratios by

$$r_{T+\tau} = \frac{\hat{r}_{T+\tau}}{\bar{r}} \qquad \tau = 1, 2, \ldots, 12$$

Having carried out these five steps, the initialization phase is complete. At this time the first set of forecasts can be generated.

The forecast for the τth future time period is

$$\hat{x}_T(\tau) = \hat{a}_T r_{T+\tau}$$

Note that because there is one seasonal ratio for each calendar month, the forecasts repeat themselves after $\tau = 12$, i.e.,

$$\hat{x}_T(13) = \hat{x}_T(1)$$
$$\hat{x}_T(14) = \hat{x}_T(2)$$

and so forth.

Example 9.1

Consider an item where the forecaster has the following 24 demand entries available in the past history: 9, 8, 10, 10, 14, 15, 15, 14, 8, 6, 3, 10, 7, 10, 9, 13, 12, 16, 13, 11, 11, 9, 7, and 6. In the above, $x_1 = 9$ and $x_{24} = 6$. The horizontal seasonal model is to be used with $\alpha = 0.1$ and $\gamma = 0.2$. Carry out the initialization phase to arrive at estimates $\hat{a}_{24}, r_{25}, r_{26}, \ldots, r_{36}$.

Table 9-1 is used as a worksheet for this example. The column denoted by (1) represents the results from the first step in the initialization. In a like manner, (2) through (5) give results for the second through fifth steps. A review of these steps follows.

In step 1 the average demands for years 1 and 2 are found, yielding 10.17 and 10.33, respectively. In step 2 the seasonal ratios for months 1 to 24 are found, i.e.,

$$\tilde{r}_1 = \frac{9}{10.17} = 0.88, \ldots, \tilde{r}_{12} = \frac{10}{10.17} = 0.98$$

$$\tilde{r}_{13} = \frac{7}{10.33} = 0.68, \ldots, \tilde{r}_{24} = \frac{6}{10.33} = 0.58$$

The average seasonal ratios for each calendar month are calculated in step 3. For example,

$$\hat{r}_1 = \tfrac{1}{2}(0.88 + 0.68) = 0.78$$

Table 9-1. WORKSHEET FOR EXAMPLE 9.1 INITIALIZING WITH THE HORIZONTAL SEASONAL MODEL

		(1)	(2)	(3)	(4)		(5)
t	x_t		$\bar{r}_t$	$\hat{r}_t$	$\hat{a}_t$	$\hat{r}_{t+12}$	r_{t+12}
0					10.17		
1	9		0.88	0.78	10.31	0.80	
2	8		0.79	0.88	10.19	0.86	
3	10		0.98	0.93	10.24	0.94	
4	10		0.98	1.12	10.11	1.09	
5	14		1.38	1.27	10.20	1.29	
6	15	$\bar{x}_{(1)} = 10.17$	1.47	1.51	10.18	1.50	
7	15		1.47	1.37	10.25	1.39	
8	14		1.38	1.22	10.37	1.25	
9	8		0.79	0.93	10.20	0.90	
10	6		0.59	0.73	10.00	0.70	
11	3		0.29	0.49	9.61	0.45	
12	10		0.98	0.78	9.93	0.83	
13	7		0.68		9.81	0.78	0.78
14	10		0.97		10.00	0.89	0.89
15	9		0.87		9.95	0.93	0.93
16	13		1.26		10.15	1.13	1.13
17	12		1.16		10.07	1.27	1.26
18	16	$\bar{x}_{(2)} = 10.33$	1.55		10.13	1.52	1.51
19	13		1.26		10.05	1.37	1.36
20	11		1.06		9.92	1.22	1.21
21	11		1.06		10.15	0.94	0.94
22	9		0.87		10.42	0.73	0.73
23	7		0.68		10.94	0.49	0.49
24	6		0.58		10.57	0.78	0.78

Now in step 4, $\hat{a}_0 = 10.17$ and the recursive formulas are applied at this time. For $t = 1$,

$$\hat{a}_1 = 0.1\left(\frac{9}{0.78}\right) + 0.9(10.17) = 10.31$$

$$\hat{r}_{13} = 0.2\left(\frac{9}{10.31}\right) + 0.8(0.78) = 0.80$$

This process continues until $t = 24$, where $\hat{a}_{24} = 10.57$, $\hat{r}_{25} = 0.78$, $\hat{r}_{26} = 0.89$, $\ldots$, $\hat{r}_{36} = 0.78$.

Applying step 5, the seasonal ratios $\hat{r}_{25}$ to $\hat{r}_{36}$ are normalized so that their average is 1. This is done by first taking

$$\bar{r} = \tfrac{1}{12}(0.78 + 0.89 + \ldots + 0.78) = 1.004$$

and then normalizing by

$$r_1 = \frac{1}{1.004}(0.78) = 0.78$$

$$r_2 = \frac{1}{1.004}(0.89) = 0.89$$

$$\vdots$$

$$r_{12} = \frac{1}{1.004}(0.78) = 0.78$$

The seasonal ratios listed are carried out to two decimal places.

Example 9.2

Use the results from Example 9.1 to find the forecasts for the following 12 months (i.e., $t = 25$ to 36).

The forecasts become

$$\hat{x}_{24}(1) = 10.57(0.78) = 8.24$$
$$\hat{x}_{24}(2) = 10.57(0.89) = 9.41$$
$$\hat{x}_{24}(3) = 10.57(0.93) = 9.83$$
$$\hat{x}_{24}(4) = 10.57(1.13) = 11.94$$
$$\hat{x}_{24}(5) = 10.57(1.26) = 13.32$$
$$\hat{x}_{24}(6) = 10.57(1.51) = 15.96$$
$$\hat{x}_{24}(7) = 10.57(1.36) = 14.38$$
$$\hat{x}_{24}(8) = 10.57(1.21) = 12.79$$
$$\hat{x}_{24}(9) = 10.57(0.94) = 9.94$$
$$\hat{x}_{24}(10) = 10.57(0.73) = 7.22$$
$$\hat{x}_{24}(11) = 10.57(0.49) = 5.18$$
$$\hat{x}_{24}(12) = 10.57(0.78) = 8.24$$

A plot of the 24 prior demands and the following 12 forecasts is shown in Figure 9-1.

Updating

As each new demand entry becomes available to the forecaster, an updating scheme is carried forward to yield the current estimates of the level and the seasonal ratios. Calling the current time period T, the new observation is x_T, and the updating relations are the following:

$$\hat{a}_T = \alpha\left(\frac{x_T}{r_T}\right) + (1 - \alpha)\hat{a}_{T-1}$$

$$r_{T+12} = \gamma\left(\frac{x_T}{\hat{a}_T}\right) + (1 - \gamma)r_T$$

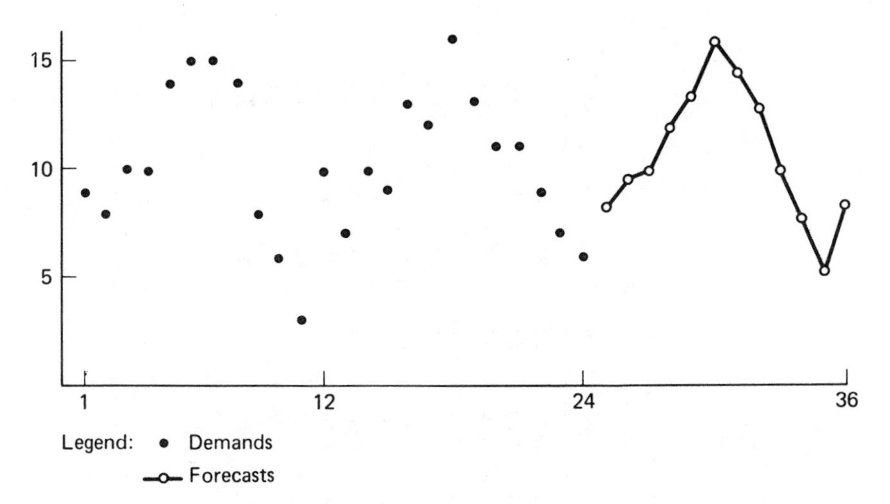

Figure 9-1. Forecasts from Example 9.2 using initialization from the horizontal seasonal model.

As before, the seasonal ratios are normalized so that their average is 1. Three steps are required for this purpose:

1. $\bar{r} = \frac{1}{12}(r_{T+1} + \ldots + r_{T+12})$.
2. $\hat{r}_{T+\tau} = \frac{r_{T+\tau}}{\bar{r}}$ for $\tau = 1, 2, \ldots, 12$.
3. $r_{T+\tau} = \hat{r}_{T+\tau}$[1] for $\tau = 1, 2, \ldots, 12$.

With the updating of the estimates completed, the forecasts for the τth future time period is now found. This is

$$\hat{x}_T(\tau) = \hat{a}_T r_{T+\tau}$$

Example 9.3

Suppose that $x_{25} = 7$. Now use the results from Example 9.1 to find the updated forecasts for $\tau = 1$ and 2.

Since $\hat{a}_{24} = 10.57$ and $r_{25} = 0.78$, then

$$\hat{a}_{25} = 0.1\left(\frac{7}{0.78}\right) + 0.9(10.57) = 10.41$$

and

$$r_{37} = 0.2\left(\frac{7}{10.41}\right) + 0.8(0.78) = 0.76$$

[1]The results from steps 2 and 3 are identical. Step 3 is used to avoid confusion in the notation.

Table 9-2. Worksheet for Example 9.4—Updating with the Horizontal Seasonal Model

t	x_t	$\hat{a}_t$	r_{t+1}	r_{t+2}	r_{t+3}	r_{t+4}	r_{t+5}	r_{t+6}	r_{t+7}	r_{t+8}	r_{t+9}	r_{t+10}	r_{t+11}	r_{t+12}	$\hat{x}_t(1)$
25	7	10.41	0.89	0.93	1.13	1.26	1.51	1.36	1.21	0.94	0.73	0.49	0.78	0.76	9.26
26	9	10.38	0.93	1.13	1.26	1.51	1.36	1.21	0.94	0.73	0.49	0.78	0.76	0.89	9.68
27	11	10.52	1.13	1.26	1.51	1.36	1.21	0.94	0.73	0.49	0.78	0.76	0.88	0.95	11.89
28	10	10.36	1.26	1.51	1.36	1.21	0.94	0.73	0.49	0.78	0.76	0.89	0.95	1.10	13.05
29	13	10.35	1.51	1.36	1.21	0.94	0.73	0.49	0.78	0.76	0.89	0.95	1.10	1.26	15.62
30	16	10.37	1.36	1.21	0.94	0.73	0.49	0.78	0.76	0.89	0.95	1.10	1.26	1.52	14.10
31	14	10.36	1.21	0.94	0.73	0.49	0.78	0.76	0.89	0.95	1.10	1.26	1.52	1.36	12.54
32	13	10.40	0.94	0.73	0.49	0.78	0.76	0.89	0.95	1.10	1.26	1.52	1.36	1.22	9.78
33	9	10.31	0.73	0.49	0.78	0.76	0.89	0.96	1.10	1.26	1.52	1.36	1.22	0.93	7.53
34	4	9.83	0.49	0.79	0.76	0.89	0.96	1.11	1.27	1.53	1.37	1.23	0.93	0.67	4.82
35	8	10.47	0.78	0.76	0.89	0.96	1.10	1.26	1.52	1.36	1.22	0.93	0.67	0.55	8.17
36	8	10.44	0.76	0.89	0.96	1.10	1.26	1.52	1.36	1.22	0.93	0.67	0.55	0.78	7.93

Also, since

$$\bar{r} = \tfrac{1}{12}(r_{26} + \ldots + r_{37})$$
$$= \tfrac{1}{12}(0.89 + \ldots + 0.76) = 1.00$$

the seasonal ratios remain unchanged. The forecasts needed are

$$\hat{x}_{25}(1) = 10.41(0.89) = 9.26$$
$$\hat{x}_{25}(2) = 10.41(0.93) = 9.68$$

Example 9.4

Continue with Example 9.3 for the demands listed in Table 9-2 and find the one-period-ahead forecasts.

The results are listed in Table 9-2 and a plot of the one-period-ahead forecasts is shown in Figure 9-2.

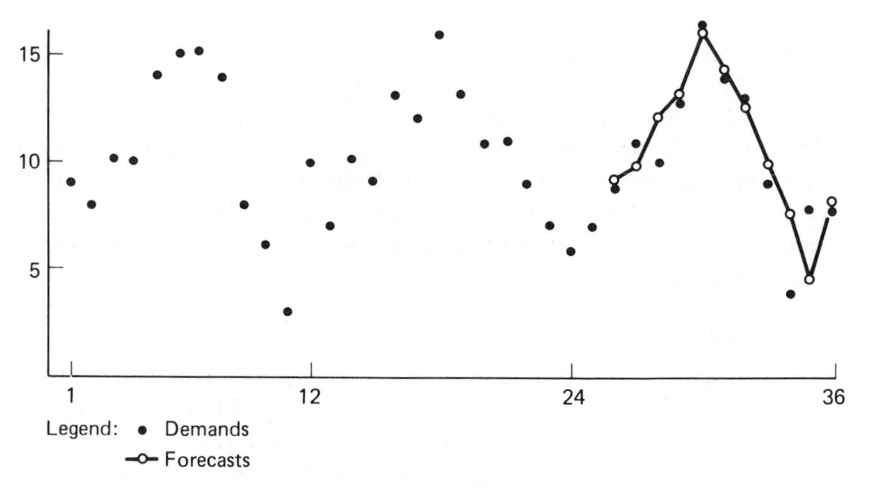

Figure 9-2. One-period-ahead forecasts using the horizontal seasonal model.

9-2 MULTIPLICATIVE TREND SEASONAL MODEL

In the more general case, an item may follow a demand pattern that has both trend and seasonal influences. The trend is either rising or falling at a steady rate, and the seasonal influence is the same as shown in the horizontal seasonal demand pattern. These two factors coupled together give a pattern where the seasonal swings are wider for the higher levels of the trend than for the lower levels (see Figure 3-5). A forecasting technique that applies for this type of demand pattern is the multiplicative trend seasonal model.

The expected demand at time $(t + \tau)$ is defined by

$$\mu_{t+\tau} = (a_t + b\tau)\rho_{t+\tau}$$

where a_t is the level at time t, b is the slope, and $\rho_{t+\tau}$ is the seasonal ratio at $(t + \tau)$. The seasonal ratios have the same properties listed in Section 9-1 for the horizontal seasonal demand pattern.

The role of the forecasting model is to use the history of demands to estimate 14 unknown coefficients in the demand pattern. At the current time T, these estimates are

$$\hat{a}_T \text{ for } a_T$$

$$\hat{b}_T \text{ for } b$$

and

$$r_{T+\tau} \text{ for } \rho_{T+\tau} \qquad \tau = 1, 2, \ldots, 12$$

Three smoothing parameters (α, β, and γ) are used for this purpose, where each lie in the interval (0, 1). Parameter α is a smoothing parameter for $\hat{a}_T$, β is used to find $\hat{b}_T$, and γ is for $r_{T+\tau}$.

As in the horizontal seasonal model, two phases are followed in applying the model. The first is the initialization phase, which uses past demands to get the system started. The second is the updating phase, where the forecasts are altered as each new demand entry becomes available. Descriptions of each of these phases follow.

Initialization

The purpose of the initialization phase is to start the system off with good estimates of a_T, b, and $\rho_{T+\tau}$ $(\tau = 1, 2, \ldots, 12)$. These estimates are found through the use of the past demand entries that are available to the forecaster. The past demands are again conveniently grouped into J years:

$$x_1, x_2, \ldots, x_{12}$$

$$x_{13}, x_{14}, \ldots, x_{24}$$

$$\vdots$$

$$x_{T-11}, x_{T-10}, \ldots, x_T$$

where

$$J = \frac{T}{12}$$

Various initializing schemes can be employed. The method presented here is a slight modification to Winters [1] original description for this purpose. In order to carry out the initialization, the following nine steps are followed:

1. Find the average monthly demand for the first and last year (year J). These are

$$\bar{x}_{(1)} = \tfrac{1}{12}(x_1 + \ldots + x_{12})$$
$$\bar{x}_{(J)} = \tfrac{1}{12}(x_{T-11} + \ldots + x_T)$$

2. Estimate the slope using

$$\bar{b}_0 = \frac{\bar{x}_{(J)} - \bar{x}_{(1)}}{T - 12}$$

   Note that the change from $\bar{x}_{(1)}$ to $\bar{x}_{(J)}$ takes place over $(T - 12)$ months. This is because $\bar{x}_{(1)}$ is the average of the first year and is representative of time period $t = 6.5$, and in a like way $\bar{x}_{(J)}$ is representative of $t = T - 5.5$. Hence, the time span separating these time periods is $T - 12$ months.
3. The deseasonalized level at $t = 0$ is now estimated by

$$\bar{a}_0 = \bar{x}_{(1)} - 6.5\bar{b}_0$$

4. Carrying forward this trend, the deseasonalized level at time t becomes

$$\bar{a}_t = \bar{a}_0 + \bar{b}_0 t$$

5. A seasonal ratio for each month is now found. For month t this is

$$\tilde{r}_t = \frac{x_t}{\bar{a}_t} \qquad t = 1, 2, \ldots, T$$

6. The seasonal ratios are grouped by calendar months to find the average for each such month, i.e.,

$$\bar{r}_1 = \frac{1}{J}(\tilde{r}_1 + \tilde{r}_{13} + \ldots + \tilde{r}_{T-11})$$

$$\bar{r}_2 = \frac{1}{J}(\tilde{r}_2 + \tilde{r}_{14} + \ldots + \tilde{r}_{T-10})$$

$$\vdots$$

$$\bar{r}_{12} = \frac{1}{J}(\tilde{r}_{12} + \tilde{r}_{24} + \ldots + \tilde{r}_T)$$

7. The seasonal ratios ($\bar{r}_1, \ldots, \bar{r}_{12}$) are normalized so that their average value is 1. This is by

$$\bar{r} = \tfrac{1}{12}(\bar{r}_1 + \ldots + \bar{r}_{12})$$
$$\hat{r}_t = \bar{r}_t/\bar{r} \qquad \text{for } t = 1, 2, \ldots, 12$$

8. Starting at $t = 1$ and continuing until $t = T$, the following three recursive relations are carried forward:

$$\hat{a}_t = \alpha\left(\frac{x_t}{\hat{r}_t}\right) + (1 - \alpha)(\hat{a}_{t-1} + \hat{b}_{t-1})$$
$$\hat{b}_t = \beta(\hat{a}_t - \hat{a}_{t-1}) + (1 - \beta)\hat{b}_{t-1}$$
$$\hat{r}_{t+12} = \gamma\left(\frac{x_t}{\hat{a}_t}\right) + (1 - \gamma)\hat{r}_t$$

9. The 12 most current estimates of the seasonal ratios are now normalized so that their average is 1. This step is performed as follows:

$$\bar{r} = \tfrac{1}{12}(\hat{r}_{T+1} + \ldots + \hat{r}_{T+12})$$
$$r_{T+\tau} = \frac{1}{\bar{r}}\hat{r}_{T+\tau} \qquad \text{for } \tau = 1, 2, \ldots, 12$$

Having performed the preceding nine steps, the initialization process is complete. The estimates that are carried forward are $\hat{a}_T$, $\hat{b}_T$, and r_{T+1}, $r_{T+2}, \ldots, r_{T+12}$. These may also be used to generate forecasts as of $t = T$. The forecast for the τth future time period is

$$\hat{x}_T(\tau) = (\hat{a}_T + \hat{b}_T\tau)r_{T+\tau}$$

Note that a seasonal ratio is estimated for each of the 12 calendar months. In this way

$$r_{T+13} = r_{T+1}$$
$$r_{T+14} = r_{T+2}$$
$$\vdots$$
$$r_{T+24} = r_{T+12}$$

can be used should forecasts be required for $\tau > 12$.

Example 9.5

Consider an item with the following 24 demand entries available to the forecaster: 5, 3, 12, 9, 13, 20, 37, 44, 26, 14, 5, 1, 3, 13, 18, 19, 31, 34, 60, 62, 56, 24, 8, and 2. Initialize the system using the multiplicative trend seasonal model with $\alpha = 0.1$, $\beta = 0.2$, and $\gamma = 0.3$.

A worksheet that carries out the steps of the initialization is given in Table 9-3.

Table 9-3. WORKSHEET FOR EXAMPLE 9.5—INITIALIZING WITH THE MULTIPLICATIVE TREND SEASONAL MODEL

t	x_t	(1)	(4) $\bar{a}_t$	(5) $\hat{r}_t$	(6) $\bar{r}_t$	(7) $\hat{r}_t$	(8) $\hat{a}_t$	(8) $\hat{b}_t$	(8) $\hat{r}_{t+12}$	(9) r_{t+12}
0			9.38				9.38	0.98		
1	5		10.36	0.48	0.31	0.31	10.94	1.10	0.35	
2	3		11.34	0.26	0.41	0.41	11.56	1.00	0.36	
3	12		12.32	0.97	0.86	0.87	12.68	1.03	0.89	
4	9		13.30	0.68	0.72	0.73	13.57	1.00	0.71	
5	13		14.28	0.91	1.05	1.06	14.34	0.95	1.01	
6	20	$\bar{x}_{(1)} =$	15.26	1.31	1.28	1.29	15.31	0.96	1.29	
7	37	15.75	16.24	2.28	2.21	2.23	16.30	0.96	2.24	
8	44		17.22	2.56	2.35	2.37	17.39	0.99	2.42	
9	26		18.20	1.43	1.65	1.66	18.11	0.93	1.59	
10	14		19.18	0.73	0.76	0.77	18.96	0.92	0.76	
11	5		20.16	0.25	0.25	0.25	19.89	0.92	0.25	
12	1		21.14	0.05	0.06	0.06	20.39	0.84	0.06	
13	3		22.12	0.14			19.96	0.59	0.29	0.29
14	13		23.10	0.56			22.11	0.90	0.43	0.43
15	18		24.08	0.75			22.73	0.84	0.86	0.86
16	19		25.06	0.76			23.89	0.91	0.74	0.74
17	31		26.04	1.19			25.38	1.02	1.07	1.07
18	34		27.02	1.26			26.40	1.02	1.29	1.29
19	60	$\bar{x}_{(2)} =$	28.00	2.14			27.36	1.01	2.23	2.22
20	62	27.50	28.98	2.14			28.09	0.95	2.36	2.35
21	56		29.96	1.87			29.66	1.08	1.68	1.68
22	24		30.94	0.78			30.82	1.09	0.77	0.77
23	8		31.92	0.25			31.92	1.09	0.25	0.25
24	2		32.90	0.06			33.05	1.10	0.06	0.06

The parenthesized column headings identify results from the corresponding step. A summary of each step is listed below.

1. The yearly averages are found:

$$\bar{x}_{(1)} = 15.75 \text{ and } \bar{x}_{(2)} = 27.50$$

2. The slope is estimated as

$$\bar{b}_0 = \frac{27.50 - 15.75}{12} = 0.98$$

3. Next, the level at $t = 0$ is

$$\bar{a}_0 = 15.75 - 6.5(0.98) = 9.38$$

4. The level at $t = 1$ to 24 is found by

$$\bar{a}_t = 9.38 + 0.98t$$

5. Estimates of the seasonal ratios for $t = 1$ to 24 are

$$\tilde{r}_t = \frac{x_t}{\bar{a}_t}$$

6. The seasonal ratios are averaged for each calendar month, i.e.,

$$\bar{r}_1 = \tfrac{1}{2}(\tilde{r}_1 + \tilde{r}_{13}) = \tfrac{1}{2}(0.48 + 0.14) = 0.31$$

$$\vdots$$

$$\bar{r}_{12} = \tfrac{1}{2}(\tilde{r}_{12} + \tilde{r}_{24}) = \tfrac{1}{2}(0.05 + 0.06) = 0.06$$

7. The seasonal ratios are now normalized so that their average over the 12 calendar months is 1.
8. Using $\hat{a}_0 = \bar{a}_0 = 9.38$, $\hat{b}_0 = \bar{b}_0 = 0.98$, the estimates $\hat{a}_t$, $\hat{b}_t$, and $\hat{r}_{t+12}$ are now found for all time periods from $t = 1$ to 24.
9. The seasonal ratios (r_t) for $t = 25$ to 36 are found by normalizing the corresponding seasonal ratios $(\hat{r}_t)$ from step 8.

Having completed the initialization, the following 14 coefficients are saved for further processing: $\hat{a}_{24}$, $\hat{b}_{24}$, and r_t for $t = 25$ to 36.

Example 9.6

Use the results from Example 9.5 to generate the forecasts for months 25 to 36. Note these forecasts are based on the demands from months 1 to 24.

Since $\hat{a}_{24} = 33.05$, $\hat{b}_{24} = 1.10$, and the seasonal ratios are as shown in Table 9-3, the forecasts become

$$\hat{x}_{24}(\tau) = (33.05 + 1.10\tau)r_{24+\tau}$$

With $\tau = 1, 2, \ldots, 12$, the results are

$$\hat{x}_{24}(1) = 34.15(0.29) = 9.90$$
$$\hat{x}_{24}(2) = 35.25(0.43) = 15.16$$
$$\hat{x}_{24}(3) = 36.35(0.86) = 31.26$$
$$\hat{x}_{24}(4) = 37.45(0.74) = 27.71$$
$$\hat{x}_{24}(5) = 38.55(1.07) = 41.25$$
$$\hat{x}_{24}(6) = 39.65(1.29) = 51.15$$
$$\hat{x}_{24}(7) = 40.75(2.22) = 90.47$$
$$\hat{x}_{24}(8) = 41.85(2.35) = 98.35$$
$$\hat{x}_{24}(9) = 42.95(1.68) = 72.16$$
$$\hat{x}_{24}(10) = 43.05(0.77) = 33.15$$
$$\hat{x}_{24}(11) = 44.15(0.25) = 11.04$$
$$\hat{x}_{24}(12) = 45.25(0.06) = 2.72$$

A plot of these forecasts is shown in Figure 9-3.

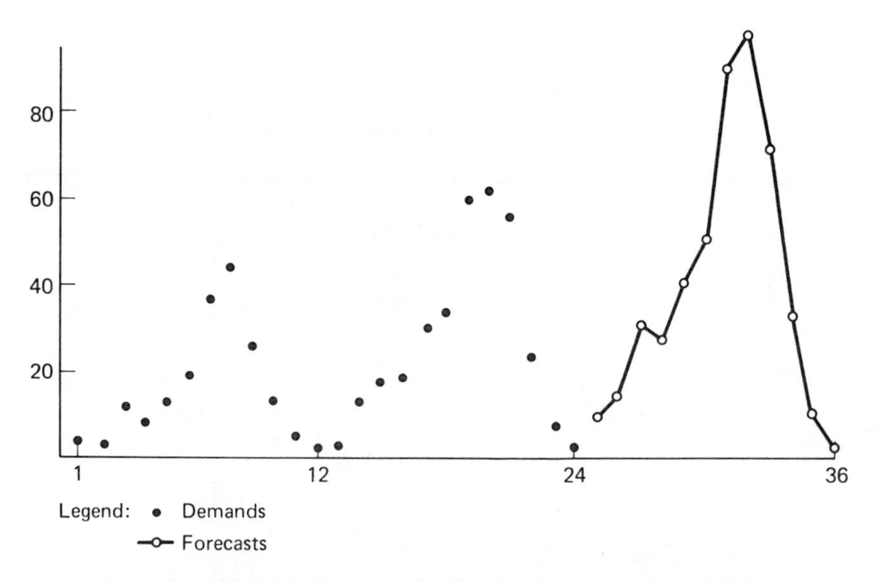

Figure 9-3. Monthly forecasts for the coming year using initialization results of the multiplicative trend seasonal model.

Updating

With each passing time period, a new demand entry becomes available and is used to update the 14 coefficients. Again using x_T as the current demand, then

$$\hat{a}_T = \alpha\left(\frac{x_T}{r_T}\right) + (1 - \alpha)(\hat{a}_{T-1} + \hat{b}_{T-1})$$

$$\hat{b}_T = \beta(\hat{a}_T - \hat{a}_{T-1}) + (1 - \beta)\hat{b}_{T-1}$$

$$r_{T+12} = \gamma\left(\frac{x_T}{\hat{a}_T}\right) + (1 - \gamma)r_T$$

Now r_{T+1} to r_{T+12} are normalized so that their average is 1.

The updated forecasts are generated at this time. For the τth future time period, the forecast is

$$\hat{x}_T(\tau) = (\hat{a}_T + \hat{b}_T\tau)r_{T+\tau}$$

As before, should τ exceed 12, the seasonal ratios are repeated. For example, $r_{T+13} = r_{T+1}$, $r_{T+14} = r_{T+2}$, and so forth.

Example 9.7

Applying the initialization results from Example 9.5, find the updated coefficients using the demands listed in Table 9-4. Generate the one-period-ahead forecasts for these months.

Table 9-4. WORKSHEET FOR EXAMPLE 9.7—UPDATING WITH THE MULTIPLICATIVE TREND SEASONAL MODEL

t	x_t	$\hat{a}_t$	$\hat{b}_t$	r_{t+1}	r_{t+2}	r_{t+3}	r_{t+4}	r_{t+5}	r_{t+6}	r_{t+7}	r_{t+8}	r_{t+9}	r_{t+10}	r_{t+11}	r_{t+12}	$\hat{x}_t(1)$
25	9	33.84	1.04	0.43	0.86	0.74	1.07	1.29	2.22	2.35	1.68	0.77	0.25	0.06	0.28	15.00
26	8	33.25	0.71	0.86	0.74	1.08	1.30	2.23	2.36	1.69	0.77	0.25	0.06	0.28	0.37	29.21
27	13	32.08	0.34	0.75	1.09	1.32	2.26	2.39	1.72	0.78	0.25	0.06	0.28	0.37	0.72	24.32
28	35	33.84	0.62	1.08	1.31	2.24	2.37	1.70	0.77	0.25	0.06	0.28	0.37	0.72	0.84	37.22
29	48	35.46	0.82	1.30	2.23	2.35	1.69	0.76	0.25	0.06	0.28	0.37	0.71	0.83	1.16	47.16
30	55	36.88	0.94	2.22	2.34	1.68	0.76	0.25	0.06	0.28	0.37	0.71	0.83	1.15	1.35	83.96
31	73	37.33	0.84	2.36	1.69	0.76	0.25	0.06	0.28	0.37	0.72	0.83	1.16	1.36	2.16	90.08
32	78	37.66	0.74	1.71	0.78	0.25	0.06	0.28	0.37	0.72	0.84	1.17	1.38	2.16	2.29	65.66
33	83	39.46	0.94	0.76	0.25	0.06	0.28	0.37	0.72	0.83	1.16	1.35	2.14	2.27	1.81	30.70
34	34	40.79	1.03	0.25	0.06	0.28	0.37	0.72	0.83	1.16	1.35	2.13	2.26	1.81	0.78	10.46
35	12	42.44	1.15	0.06	0.28	0.37	0.72	0.82	1.16	1.35	2.13	2.26	1.81	0.78	0.26	2.62
36	4	45.90	1.61	0.28	0.37	0.72	0.82	1.16	1.35	2.13	2.25	1.81	0.78	0.26	0.07	13.30

At $T = 25$, the updated coefficients become

$$\hat{a}_{25} = 0.1\left(\frac{9}{0.29}\right) + 0.9(33.05 + 1.10) = 33.84$$

$$\hat{b}_{25} = 0.2(33.84 - 33.05) + 0.8(1.10) = 1.04$$

$$r_{37} = 0.3\left(\frac{9}{33.84}\right) + 0.7(0.29) = 0.28$$

Now since

$$\bar{r} = \tfrac{1}{12}(r_{26} + \ldots + r_{37}) = 1.00$$

no adjustments are needed in the seasonal ratios. With $\tau = 1$, the forecast is

$$\hat{x}_{25}(1) = (33.84 + 1.04)0.43 = 15.00$$

The example continues and the results are listed in Table 9-4 and are plotted in Figure 9-4.

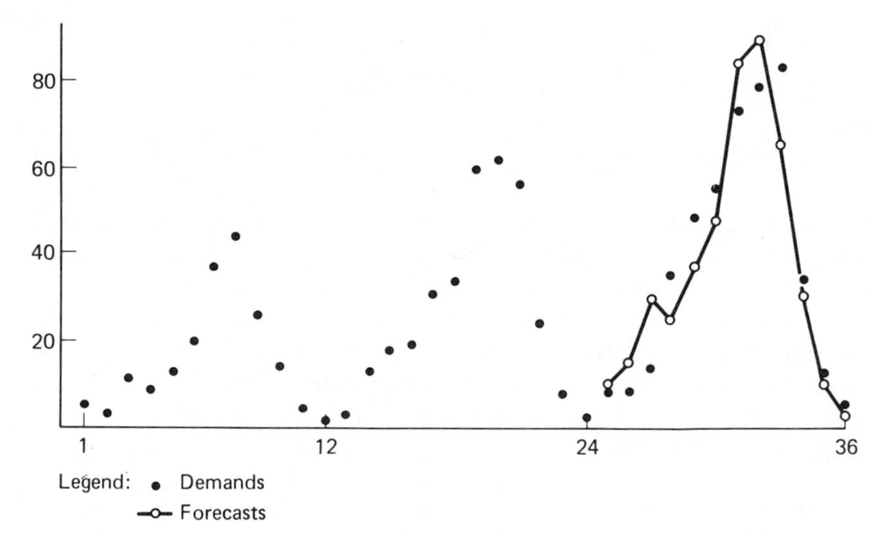

Figure 9-4. One-period-ahead forecasts using the multiplicative trend seasonal model.

9-3 ADDITIVE TREND SEASONAL MODEL

Another form of the trend seasonal pattern is where the seasonal influences are additive rather than multiplicative (see Figure 3-6). Here the swings in the demands that are due to the seasonal influences are of the same magnitude from year to year. The forecasting technique that is used with items following this pattern is the additive trend seasonal model.

The expected demand at $(t + \tau)$ is

$$\mu_{t+\tau} = a_t + b\tau + \delta_{t+\tau}$$

where a_t is the level at time t, b is the slope, and $\delta_{t+\tau}$ is the seasonal increment at time $t + \tau$. When $\delta_{t+\tau} = 0$, then month $t + \tau$ has no seasonal influence. With $\delta_{t+\tau} > 0$, month $t + \tau$ has a higher expected demand than an average month, and with $\delta_{t+\tau} < 0$, the expected demand is less. Since the seasonal increments must balance out over a year, their sum over a calendar year is zero, i.e.,

$$\sum_{\tau=1}^{12} \delta_{t+\tau} = 0$$

As in the prior trend seasonal model, 14 coefficients are estimated at each time period. These estimates are

$$\hat{a}_t \text{ for } a_t$$

$$\hat{b}_t \text{ for } b$$

and

$$d_{t+\tau} \text{ for } \delta_{t+\tau} \qquad \tau = 1, 2, \ldots, 12$$

The model uses three smoothing parameters in this process. These are α, β, and γ, where each lies in the interval (0, 1). The smoothing parameter α is used to find $\hat{a}_t$, β is used in seeking $\hat{b}_t$, and γ is needed in obtaining $d_{t+\tau}$.

Again two phases are carried out in the implementation of this model. The first is an initialization phase, where past demands $(x_1, x_2, \ldots, x_T)$ are used to seek the estimates that are needed to get started. The second is an updating phase, whereby the coefficients are modified as each new demand entry becomes available.

Initialization

To get started, the forecaster assembles his/her prior demand entries $(x_1, x_2, \ldots, x_T)$. These are conveniently listed into yearly clusters:

$$x_1, \ldots, x_{12}$$

$$x_{13}, \ldots, x_{24}$$

$$\vdots$$

$$x_{T-11}, \ldots, x_T$$

Let $J = T/12$ represent the number of full years of demand entries available. With these data, the initialization is carried out using the following nine steps:

1. The average demand per month is found for years 1 and J. These are

$$\bar{x}_{(1)} = \tfrac{1}{12}(x_1 + \ldots + x_{12})$$
$$\bar{x}_{(J)} = \tfrac{1}{12}(x_{T-11} + \ldots + x_T)$$

2. The slope is estimated from

$$\bar{b}_0 = \frac{1}{T-12}(\bar{x}_{(J)} - \bar{x}_{(1)})$$

3. Now the level at $t = 0$ is estimated using

$$\bar{a}_0 = \bar{x}_{(1)} - 6.5b_0$$

4. The level for each month from $t = 1$ to T is found by

$$\bar{a}_t = \bar{a}_0 + \bar{b}_0 t$$

5. The seasonal increments for months $t = 1$ to T are estimated from

$$\tilde{d}_t = x_t - \bar{a}_t$$

6. An average value of the seasonal increments for each of the 12 calendar months is obtained. These are

$$\bar{d}_1 = \frac{1}{J}(\tilde{d}_1 + \tilde{d}_{13} + \ldots + \tilde{d}_{T-11})$$
$$\vdots$$
$$\bar{d}_{12} = \frac{1}{J}(\tilde{d}_{12} + \tilde{d}_{24} + \ldots + \tilde{d}_T)$$

7. The 12 seasonal increments are normalized so that they sum to zero. This step is carried out with the following procedure:

$$\bar{d} = \tfrac{1}{12}\sum_{t=1}^{12} \bar{d}_t$$
$$\hat{d}_t = \bar{d}_t - \bar{d} \qquad \text{for } t = 1, 2, \ldots, 12$$

8. Using $\hat{a}_0 = \bar{a}_0$ and $\hat{b}_0 = \bar{b}_0$, the following three estimates are calculated at each time period from $t = 1$ to T:

$$\hat{a}_t = \alpha(x_t - \hat{d}_t) + (1 - \alpha)(\hat{a}_{t-1} + \hat{b}_{t-1})$$
$$\hat{b}_t = \beta(\hat{a}_t - \hat{a}_{t-1}) + (1 - \beta)\hat{b}_{t-1}$$
$$\hat{d}_{t+12} = \gamma(x_t - \hat{a}_t) + (1 - \gamma)\hat{d}_t$$

9. Finally, the most current seasonal increments $(\hat{d}_{T+1}, \ldots, \hat{d}_{T+12})$ are normalized so that they sum to zero. This is carried out by the following procedure:

$$\bar{d} = \tfrac{1}{12}(\hat{d}_{T+1} + \ldots + \hat{d}_{T+12})$$

$$d_{T+\tau} = \hat{d}_{T+\tau} - \bar{d} \qquad \text{for } \tau = 1, 2, \ldots, 12$$

Having completed the nine steps above, the initialization process is completed. It is now possible to generate forecasts as of time T. For the τth future time period, the forecast is

$$\hat{x}_T(\tau) = \hat{a}_T + \hat{b}_T\tau + d_{T+\tau}$$

In the event that $\tau > 12$ is needed, the seasonal increments that are used for this purpose are

$$d_{T+13} = d_{T+1}$$

$$d_{T+14} = d_{T+2}$$

and so forth.

Example 9.8

Consider an item with the following 24 demand entries: 122, 118, 115, 100, 94, 82, 74, 77, 80, 79, 88, 97, 116, 118, 99, 93, 76, 77, 71, 69, 62, 72, 72, and 90, where $x_1 = 122$ and $x_{24} = 90$. Apply the initializing scheme using the additive trend seasonal model with $\alpha = 0.1$, $\beta = 0.2$, and $\gamma = 0.3$.

The results from the nine steps in the initialization are shown in Table 9-5 and a review of the steps is listed below.

1. The average demands for years 1 and 2 are

$$\bar{x}_{(1)} = 93.83 \quad \text{and} \quad \bar{x}_{(2)} = 84.58$$

2. The slope at $t = 0$ is

$$\bar{b}_0 = \frac{84.58 - 93.83}{12} = -0.77$$

3. The level at $t = 0$ becomes

$$\bar{a}_0 = 93.83 - 6.5(-0.77) = 98.83$$

4. Now the level at $t = 1$ to 24 are found. These are found by

$$\bar{a}_t = 98.83 - 0.77t$$

5. Seasonal increments for $t = 1$ to 24 are calculated using

$$\bar{d}_t = x_t - \bar{a}_t$$

Table 9-5. WORKSHEET FOR EXAMPLE 9.8—INITIALIZING WITH THE ADDITIVE TREND SEASONAL MODEL

t	x_t	(1)	(4) $\bar{a}_t$	(5) $\tilde{d}_t$	(6) $\bar{d}_t$	(7) $\dot{d}_t$	(8) $\hat{a}_t$	(8) $\hat{b}_t$	(8) $\hat{d}_{t+12}$	(9) d_{t+12}
0			98.83				98.83	−0.77		
1	122		98.06	23.94	25.56	25.55	97.90	−0.80	25.12	
2	118		97.29	20.71	25.33	25.32	96.66	−0.89	24.13	
3	115		96.52	18.48	15.10	15.10	96.18	−0.81	16.22	
4	100		95.75	4.25	5.37	5.37	95.30	−0.82	5.17	
5	94		94.98	−0.98	−5.36	−5.36	94.96	−0.72	−4.04	
6	82	$\bar{x}_{(1)} =$	94.21	−12.21	−10.09	−10.09	94.02	−0.77	−10.67	
7	74	93.83	93.44	−19.44	−16.32	−16.33	92.96	−0.83	−17.12	
8	77		92.67	−15.67	−15.05	−15.05	92.13	−0.83	−15.07	
9	80		91.90	−11.90	−16.28	−16.29	91.80	−0.73	−14.94	
10	79		91.13	−12.13	−11.01	−11.01	90.96	−0.75	−11.30	
11	88		90.36	−2.36	−5.74	−5.74	90.56	−0.68	−4.79	
12	97		89.59	7.41	8.53	8.53	89.74	−0.71	8.15	
13	116		88.82	27.18			89.22	−0.67	25.62	25.72
14	118		88.05	29.95			89.08	−0.57	25.57	25.67
15	99		87.05	11.72			87.94	−0.68	14.57	14.77
16	93		86.51	6.49			87.32	−0.67	5.32	5.42
17	76		85.74	−9.74			85.99	−0.80	−5.82	−5.72
18	77	$\bar{x}_{(2)} =$	84.97	−7.97			85.43	−0.75	−10.00	−9.90
19	71	84.58	84.20	−13.20			85.03	−0.68	−16.19	−16.09
20	69		83.43	−14.43			84.32	−0.69	−15.14	−15.04
21	62		82.66	−20.66			82.96	−0.82	−16.75	−16.65
22	72		81.89	−9.89			82.25	−0.80	−10.99	−10.89
23	72		81.12	−9.12			80.99	−0.89	−6.05	−5.95
24	90		80.35	9.65			80.27	−0.86	8.62	8.72

6. An average seasonal increment is found for each calendar month. These are:

$$\bar{d}_1 = \tfrac{1}{2}(23.94 + 27.18) = 25.56$$

$$\vdots$$

$$\bar{d}_{12} = \tfrac{1}{2}(7.41 + 9.65) = 8.53$$

7. The seasonal increments $(\bar{d}_1, \ldots, \dot{d}_{12})$ are used to find normalized seasonal increments $(\hat{d}_1, \ldots, \hat{d}_{12})$.
8. Using $\hat{a}_0 = 98.83$ and $\hat{b}_0 = -0.77$, the estimates for $\hat{a}_t$, $\hat{b}_t$, and $\hat{d}_{t+12}$ are generated for $t = 1$ to 24.
9. The seasonal increments $(\hat{d}_{25}, \ldots, \hat{d}_{36})$ are normalized to give $(d_{25}, \ldots, d_{36})$.

The initialization phase is now complete. The forecaster saves the coefficient estimates $\hat{a}_{24} = 80.27$, $\hat{b}_{24} = -0.86$, and $d_{25}, \ldots, d_{36}$ (as listed in Table 9-5).

Example 9.9

Using the results from Example 9.8, find the forecasts for months 25 to 36.

The forecast equation is

$$\hat{x}_{24}(\tau) = 80.27 - 0.86\tau + d_{24+\tau}$$

where d_{25} to d_{36} are as listed in Table 9-5. Hence, the forecasts are

$$\hat{x}_{24}(1) = 79.41 + 25.72 = 105.13$$
$$\hat{x}_{24}(2) = 78.55 + 25.67 = 104.22$$
$$\hat{x}_{24}(3) = 77.69 + 14.77 = 92.46$$
$$\hat{x}_{24}(4) = 76.83 + 5.42 = 82.25$$
$$\hat{x}_{24}(5) = 75.97 - 5.72 = 70.25$$
$$\hat{x}_{24}(6) = 75.11 - 9.90 = 65.21$$
$$\hat{x}_{24}(7) = 74.25 - 16.09 = 58.16$$
$$\hat{x}_{24}(8) = 73.39 - 15.04 = 58.35$$
$$\hat{x}_{24}(9) = 72.53 - 16.65 = 55.88$$
$$\hat{x}_{24}(10) = 71.67 - 10.89 = 60.78$$
$$\hat{x}_{24}(11) = 70.81 - 5.95 = 64.86$$
$$\hat{x}_{24}(12) = 69.95 + 8.72 = 78.67$$

These are plotted in Figure 9-5.

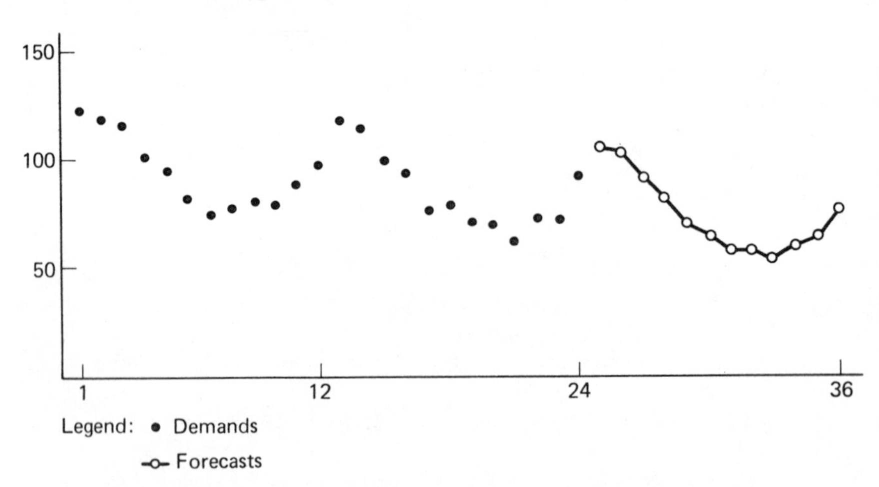

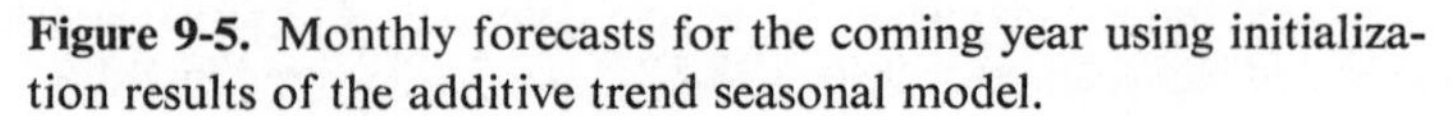

Figure 9-5. Monthly forecasts for the coming year using initialization results of the additive trend seasonal model.

Updating

For each time period beyond the date of the initialization phase, the 14 coefficients are updated as each new demand entry (x_T) becomes available. This process is carried out with the following recursive relations:

$$\hat{a}_T = \alpha(x_T - d_T) + (1 - \alpha)(\hat{a}_{T-1} + \hat{b}_{T-1})$$
$$\hat{b}_T = \beta(\hat{a}_T - \hat{a}_{T-1}) + (1 - \beta)\hat{b}_{T-1}$$
$$d_{T+12} = \gamma(x_T - \hat{a}_T) + (1 - \gamma)d_T$$

The most current 12 seasonal increments ($d_{T+1}, d_{T+2}, \ldots, d_{T+12}$) are normalized to sum to zero.

Having updated the coefficients, the current forecasts can be generated. For the τth future time period, the forecast is

$$\hat{x}_T(\tau) = \hat{a}_T + \hat{b}_T\tau + d_{T+\tau}$$

Example 9.10

Use the results of Example 9.8 along with the following demands: 100, 92, 94, 79, 88, 55, 58, 51, 59, 54, 62, and 76, to update the one-period-ahead forecasts for months $t = 25$ to 36.

At $T = 25$, the demand entry $x_{25} = 100$ is used to update the coefficients. The results yield

$$\hat{a}_{25} = 0.1(100 - 25.72) + 0.9(80.27 - 0.86) = 78.90$$
$$\hat{b}_{25} = 0.2(78.90 - 80.27) + 0.8(-0.86) = -0.96$$
$$d_{37} = 0.3(100 - 78.90) + 0.7(25.72) = 24.33$$

Now since

$$d_{26} + d_{27} + \ldots + d_{37} = -1.33$$

and

$$\bar{d} = \frac{-1.33}{12} = -0.11$$

the seasonal increments from Table 9-5 are normalized as follows:

$$d_{26} = 25.67 + 0.11 = 25.78$$
$$\vdots$$
$$d_{37} = 24.33 + 0.11 = 24.44$$

With $\tau = 1$, the forecast becomes

$$\hat{x}_{25}(1) = 78.90 - 0.96 + 25.78 = 103.72$$

Table 9-6. UPDATING PROCEDURE IN EXAMPLE 9.10 USING THE ADDITIVE TREND SEASONAL MODEL

t	x_t	$\hat{a}_t$	$\hat{b}_t$	d_{t+1}	d_{t+2}	d_{t+3}	d_{t+4}	d_{t+5}	d_{t+6}	d_{t+7}	d_{t+8}	d_{t+9}	d_{t+10}	d_{t+11}	d_{t+12}	$\hat{x}_t(1)$
25	100	78.90	−0.96	25.78	14.88	5.53	−5.61	−9.79	−15.98	−14.93	−16.54	−10.78	−5.84	8.83	24.44	103.71
26	92	76.76	−1.20	15.14	5.79	−5.35	−9.53	−15.72	−14.67	−16.28	−10.52	−5.58	9.09	24.71	22.88	90.71
27	94	75.90	−1.13	5.72	−5.42	−9.60	−15.79	−14.74	−16.35	−10.59	−5.65	9.02	24.63	22.81	15.96	80.48
28	79	74.62	−1.16	−5.39	−9.57	−15.76	−14.71	−16.32	−10.56	−5.62	9.05	24.67	22.84	15.99	5.35	68.07
29	68	73.45	−1.16	−9.57	−15.76	−14.71	−16.32	−10.56	−5.62	9.05	24.67	22.84	15.99	5.35	−5.40	62.72
30	55	71.51	−1.32	−15.58	−14.53	−16.14	−10.38	−5.44	9.32	24.84	23.02	16.17	5.53	−5.23	−11.48	54.62
31	58	70.54	−1.25	−14.61	−16.22	−10.46	−5.52	9.15	24.77	22.94	16.09	5.45	−5.31	−11.55	−14.74	54.68
32	51	68.92	−1.32	−16.13	−10.37	−5.43	9.24	24.85	23.02	16.17	5.53	−5.22	−11.47	−14.66	−15.52	51.46
33	59	68.35	−1.17	−10.54	−5.60	9.07	24.68	22.85	16.00	5.36	−5.39	−11.64	−14.83	−15.69	−14.27	56.63
34	54	66.92	−1.22	−5.55	9.12	24.74	22.91	16.06	5.42	−5.33	−11.58	−14.77	−15.63	−14.21	−11.20	60.15
35	62	65.88	−1.19	9.08	24.70	22.87	16.02	5.38	−5.38	−11.62	−14.81	−15.67	−14.25	−11.24	−5.09	73.77
36	76	64.91	−1.14	24.65	22.82	15.97	5.33	−5.43	−11.67	−14.86	−15.72	−14.30	−11.29	−5.14	9.63	88.42

Continuing in this manner, the coefficients are updated as each new demand entry becomes available and the results are listed in Table 9-6. A plot of the one-period-ahead forecasts is shown in Figure 9-6.

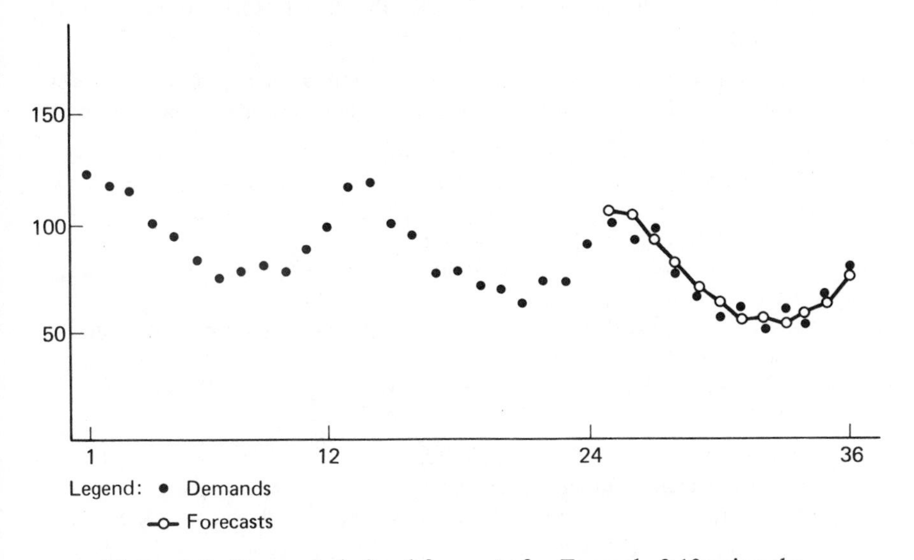

Figure 9-6. One-period-ahead forecasts for Example 9.10 using the additive trend seasonal model.

PROBLEMS

9-1. Carry out the initialization phase in Example 9.1 using $\alpha = 0.2$ and $\gamma = 0.1$. List the forecasts for the next three time periods that are obtained from the initialization.

9-2. Using the results of Problem 9-1, find the forecasts for the next three time periods when at $t = 25$, the demand is $x_{25} = 10$.

9-3. Suppose that $M = 4$, and the following 12 observations are known: 6, 12, 10, 4, 5, 13, 9, 3, 7, 15, 8, and 6. Carry out the initialization phase of the horizontal seasonal model with $\alpha = 0.1$ and $\gamma = 0.2$.

9-4. Use the results of Problem 9-3 as of $T = 12$ to find the forecasts for the next four time periods. If $x_{13} = 7$, find the updated forecasts for the following four time periods as of $T = 13$.

9-5. Suppose that $\alpha = 0.2$, $\gamma = 0.3$, and the horizontal seasonal model with $M = 6$ is in use for an item. Also, assume that the following is known as of time $T = 12$: $\hat{a}_{12} = 50$, $r_{13} = 0.6$, $r_{14} = 0.8$, $r_{15} = 0.9$, $r_{16} = 1.2$, $r_{17} = 1.4$, and $r_{18} = 1.1$.

a. Find the forecasts for each of the next 12 time periods.
b. If at $t = 13$, the demand is $x_{13} = 20$, find the updated estimates of the level and the seasonal ratios.
c. Use the results of part b to forecast the demands for the next two time periods.

9-6. Carry out the initialization phase in Example 9.5 using $\alpha = 0.2$, $\beta = 0.3$, and $\gamma = 0.4$. Use the results to forecast the demand for the next time period.

9-7. Suppose that the multiplicative trend seasonal model with $M = 4$, $\alpha = 0.1$, $\beta = 0.2$, and $\gamma = 0.3$ is in use for an item. Also, assume that the following eight demand entries are known: 3, 4, 11, 6, 4, 6, 18, and 8.
a. Carry out the initialization as of $T = 8$.
b. Use the results of part a to forecast the demands for the next eight time periods.
c. If at $t = 9$, the demand is $x_9 = 6$, find the updated level, slope, and seasonal ratios.
d. Use the results of part c to forecast the demands for the next four time periods.

9-8. Suppose that the multiplicative model with $M = 6$, $\alpha = 0.2$, $\gamma = 0.3$, and $\beta = 0.4$ is in use for an item. Also assume that the following is known as of $T = 12$: $\hat{a}_{12} = 60$, $\hat{b}_{12} = 2$, $r_{13} = 0.7$, $r_{14} = 0.9$, $r_{15} = 1.3$, $r_{16} = 1.2$, $r_{17} = 1.0$, and $r_{18} = 0.9$.
a. Now if at $t = 13$, the demand entry is $x_{13} = 38$, find the updated estimates of the level, slope, and the seasonal ratios.
b. Use the results of part a to find the forecasts of the demands for the next six time periods.

9-9. Carry out the initialization phase of Example 9.8 using $\alpha = 0.2$, $\beta = 0.3$, and $\gamma = 0.4$. Use these results to find the forecasts for the next two time periods.

9-10. Suppose that the additive trend seasonal model with $M = 4$, $\alpha = 0.2$, $\beta = 0.4$, and $\gamma = 0.5$ is in use for an item. Also, assume that the following 12 demand entries are known: 3, 7, 9, 5, 4, 10, 12, 6, 8, 15, 18, and 7.
a. Carry out the initialization phase as of $T = 12$.
b. Use the results of part a to forecast the demands for the next eight time periods.
c. If at $t = 13$, the demand is $x_{13} = 10$, find the updated estimates of the level, slope, and the seasonal increments.

9-11. Suppose that the additive trend seasonal model with $M = 6$, $\alpha = 0.2$, $\beta = 0.1$ and $\gamma = 0.4$ is in use. Assume that as of $T = 18$, the following is known: $\hat{a}_{18} = 100$, $\hat{b}_{18} = -2$, $d_{19} = -10$, $d_{20} = -20$, $d_{21} = -5$, $d_{22} = 0$, $d_{23} = 10$, and $d_{24} = 25$.
a. If at $T = 19$, the demand is $x_{19} = 80$, find the updated estimates of the level, slope, and seasonal increments.
b. Use the results of part a to find the forecasts for each of the next six time periods.

REFERENCE

[1] WINTERS, P. R., "Forecasting Sales by Exponentially Weighted Moving Averages." *Management Science*, April 1960, pp. 324–342.

REFERENCES FOR FURTHER STUDY

HOLT, C. C., "Forecasting Seasonal and Trends by Exponentially Weighted Moving Averages." Carnegie Institute of Technology, Pittsburgh, Pa., 1957.

HOLT, C. C., F. MODIGLIANI, J. F. MUTH, AND H. A. SIMON, *Planning Production, Inventories and Work Force*. Englewood Cliffs, N.J.: Prentice-Hall, Inc., 1960, Chap. 14.

10

ADAPTIVE CONTROL MODELS

This chapter presents various techniques in forecasting that have been developed to modify forecasting parameters in a dynamic manner. In this sense, the parameters may be altered at each time period or at intermittent time periods, depending on the flow of the forecast errors. Such methods are generally classified as adaptive control models.

The ability to monitor the forecast errors and modify the parameter values can be highly useful to the forecaster. In this way there is no need to commit an item to be forecast with a fixed set of parameter values. Instead, the system is allowed to seek the appropriate value for each parameter in the model.

Six models are presented in all. These include models where one or more forecasting parameters are monitored in an adaptive control manner.

10-1 EXPONENTIAL SMOOTHING WITH AN ADAPTIVE RESPONSE RATE

Trigg and Leach [1] have developed a technique that is a modification to forecasting models that employ exponential smoothing. Recall in these models that a smoothing or discounting parameter is selected by the forecaster to assign weights to the past demand entries. With this modification the parameter is allowed to vary at each time period based on the flow of the forecasting errors. The objective is to allow the forecasting model to react faster to sudden changes in the level or a shift in the demand pattern.

The technique is described with two forecasting models. These are the single smoothing model (Section 4-3) and the trend model using adaptive smoothing (Examples 7.6 and 7.9). In the single smoothing model, the smoothing parameter (α) is reset at each time period, while in the trend model, the discounting parameter (β) is altered.

In the Trigg and Leach approach, two smoothed errors are calculated at each time period. For time period t these are:

$$E_t = \gamma e_t + (1 - \gamma)E_{t-1}$$
$$A_t = \gamma |e_t| + (1 - \gamma)A_{t-1}$$

where γ = smoothing parameter applied to the errors and lies within (0, 1)
$e_t = x_t - \hat{x}_{t-1}(1)$ = current one-period-ahead forecast error
E_t = smoothed average error
A_t = smoothed absolute error

The two smoothed errors above are combined to yield another measure of the error, called a *tracking signal*. The tracking signal at time t is measured by

$$TS_t = \frac{E_t}{A_t}$$

Since E_t can be a positive or negative value and A_t is always a positive value, the tracking signal will always lie within the interval $(-1, 1)$. When the forecasts are unbiased, the tracking signal will be close to zero. This is because (for unbiased forecasts) E_t fluctuates near zero while A_t does not. When the forecasts are biased, the tracking signal will approach either -1 or $+1$, depending on the direction of the bias. This is true since E_t and A_t will differ little in magnitude for biased forecasts.

With the characteristics above, the tracking signal acts as a normalized measure of the forecast error. It is close to zero when all is going well, and is approaching -1 or $+1$ when the forecasts are consistently high or low.

In the Trigg and Leach technique, the tracking signal as described above is used as the basis of modifying the forecasting parameter.

Horizontal Model

Recall in the single smoothing model that the smoothed average demand at time t is

$$S_t = \alpha x_t + (1 - \alpha)S_{t-1}$$

and the τ-period-ahead forecast becomes

$$\hat{x}_t(\tau) = S_t$$

In the standard model, the smoothing parameter α has a fixed value for all time periods.

In the modified model, the smoothing parameter is varied at each time period and is conveniently labeled as α_t. The value for α_t is found from

$$\alpha_t = |TS_t|$$

which is the modulus (or absolute value) of the tracking signal.

Consider the current time period $t = T$, where α_T is found as above. The smoothed average of demands becomes

$$S_T = \alpha_T x_T + (1 - \alpha_T)S_{T-1}$$

and the τ-period-ahead forecast is

$$\hat{x}_T(\tau) = S_T$$

Now when $t = T + 1$, then x_{T+1} is known and

$$e_{T+1} = x_{T+1} - \hat{x}_T(1)$$
$$E_{T+1} = \gamma e_{T+1} + (1 - \gamma)E_T$$
$$A_{T+1} = \gamma |e_{T+1}| + (1 - \gamma)A_T$$
$$TS_{T+1} = \frac{E_{T+1}}{A_{T+1}}$$

and

$$\alpha_{T+1} = |TS_{T+1}|$$

With these calculations the cycle is completed and the system carries on in this manner.

Example 10.1

Table 10-1 shows how the method applies to the demand entries listed. Here $\gamma = 0.2$ is used to smooth the forecast errors and $\alpha = 0.1$ is used as the starting value of the smoothing parameter.

At $t = 1$, the first forecast is

$$\hat{x}_1(\tau) = x_1 = 12$$

Also, $E_1 = A_1 = TS_1 = 0$, since no other information is available.

Table 10-1. EXAMPLE 10.1—WORKSHEET FOR THE TRIGG AND LEACH TECHNIQUE WITH SINGLE SMOOTHING

t	x_t	e_t	E_t	A_t	TS_t	α_t	$\hat{x}_t(\tau)$
1	12	—	0.00	0.00	0.00	—	12.00
2	11	−1.00	−1.00	1.00	−1.00	0.10	11.90
3	13	1.10	−0.58	1.02	−0.57	0.57	12.53
4	10	−1.90	−0.97	1.32	−0.73	0.73	10.67
5	12	−0.53	−0.51	1.32	−0.39	0.39	11.18
6	9	−2.18	−0.84	1.49	−0.57	0.57	9.95
7	13	3.05	−0.07	1.81	−0.04	0.04	10.06
8	10	−0.06	−0.06	1.46	−0.04	0.04	10.06
9	14	3.94	0.74	1.95	0.38	0.38	11.54
10	17	5.46	1.68	2.65	0.63	0.63	15.00
11	15	0.00	1.34	2.12	0.63	0.63	15.00
12	20	5.00	2.08	2.79	0.77	0.77	18.85
13	18	−0.85	1.89	2.39	0.79	0.79	19.76
14	17	−2.76	1.56	1.96	0.80	0.80	19.95
15	20	0.05	1.26	1.58	0.80	0.80	19.99
16	26	6.01	2.21	2.46	0.90	0.90	25.38
17	23	−2.38	1.29	2.45	0.53	0.53	24.12
18	24	−0.12	1.01	1.98	0.51	0.51	24.06
19	23	−1.06	0.60	1.80	0.33	0.33	23.71
20	22	−1.71	0.13	1.78	0.08	0.08	23.58
21	25	1.42	0.39	1.71	0.23	0.23	23.91
22	22	−1.91	−0.07	1.75	−0.04	0.04	23.83
23	23	−0.83	−0.22	1.57	−0.14	0.14	23.72
24	26	2.28	0.28	1.71	0.16	0.16	24.09

Now when $t = 2$, then

$$e_2 = 11 - 12 = -1$$

$$E_2 = -1$$

$$A_2 = 1$$

and

$$TS_2 = -1$$

However, since E_2 has not really been smoothed yet, the modified technique is not yet applied and $\alpha_2 = 0.1$. Hence,

$$S_2 = 0.1(11) + 0.9(12) = 11.90$$

$$\hat{x}_2(\tau) = 11.90$$

At $t = 3$,

$$e_3 = 13 - 11.90 = 1.1$$

$$E_3 = 0.2(1.1) + 0.8(-1) = -0.58$$

$$A_3 = 0.2|1.1| + 0.8(1) = 1.02$$
$$TS_3 = \frac{-0.58}{1.02} = -0.57$$
$$\alpha_3 = 0.57$$
$$S_3 = 0.57(13) + 0.43(11.90) = 12.53$$
$$\hat{x}_3(\tau) = 12.53$$

The method continues in this manner and the results are listed in the table. A plot of the demands and the one-period-ahead forecasts is shown in Figure 10-1.

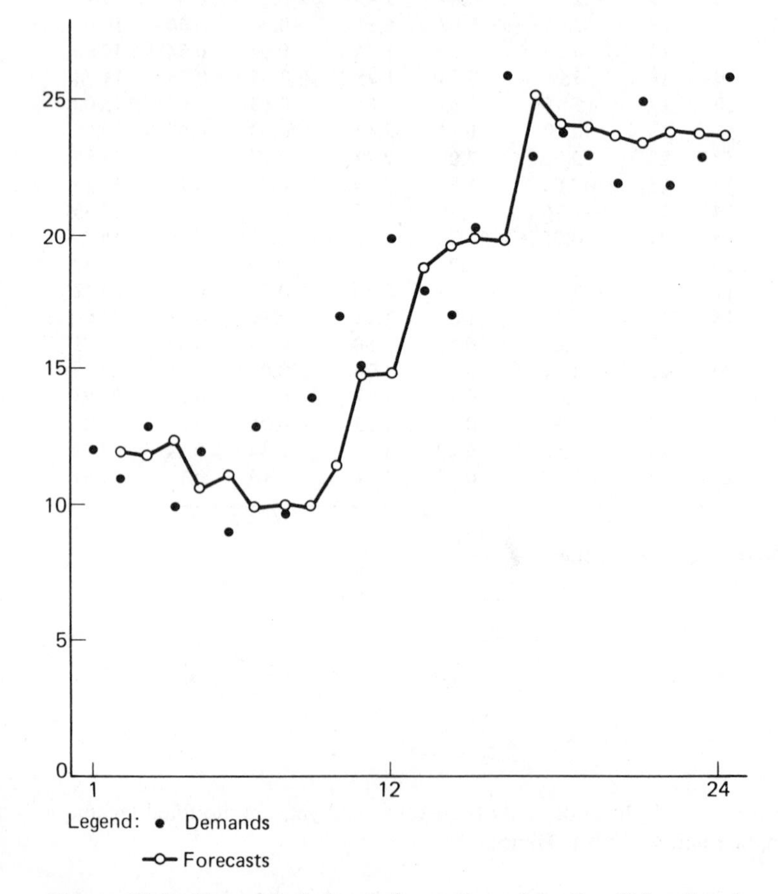

Figure 10-1. One-period-ahead forecasts using the Trigg–Leach technique.

Trend Model

Consider the trend model from the adaptive smoothing method. Here two coefficients a_1 and a_2 are estimated at each time period to yield the forecasts

$$\hat{x}_T(\tau) = \hat{a}_1(T) + \hat{a}_2(T)\tau$$

where $\hat{a}_1(T)$ and $\hat{a}_2(T)$ are estimates of a_1 and a_2, respectively, at time T.

The coefficients estimates are found from

$$\hat{a}_1(T) = \hat{a}_1(T-1) + \hat{a}_2(T-1) + h_1 e_T$$
$$\hat{a}_2(T) = \hat{a}_2(T-1) + h_2 e_T$$

Recall that the smoothing coefficients (h_1, h_2) are obtained for a particular discounting parameter, β. In the trend model these become

$$h_1 = 1 - \beta^2$$
$$h_2 = (1 - \beta)^2$$

In the standard approach β is set for all time periods, and h_1 and h_2 remain fixed. Now in the modified approach, h_1 is varied, depending on the tracking signal. Only h_1 is varied since a change in both h_1 and h_2 will result in too sharp a change in the forecasts. In this way, only the smoothing parameter (h_1) affecting the level is affected, whereas the parameter that influences the slope (h_2) is smoothed in the normal manner.

For the reasons above, the notation h_{1_t} is used, where

$$h_{1_t} = 1 - \beta_{1_t}^2$$

and

$$\beta_{1_t} = \sqrt{1 - |TS_t|}$$

Note the simpler relation

$$h_{1_t} = |TS_t|$$

Now as stated earlier,

$$h_2 = (1 - \beta)^2$$

where β is the standard discounting parameter selected by the forecaster.

Summarizing, at time T, the one-period-ahead error is

$$e_T = x_T - \hat{x}_{T-1}(1)$$

This yields

$$E_T = \gamma e_T + (1 - \gamma)E_{T-1}$$
$$A_T = \gamma |e_T| + (1 - \gamma)A_{T-1}$$
$$TS_T = \frac{E_T}{A_T}$$
$$h_{1T} = |TS_T|$$

$$\hat{a}_1(T) = \hat{a}_1(T-1) + \hat{a}_2(T-1) + h_{1T}e_T$$
$$\hat{a}_2(T) = \hat{a}_2(T-1) + h_2e_T$$

and

$$\hat{x}_T(\tau) = \hat{a}_1(T) + \hat{a}_2(T)\tau$$

Example 10.2

Table 10-2 gives an example where the technique above is used with $\gamma = 0.2$ and $\beta = 0.9$ ($h_2 = 0.01$). Suppose that at the outset, the forecaster estimates the coefficients as $\hat{a}_1(0) = 10$ and $\hat{a}_2(0) = 1$. Hence $\hat{x}_0(1) = 10 + 1 = 11$.

Table 10-2. WORKSHEET USING THE TRIGG LEACH TECHNIQUE WITH A TREND MODEL

t	x_t	e_t	E_t	A_t	TS_t	h_{1t}	$\hat{a}_1(t)$	$\hat{a}_2(t)$	$\hat{x}_t(1)$
1	10	−1.00	−1.00	1.00	−1.00	0.19	10.81	0.99	11.80
2	12	0.20	−0.76	0.84	−0.90	0.90	11.98	0.99	12.97
3	11	−1.97	−1.00	1.07	−0.94	0.94	11.11	0.97	12.08
4	14	1.92	−0.42	1.24	−0.34	0.34	12.73	0.99	13.72
5	16	2.28	0.12	1.45	0.08	0.08	13.91	1.01	14.92
6	15	0.08	0.11	1.17	0.09	0.09	14.94	1.01	15.95
7	28	12.05	2.50	3.35	0.75	0.75	24.95	1.13	26.08
8	27	0.92	2.18	2.86	0.76	0.76	26.78	1.14	27.92
9	31	3.08	2.36	2.90	0.81	0.81	30.42	1.17	31.59
10	33	1.41	2.17	2.61	0.83	0.83	32.77	1.18	33.95
11	32	−1.95	1.35	2.48	0.54	0.54	32.90	1.17	34.07
12	35	0.97	1.26	2.17	0.58	0.58	34.61	1.17	35.78
13	34	−1.78	0.66	2.09	0.31	0.31	35.22	1.16	36.38
14	36	−0.38	0.45	1.75	0.26	0.26	36.28	1.15	37.43
15	35	−2.43	−0.13	1.89	−0.07	0.07	37.27	1.13	38.40
16	39	0.60	0.02	1.63	0.01	0.01	38.40	1.13	39.53
17	38	−1.53	−0.29	1.61	−0.18	0.18	39.26	1.12	40.38
18	42	1.62	0.09	1.61	0.06	0.06	40.47	1.14	41.61

Now at $t = 1$, the forecast error is

$$e_1 = 10 - \hat{x}_0(1) = 10 - 11 = -1$$

and

$$E_1 = -1$$
$$A_1 = 1$$
$$TS_1 = -1$$

The value for β_1 is the same as the standard parameter at this time since neither E_1 nor A_1 has yet been smoothed. So using $h_{11} = (1 - 0.9^2) = 0.19$ and $h_2 = (1 - 0.9)^2 = 0.01$, then

$$\hat{a}_1(1) = 10 + 1 + 0.19(-1) = 10.81$$
$$\hat{a}_2(1) = 1 + 0.01(-1) = 0.99$$
$$\hat{x}_1(\tau) = 10.81 + 0.99\tau$$

When $t = 2$, then

$$e_2 = 12 - 11.80 = 0.20$$
$$E_2 = 0.2(0.20) + 0.8(-1.00) = -0.76$$
$$A_2 = 0.2(0.20) + 0.8(1.00) = 0.84$$
$$TS_2 = \frac{-0.76}{0.84} = -0.90$$
$$h_{12} = |TS_2| = 0.90$$
$$\hat{a}_1(2) = 10.81 + 0.99 + 0.90(0.20) = 11.98$$
$$\hat{a}_2(2) = 0.99 + 0.01(0.20) = 0.99$$
$$\hat{x}_2(\tau) = 11.98 + 0.99\tau$$

A plot of the one-period-ahead forecasts for $t = 1$ to 18 is shown in Figure 10-2.

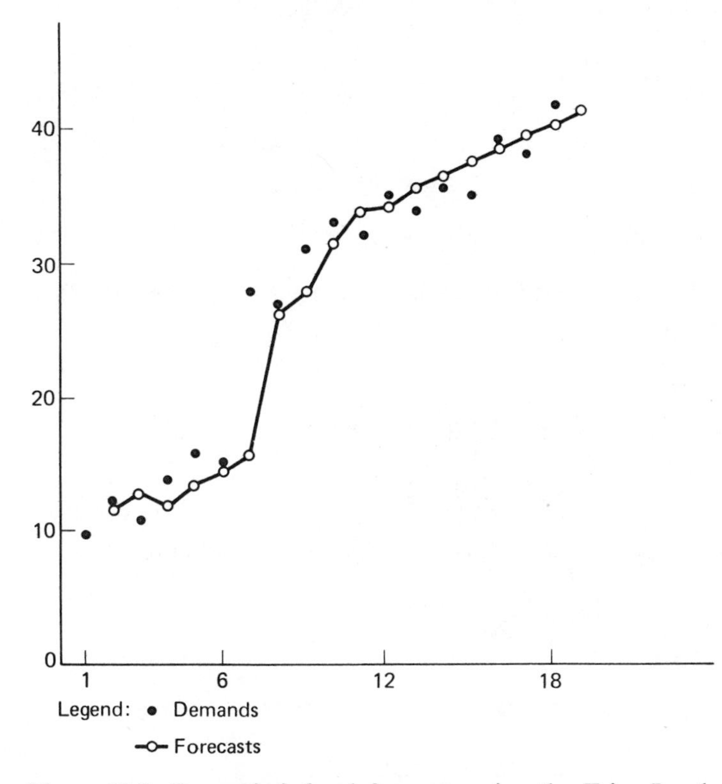

Figure 10-2. One-period-ahead forecasts using the Trigg–Leach technique with a trend model.

10-2 COMBINATION FORECASTS

Bates and Granger [2] propose a technique of generating a forecast by combining the forecasts from two alternative forecasting methods. Let the forecasting methods be identified as F_1 and F_2, and let the τ period ahead forecasts from these methods be denoted as $\hat{x}_{1T}(\tau)$ and $\hat{x}_{2T}(\tau)$, respectively. The combined forecast is

$$\hat{x}_T(\tau) = w_T\hat{x}_{1T}(\tau) + (1 - w_T)\hat{x}_{2T}(\tau)$$

where w_T is a combining parameter that lies within (0, 1) and is varied at each time period.

The parameter w_T is chosen in such a way that the forecast error of the combined forecast is minimized. Bates and Granger list five options in seeking w_T. These are

1. $w_T = \dfrac{E_2}{E_1 + E_2}$

   where

$$E_1 = \sum_{t=T-5}^{T} e_{1t}^2$$

$$E_2 = \sum_{t=T-5}^{T} e_{2t}^2$$

$$e_{1t} = \hat{x}_{1,t-1}(1) - x_t$$

$$e_{2t} = \hat{x}_{2,t-1}(1) - x_t$$

2. $w_T = \theta w_{T-1} + (1 - \theta)\dfrac{E_2}{E_1 + E_2}$

   where θ is a fixed smoothing parameter lying within (0, 1).

3. $w_T = \dfrac{S_2^2}{S_1^2 + S_2^2}$

   where

$$S_1^2 = \sum_{t=1}^{T} \beta^{T-t}e_{1t}^2$$

$$S_2^2 = \sum_{t=1}^{T} \beta^{T-t}e_{2t}^2$$

$$\beta = \text{discounting parameter within } (0, 1)$$

4. $w_T = \dfrac{S_2^2 - S_{12}}{S_1^2 + S_2^2 - 2S_{12}}$

   where

$$S_{12} = \sum_{t=1}^{T} \beta^{T-t}e_{1t}e_{2t}$$

5. $w_T = \theta w_{T-1} + (1 - \theta)\dfrac{|e_{2,T}|}{|e_{1,T}| + |e_{2,T}|}$

At the outset when $T = 1$, the combining parameter (w_1) is set to $\frac{1}{2}$, i.e., $w_1 = 0.50$. The forecaster selects one of the options above to generate w_T for $T \geq 2$. From here, the two forecasts $\hat{x}_{1T}(\tau)$ and $\hat{x}_{2T}(\tau)$ are generated, and w_T is found. Next, the combined forecast

$$\hat{x}_T(\tau) = w_T\hat{x}_{1T}(\tau) + (1 - w_T)\hat{x}_{2T}(\tau)$$

is calculated and the system proceeds in this manner.

The forecasting methods F_1 and F_2 can be of any type. They may be the exact same forecast model with alternative forecasting parameters (e.g., F_1 may use $\alpha = 0.1$ and F_2 uses $\alpha = 0.3$), or they may be two completely different forecast models (e.g., F_1 may be a single smoothing model and F_2 a double smoothing model). The purpose of the combining technique is to give more weight to the forecast model (F_1 or F_2) that produces the better results.

Example 10.3

Table 10-3 shows an example of how the technique applies using option 1 when F_1 is the single smoothing model with $\alpha = 0.1$ and F_2 is the double smoothing model

Table 10-3. Worksheet for Example 10.3 Using Combined Forecasts

t	x_t	$\hat{x}_{1t}(1)$	$\hat{x}_{2t}(1)$	e_{1t}	e_{2t}	E_{1t}	E_{2t}	W_t	$\hat{x}_t(1)$
1	12	12.0	12.0	—	—	—	—	0.50	12.0
2	11	11.9	11.8	1.0	1.0	1.0	1.0	0.50	11.8
3	7	11.4	10.8	4.9	4.8	25.0	24.0	0.49	11.1
4	13	11.6	11.2	−1.6	−2.2	27.5	28.7	0.51	11.4
5	8	11.2	10.5	3.6	3.2	40.3	39.0	0.49	10.9
6	9	11.0	10.1	2.2	1.5	44.2	40.4	0.48	10.5
7	13	11.2	10.6	−2.0	−2.8	24.2	25.4	0.51	10.9
8	10	11.1	10.5	1.2	0.6	23.1	21.1	0.48	10.7
9	14	11.4	11.1	−2.9	−3.5	18.9	23.4	0.55	11.2
10	17	11.9	12.3	−5.6	−5.9	45.8	55.8	0.55	12.1
11	12	11.9	12.2	−0.1	0.3	41.7	47.8	0.53	12.1
12	15	12.2	12.8	−3.1	−2.8	49.7	55.0	0.53	12.5
13	20	13.0	14.3	−7.8	−7.1	101.3	94.0	0.48	13.7
14	17	13.4	15.0	−4.0	−2.7	85.4	66.4	0.44	14.3
15	20	14.1	16.1	−6.6	−5.0	128.8	91.6	0.42	15.3
16	26	15.3	18.3	−11.9	−9.9	261.6	181.1	0.41	17.1
17	19	15.6	18.8	−3.7	−0.7	215.3	130.0	0.38	17.6
18	21	16.2	19.5	−5.4	−2.2	228.2	127.9	0.36	18.3
19	20	16.6	19.9	−3.8	−0.5	199.5	102.9	0.34	18.8
20	28	17.7	21.9	−11.4	−8.1	188.2	70.5	0.27	20.8
21	31	19.0	24.1	−13.3	−9.1	351.1	152.7	0.30	22.6
22	27	19.8	25.2	−8.0	−2.9	385.8	156.0	0.29	23.7
23	34	21.2	27.5	−14.1	−8.8	572.8	232.8	0.29	25.7
24	26	21.7	27.8	−4.7	1.5	463.7	170.3	0.27	26.2

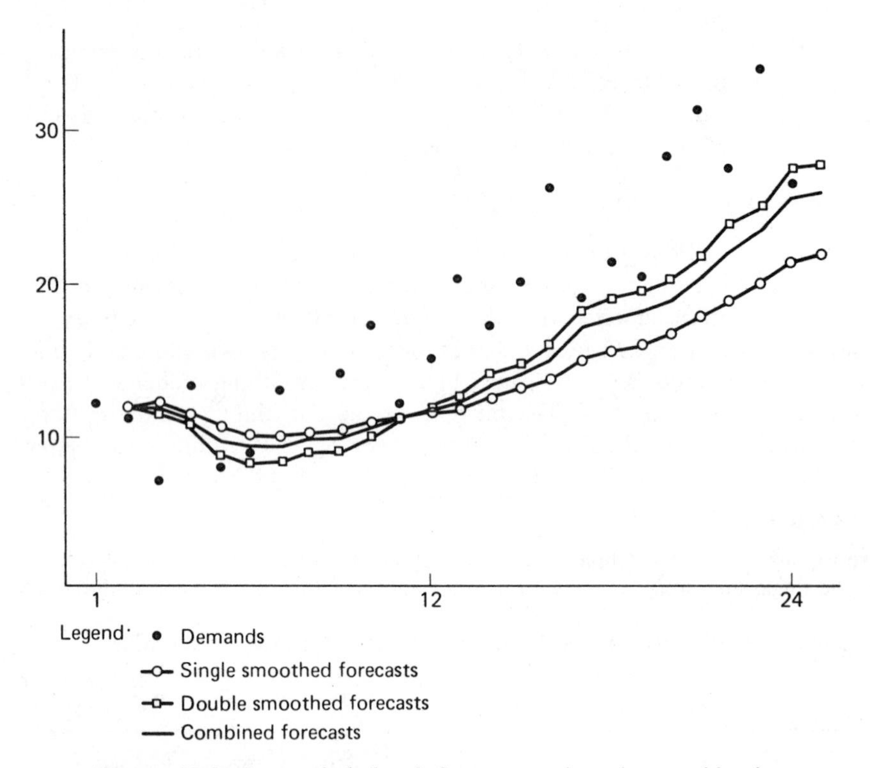

Figure 10-3. One-period-ahead forecasts using the combination forecast technique (Example 10.3).

using $\alpha = 0.1$ (Figure 10-3). At the outset the forecaster initializes F_1 by setting $S_1 = x_1 = 12$. He/she initializes F_2 using $S_1 = \hat{a}_1 = x_1 = 12$ and $\hat{b}_1 = 0$.

Recall that in F_1,

$$S_t = \alpha x_t + (1 - \alpha)S_{t-1}$$

$$\hat{x}_{1t}(\tau) = S_t$$

and in F_2,

$$e_t = \hat{x}_{2,t-1}(1) - x_t$$

$$\hat{a}_t = x_t + (1 - \alpha)^2 e_t$$

$$\hat{b}_t = \hat{b}_{t-1} - \alpha^2 e_t$$

$$\hat{x}_{2t}(\tau) = \hat{a}_t + \hat{b}_t \tau$$

So at $t = 1$, $\hat{x}_{11}(1) = \hat{x}_{21}(1) = \hat{x}_1(1) = 12$. Now when $t = 2$, $e_{11} = e_{12} = 1$, $E_{11} = E_{21} = 1.0$, $w_1 = 0.50$, $\hat{x}_{1t}(1) = 11.9$, $\hat{x}_{2t}(1) = 11.8$, and

$$\hat{x}_t(1) = 0.5(11.9) + 0.5(11.8) = 11.85$$

For $t = 3$, Table 10-3 shows that $e_{13} = 4.9$, $e_{23} = 4.8$, and hence

$$E_{12} = (1)^2 + (4.9)^2 = 25.01$$

$$E_{22} = (1)^2 + (4.8)^2 = 24.02$$

$$w_2 = \frac{24.02}{25.01 + 24.02} = 0.49$$

$$\hat{x}_t(\tau) = 0.49(11.4) + 0.51(10.8) = 11.1$$

The system continues in this manner. Recall that E_{1t} and E_{2t} includes the sum of squares of the corresponding six most recent forecast errors.

10-3 ADAPTIVE CONTROL OF THE SMOOTHING PARAMETER

An adaptive control method that is applicable for models with one smoothing parameter has been developed by Chow [3]. These include the single, double, and triple smoothing models of Chapters 4, 5, and 6. Letting α designate the smoothing parameter, the method seeks the best value of α at each time period. The procedure is easy to use, and Chow shows the results from experimental trials.

The method is described in the following two steps:

1. At the outset, the forecaster chooses three values of α: α_0, a nominal value; α_L, a low value; and α_H, a high value, where

$$\alpha_L = \alpha_0 - \delta$$

$$\alpha_H = \alpha_0 + \delta$$

and δ is a fixed constant. Chow uses $\delta = 0.05$. Now three forecasts are generated, using the chosen values of α. These forecasts are $\hat{x}_{0t}(\tau)$, using α_0; $\hat{x}_{Lt}(\tau)$, using α_L; and $\hat{x}_{Ht}(\tau)$, using α_H. The forecast that is employed in production is $\hat{x}_{0t}(\tau)$. With each forecast model, the smoothed absolute error is found. These are

$$A_{0t} \quad \text{for } \hat{x}_{0t}(\tau)$$

$$A_{Lt} \quad \text{for } \hat{x}_{Lt}(\tau)$$

$$A_{Ht} \quad \text{for } \hat{x}_{Ht}(\tau)$$

and are initially set to zero.

2. At the subsequent time periods, the following calculations are carried out:

The one-period-ahead forecast errors are measured:

$$e_o = x_t - \hat{x}_{o,t-1}(1)$$

$$e_L = x_t - \hat{x}_{L,t-1}(1)$$

$$e_H = x_t - \hat{x}_{H,t-1}(1)$$

The smoothed absolute averages are updated:

$$A_{ot} = \alpha_o|e_o| + (1 - \alpha_o)A_{o,t-1}$$

$$A_{Lt} = \alpha_L|e_L| + (1 - \alpha_L)A_{L,t-1}$$

$$A_{Ht} = \alpha_H|e_H| + (1 - \alpha_H)A_{H,t-1}$$

The value of α (say $\alpha_{\min}$) associated with the minimum smoothed average error is identified.

Should $\alpha_{\min} = \alpha_o$, the three smoothing parameters are unchanged and the forecasts $\hat{x}_{ot}(\tau)$, $\hat{x}_{Lt}(1)$, and $\hat{x}_{Ht}(1)$ are generated.

If $\alpha_{\min} = \alpha_L$, the nominal value becomes $\alpha_o = \alpha_L$ and $\alpha_L = \alpha_o - \delta$, $\alpha_H = \alpha_o + \delta$. Limits on α_L and α_H are set at 0.05 and 0.95. Also, the averages A_{ot}, A_{Lt}, and A_{Ht} are set to zero and the forecasts $\hat{x}_{ot}(\tau)$, $\hat{x}_{Lt}(1)$, and $\hat{x}_{Ht}(1)$ are found.

If $\alpha_{\min} = \alpha_H$, then the nominal value becomes $\alpha_o = \alpha_H$ and the process continues as in the above situation, where $\alpha_{\min} = \alpha_L$.

Example 10.4

An example is given in Table 10-4. Here $\delta = 0.05$ and $\alpha_L = 0.15$, $\alpha_o = 0.20$, and $\alpha_H = 0.25$ are used at the outset.

At $t = 1$, the three forecasts are initialized to $x_1 = 12$. When $t = 2$, the three forecasts are found with $\alpha_L = 0.15$, $\alpha_o = 0.20$, and $\alpha_H = 0.25$. Now when $t = 3$, $e_L = -4.9$, and $e_o = e_H = -4.8$. Hence,

$$A_{L3} = 0.2(4.9) + 0.8(1.0) = 1.8$$

Also, $A_{o3} = A_{H3} = 1.8$. Since, all are the same, the parameters are not changed.

Now at $t = 4$, $A_{L4} = 1.8$ is the minimum smoothed absolute error and $\alpha_{\min} = 0.15$. So the nominal value becomes $\alpha_o = 0.15$. Further, $\alpha_L = 0.10$, $\alpha_H = 0.20$, and the smoothed averages are set to zero. The forecasts are found with the new parameters, and in this way the system continues. A plot of the one-period-ahead forecasts is shown in Figure 10-4.

Table 10-4. WORKSHEET FOR EXAMPLE 10.4 USING THE ADAPTIVE CONTROL OF SMOOTHING PARAMETER METHOD

t	x_t	e_L	e_o	e_H	A_{Lt}	A_{ot}	A_{Ht}	α_L	α_o	α_H	$\hat{x}_{Lt}(1)$	$\hat{x}_{ot}(1)$	$\hat{x}_{Ht}(1)$
1	12	0.0	0.0	0.0	0.0	0.0	0.0	0.15	0.20	0.25	12.0	12.0	12.0
2	11	−1.0	−1.0	−1.0	1.0	1.0	1.0	0.15	0.20	0.25	11.9	11.8	11.8
3	7	−4.9	−4.8	−4.8	1.8	1.8	1.8	0.15	0.20	0.25	11.1	10.8	10.6
4	13	1.9	2.2	2.4	1.8	1.9	1.9	0.10	0.15	0.20	11.3	11.2	11.1
5	8	−3.3	−3.2	−3.1	3.3	3.2	3.1	0.15	0.20	0.25	10.8	10.5	10.3
6	9	−1.8	−1.5	−1.3	1.8	1.5	1.3	0.20	0.25	0.30	10.5	10.1	9.9
7	13	2.5	2.9	3.1	2.5	2.9	3.1	0.15	0.20	0.25	10.8	10.7	10.7
8	10	−0.8	−0.7	−0.7	0.8	0.7	0.7	0.15	0.20	0.25	10.7	10.6	10.5
9	14	3.3	3.4	3.5	1.3	1.3	1.3	0.15	0.20	0.25	11.1	11.2	11.3
10	17	−5.9	−5.8	−5.7	2.2	2.2	2.2	0.15	0.20	0.25	12.0	12.4	12.8
11	12	0.0	−0.4	−0.5	1.8	1.9	1.9	0.10	0.15	0.20	11.9	12.2	12.5
12	15	3.1	2.8	2.5	3.1	2.8	2.5	0.15	0.20	0.25	12.4	12.8	13.1
13	20	7.6	7.2	6.9	7.6	7.2	6.9	0.20	0.25	0.30	13.9	14.6	15.2
14	17	3.1	2.4	1.8	3.1	2.4	1.8	0.25	0.30	0.35	14.7	15.3	15.8
15	20	5.3	4.7	4.2	5.3	4.7	4.2	0.30	0.35	0.40	16.3	17.0	17.5
16	26	9.7	9.0	8.5	9.7	9.0	8.5	0.35	0.40	0.45	19.7	20.6	21.3
17	19	−0.7	−1.6	−2.3	0.7	1.6	2.3	0.30	0.35	0.40	19.5	20.0	20.4
18	21	1.5	1.0	0.6	1.5	1.0	0.6	0.35	0.40	0.45	20.0	20.4	20.7
19	20	0.0	−0.4	−0.7	0.0	0.4	0.7	0.30	0.35	0.40	20.0	20.3	20.4
20	28	8.0	7.7	7.6	8.0	7.7	7.6	0.35	0.40	0.45	22.8	23.4	23.8
21	31	8.2	7.6	7.2	8.2	7.6	7.2	0.40	0.45	0.50	26.1	26.8	27.4
22	27	0.9	0.2	−0.4	0.9	0.2	0.4	0.40	0.45	0.50	26.5	26.9	27.2
23	34	7.5	7.1	−6.8	3.9	3.3	3.3	0.40	0.45	0.50	29.5	30.1	30.6
24	26	−3.5	−4.1	−4.6	3.7	3.7	4.0	0.40	0.45	0.50	28.1	28.2	28.3

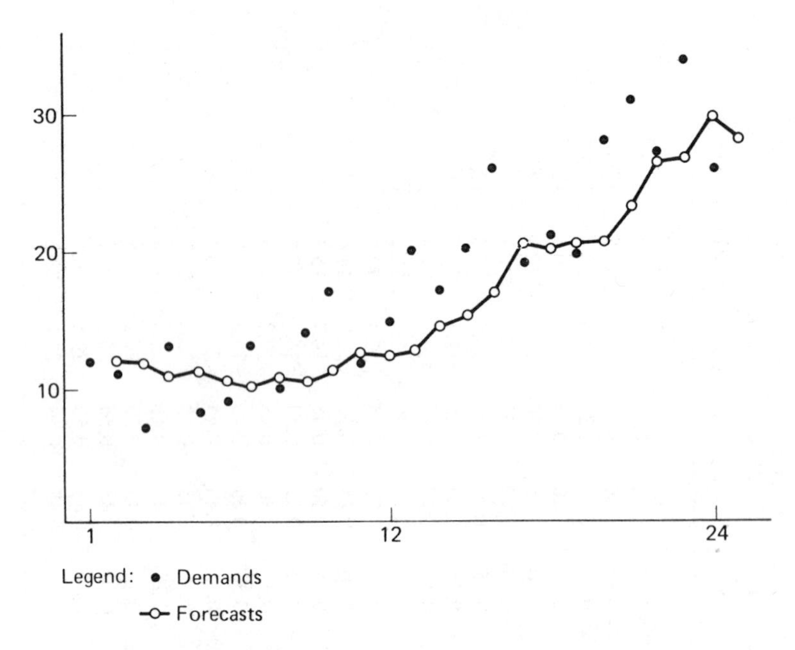

Figure 10-4. One-period-ahead forecasts using the adaptive control of the smoothing parameter method.

10-4 ADAPTIVE CONTROL FOR MODELS WITH MORE THAN ONE SMOOTHING PARAMETER

Montgomery [4] has developed a method of adaptive control for forecasting models that have more than one smoothing parameter.[1] The objective of the technique is to find the best settings for each of the parameters. A decision on the choice of parameter values is made at each time period, depending on the flow of forecast errors.

For a model with k smoothing parameters, the technique calls for $(k + 1)$ alternative design points. Each design point has a particular set of values for every smoothing parameter. At the current time T, the forecast error corresponding to each design point is measured, and through these errors a choice is made on whether or not to change one of the design points. Thereupon, the selection of a design point to use in production is made. That is, the forecasts for the coming periods are generated using the design point selected.

[1] The method is based on the simplex, which is an orthogonal first-order experimental design.

The technique is described for a forecast model with two parameters ($k = 2$). Let the smoothing parameters be identified by

$$d_1 = (\alpha_1, \beta_1)$$
$$d_2 = (\alpha_2, \beta_2)$$
$$d_3 = (\alpha_3, \beta_3)$$

The design points above are generated in such a way where the following three conditions are satisfied:

$$(\alpha_1 - \alpha_2)^2 + (\beta_1 - \beta_2)^2 = \rho^2$$
$$(\alpha_1 - \alpha_3)^2 + (\beta_1 - \beta_3)^2 = \rho^2$$
$$(\alpha_2 - \alpha_3)^2 + (\beta_2 - \beta_3)^2 = \rho^2$$

where ρ is a fixed constant set by the forecaster and is called the *edge*.

At time period $t = T - 1$, three forecasts are generated—one from each design point. These are labeled $\hat{x}_{1,T-1}(1)$, $\hat{x}_{2,T-1}(1)$, and $\hat{x}_{3,T-1}(1)$ and use d_1, d_2, and d_3, respectively.

Now at $t = T$, using the current demand entry (x_T), the absolute forecast errors corresponding to each of the three forecasts are measured. These are

$$e_1 = |x_T - \hat{x}_{1,T-1}(1)|$$
$$e_2 = |x_T - \hat{x}_{2,T-1}(1)|$$
$$e_3 = |x_T - \hat{x}_{3,T-1}(1)|$$

With these error values, a choice is now made on whether or not to alter one of the design points, and on which design point to use in production to forecast the future demands.

The following three rules are followed in the process:

1. Let d_j represent the design point which is associated with the maximum of (e_1, e_2, e_3). The design d_j is dropped and a new design d_j^* takes its place. The new values for α and β are

$$\alpha_j^* = \sum_{i \neq j} \alpha_i - \alpha_j$$
$$\beta_j^* = \sum_{i \neq j} \beta_i - \beta_j$$

   and the new design point is $d_j^* = (\alpha_j^*, \beta_j^*)$. With the three design points that remain, forecasts are generated and the forecast corresponding with d_j^* is used for production purposes.

2. Apply rule 1 unless one design point has occurred in three successive time periods without being dropped. Should this situation arise for the ith design, use the forecast generated with d_i for production purposes. None of the designs are dropped at this time. At the next time period, rule 1 is followed.
3. Should d_j give the maximum error at time $t = T - 1$ and its replacement (d_j^*) gives the maximum at $t = T$, do not replace d_j^* (this will cycle back to d_j). Instead, drop the design with the second largest absolute current error.

Example 10.5

Consider the following forecast model

$$\hat{x}_T(\tau) = \hat{a}_T + \hat{b}_T\tau$$

where

$$\hat{a}_T = \alpha x_T + (1 - \alpha)(\hat{a}_{T-1} + \hat{b}_{T-1})$$
$$\hat{b}_T = \beta(\hat{a}_T - \hat{a}_{T-1}) + (1 - \beta)\hat{b}_{T-1}$$

with α and β the smoothing parameters. Suppose that at $t = T - 1$, the three design points are

$$d_1 = (0.100, 0.075)$$
$$d_2 = (0.168, 0.093)$$
$$d_3 = (0.118, 0.143)$$

Note that the edge is $p = 0.07$, since

$$(0.100 - 0.168)^2 + (0.075 - 0.093)^2 = 0.07^2$$
$$(0.100 - 0.118)^2 + (0.075 - 0.143)^2 = 0.07^2$$
$$(0.168 - 0.118)^2 + (0.093 - 0.143)^2 = 0.07^2$$

Also, assume that

$$\hat{x}_{1,T-1}(1) = 110.54$$
$$\hat{x}_{2,T-1}(1) = 110.90$$
$$\hat{x}_{3,T-1}(1) = 110.67$$

and d_1 had been selected for use in production. In d_1, assume that $\hat{a}_{T-1} = 105.50$ and $\hat{b}_{T-1} = 5.04$. The data above are listed in Table 10-5.

Now at $t = T$, suppose that the demand entry is $x_T = 109$. This gives

$$e_1 = |109 - 110.54| = 1.54$$
$$e_2 = |109 - 110.90| = 1.90$$
$$e_3 = |109 - 110.67| = 1.67$$

Table 10-5. WORKSHEET FOR EXAMPLE 10.5

t	$T-1$	T
x_t	—	109
e_1	—	1.54
e_2	—	1.90
e_3	—	1.67
d_1	(0.100, 0.075)†	(0.100, 0.075)
d_2	(0.168, 0.093)	(0.050, 0.125)†
d_3	(0.118, 0.143)	(0.118, 0.143)
$\hat{a}_1$	105.50†	110.39
$\hat{b}_1$	5.04†	5.03
$\hat{a}_2$	—	110.46†
$\hat{b}_2$	—	5.03†
$\hat{a}_3$	—	110.36
$\hat{b}_3$	—	5.01
$\hat{x}_{1t}(1)$	110.54†	115.42
$\hat{x}_{2t}(1)$	110.90	115.49†
$\hat{x}_{3t}(1)$	110.67	115.37

†Used in production to forecast demands.

Since d_2 is associated with the maximum error, the smoothing parameters of d_2 are dropped and replaced with

$$\alpha_2^* = 0.100 + 0.118 - 0.168 = 0.050$$
$$\beta_2^* = 0.075 + 0.143 - 0.093 = 0.125$$

whereby

$$d_1 = (0.100, 0.075)$$
$$d_2 = (0.050, 0.125)$$
$$d_3 = (0.118, 0.143)$$

Note again that

$$(0.100 - 0.050)^2 + (0.075 - 0.125)^2 = 0.07^2$$
$$(0.100 - 0.118)^2 + (0.075 - 0.143)^2 = 0.07^2$$
$$(0.050 - 0.118)^2 + (0.125 - 0.143)^2 = 0.07^2$$

Now the forecasts are generated using d_1, d_2, and d_3. For d_1,

$$\hat{a}_{1T} = 0.100(109) + 0.900(110.54) = 110.39$$
$$\hat{b}_{1T} = 0.075(110.39 - 105.54) + 0.925(5.04) = 5.03$$
$$\hat{x}_{1T}(1) = 110.39 + 5.03 = 115.42$$

Also, d_2 gives

$$\hat{a}_{2T} = 0.050(109) + 0.950(110.54) = 110.46$$
$$\hat{b}_{2T} = 0.125(110.46 - 105.54) + 0.875(5.04) = 5.03$$
$$\hat{x}_{2T}(1) = 110.46 + 5.03 = 115.49$$

and for d_3

$$\hat{a}_{3T} = 0.118(109) + 0.882(110.54) = 110.36$$
$$\hat{b}_{3T} = 0.143(110.36 - 105.44) + 0.857(5.04) = 5.01$$
$$\hat{x}_{3T}(1) = 110.36 + 5.01 = 115.37$$

The forecasts used in production is the one corresponding to design d_2. This is $\hat{x}_{2T}(\tau) = 110.46 + 5.03\tau$.

10-5 ADAPTIVE CONTROL OF TWO SMOOTHING PARAMETERS USING A FACTORIAL DESIGN

Another adaptive control technique for monitoring two or more smoothing parameters has been developed by Roberts and Reed [5]. The method is based on a two-level factorial design where each smoothing parameter assumes three values—low, center, and high. At each time period, forecasts are generated with the parameters set at their center points and with all combinations of the low and high values. For a k-parameter model, this yields $(2^k + 1)$ alternative forecasts. The forecast that is used in production is the one where all the parameters assume their center values.

The forecaster selects a value n_o, whereas after n_o time periods elapse, the first analysis is performed to determine which, if any, parameter values to alter. For those parameters that require modification, the analysis shows which direction the values should be changed. In this section the technique is described for a two-parameter model.

The parameters are denoted as α and β and the three settings of each are

$$\alpha_o, \alpha_L, \alpha_H \quad \text{and} \quad \beta_o, \beta_L, \beta_H$$

Here α_o and β_o are the center values, (α_L, β_L) are the low, and (α_H, β_H) are the high. These values are related by

$$\alpha_H = \alpha_o + b_\alpha$$
$$\alpha_L = \alpha_o - b_\alpha$$
$$\beta_H = \beta_o + b_\beta$$
$$\beta_L = \beta_o - b_\beta$$

where b_α and b_β are fixed constants selected by the forecaster.

With $k = 2$, the $(2^2 + 1) = 5$ alternative design points become

$$d_1 = (\alpha_o, \beta_o)$$
$$d_2 = (\alpha_L, \beta_L)$$
$$d_3 = (\alpha_H, \beta_H)$$
$$d_4 = (\alpha_H, \beta_L)$$
$$d_5 = (\alpha_L, \beta_H)$$

So at each time period, five one-period-ahead forecasts are generated using the parameter values above. At time t, these are denoted as $\hat{x}_{it}(1)$ for $i = 1, 2, 3, 4, 5$. The forecasts used for production purposes is $\hat{x}_{1t}(\tau)$ (i.e., with $d_1 = \alpha_o, \beta_o$).

At every time period, the one-period-ahead forecast errors corresponding to the five design points are measured. These are

$$e_{it} = (x_t - \hat{x}_{i,t-1}(1)) \qquad \text{for } i = 1, 2, 3, 4, 5$$

Should $t = t_o$ be the first time period with the five design points under consideration, then n sets of these forecasts errors are found. These are

$$(e_{it_o}, e_{i,t_o+1}, \ldots, e_{i,t_o+n-1}) \qquad i = 1, 2, 3, 4, 5$$

and the average of the squared errors become

$$S_i = \frac{1}{n} \sum_{t=t_o}^{t_o+n-1} e_{it}^2 \qquad i = 1, 2, 3, 4, 5$$

The relationship between these averages can be used to determine the influence of the α and β values. Two measures are calculated and are

$$E_\alpha = \tfrac{1}{3}(S_3 + S_4 - S_2 - S_5)$$
$$E_\beta = \tfrac{1}{3}(S_3 + S_5 - S_2 - S_4)$$

E_α is called the "effect of α" and E_β is the "effect of β."

Now, approximate 99% error limits on the effects are calculated. These are

$$E_{\text{low}} = \frac{3S}{\sqrt{n}}$$

$$E_{\text{high}} = \frac{3S}{\sqrt{n}}$$

where S is an estimate of the standard deviation of the square of the errors.

Roberts and Reed suggest the use of the range test to estimate this standard deviation.[2]

The two effects E_α and E_β are compared to the limits above to decide whether or not to change the values of α and β. The procedure is as follows:

$$\text{if} \begin{cases} E_\alpha < E_{\text{low}}, \text{ then change } \alpha_o \text{ to } \alpha_o = \alpha_H \\ E_\alpha > E_{\text{high}}, \text{ then change } \alpha_o \text{ to } \alpha_o = \alpha_L \\ \text{otherwise, do not change } \alpha_o \end{cases}$$

$$\text{if} \begin{cases} E_\beta < E_{\text{low}}, \text{ then change } \beta_o \text{ to } \beta_o = \beta_H \\ E_\beta > E_{\text{high}}, \text{ then change } \beta_o \text{ to } \beta_o = \beta_L \\ \text{otherwise, do not change } \beta_o \end{cases}$$

Note that whenever a change is made to α_o and/or β_o, the appropriate changes to the corresponding low and high values are made. That is,

$$\alpha_L = \alpha_o - b_\alpha$$

$$\alpha_H = \alpha_o + b_\alpha$$

$$\beta_o = \beta_o - b_\alpha$$

$$\beta_H = \beta_o + b_\alpha$$

Roberts and Reed, however, do not allow any of the values above to lie outside the range (0.05, 0.95). Also, whenever a change in either one or both of the parameters takes place, then the S_i ($i = 1, 2, 3, 4, 5$) averages are reset to zero and the system starts anew.

Example 10.6

Consider the forecast model

$$\hat{x}_t(\tau) = \hat{a}_t + \hat{b}_t \tau$$

where

$$\hat{a}_t = \alpha x_t + (1 - \alpha)(\hat{a}_{t-1} + \hat{b}_{t-1})$$

$$\hat{b}_t = \beta(\hat{a}_t - \hat{a}_{t-1}) + (1 - \beta)\hat{b}_{t-1}$$

---

[2] The "range" of m observations is defined in [6] as the difference between the highest and lowest m values. For small samples, the range gives a reasonable estimate of the standard deviation (S) by use of

$$S = \text{range}/d_m$$

where

m	2	3	4	5	6	7	8	9	10	12	16
d_m	1.128	1.693	2.059	2.326	2.534	2.704	2.847	2.970	3.078	3.258	3.532

Suppose that the values of α and β under consideration are:

$$(\alpha_L, \alpha_o, \alpha_H) = (0.05, 0.10, 0.15)$$
$$(\beta_L, \beta_o, \beta_H) = (0.10, 0.15, 0.20)$$

whereas $b_\alpha = b_\beta = 0.05$ and

$$d_1 = (0.10, 0.15)$$
$$d_2 = (0.05, 0.10)$$
$$d_3 = (0.15, 0.20)$$
$$d_4 = (0.15, 0.10)$$
$$d_5 = (0.05, 0.20)$$

The forecasts generated with $\alpha = 0.10$ and $\beta = 0.15$ are used for production purposes. Assume, also, that at $t = 0$, the five alternative settings of the coefficients begin with $\hat{a}_o = 40$ and $\hat{b}_o = 1$. Suppose also that the forecaster selects $n_o = 2$ as the first time that he will analyze the effects to determine whether a change in the parameters is appropriate.

Table 10-6 is used as a worksheet for this example. At $t = 1$, the demand entry $x_1 = 45$ becomes available and the coefficients using d_1 become

$$\hat{a}_{11} = 0.10(45) + 0.90(41) = 41.40$$
$$\hat{b}_{11} = 0.15(41.40 - 40.00) + 0.85(1) = 1.06$$

Hence,

$$\hat{x}_{11}(1) = 41.40 + 1.06 = 42.46$$

Table 10-6. WORKSHEET FOR EXAMPLE 10.6

t	x_t	e_1^2	e_2^2	e_3^2	e_4^2	e_5^2	S_2	S_3	S_4	S_5	E_α	E_β	S	E_{low}	E_{high}
0															
1	45														
2	50	56.85	60.53	53.00	54.46	58.98									
3	51	43.82	53.73	34.22	37.45	50.41	57.13	43.61	45.95	54.69	−7.42	−1.59	11.38	−24.14	24.14
4	52	32.38	47.20	19.80	24.50	41.86	53.82	35.67	38.80	50.42	−9.92	−2.18	13.95	−24.15	24.15

$\hat{a}_{1t}$	$\hat{b}_{1t}$	$\hat{x}_{1t}(1)$	$\hat{a}_{2t}$	$\hat{b}_{2t}$	$\hat{x}_{2t}(1)$	$\hat{a}_{3t}$	$\hat{b}_{3t}$	$\hat{x}_{3t}(1)$	$\hat{a}_{4t}$	$\hat{b}_{4t}$	$\hat{x}_{4t}(1)$	$\hat{a}_{5t}$	$\hat{b}_{5t}$	$\hat{x}_{5t}(1)$
40	1	41	40	1	41	40	1	41	40	1	41	40	1	41
41.40	1.06	42.46	41.20	1.02	42.22	41.60	1.12	42.72	41.60	1.02	42.62	41.20	1.12	42.32
43.21	1.17	44.38	42.61	1.06	43.67	43.81	1.34	45.15	43.73	1.15	44.88	42.70	1.20	43.90
45.04	1.27	46.31	44.04	1.09	45.13	46.03	1.52	47.55	45.80	1.25	47.05	44.26	1.27	45.53
46.88	1.35	48.23	45.47	1.12	46.59	48.22	1.66	49.88	47.79	1.33	49.12	45.85	1.33	47.18

In the same way, like values are formed with d_2, d_3, d_4, and d_5. No forecast errors are measured at $t = 1$ since all the designs start out alike.

At $t = 2$, the demand entry is $x_2 = 50$. For d_1 the forecast error is

$$e_1 = 50 - 42.46 = 7.54$$

and the squared error is

$$e_1^2 = (7.54)^2 = 56.85$$

In the same way the corresponding squared errors are found for d_2, d_3, d_4, and d_5. Also, at this time the coefficients and forecasts for the five designs are updated.

When $t = 3$, the demand entry $x_3 = 51$ is used to find the squared errors as before. Because $n_o = 2$, the first analysis concerning the effects is performed. For d_2 the average of the squared errors is

$$S_2 = \tfrac{1}{2}(60.53 + 53.73) = 57.13$$

The averages for d_3, d_4, and d_5 are listed in the table. Now the effects are formed. These are

$$E_\alpha = \tfrac{1}{2}(43.61 + 45.95 - 57.13 - 54.69) = -7.42$$

$$E_\beta = \tfrac{1}{2}(43.61 + 54.69 - 57.13 - 45.95) = -1.59$$

Using the range method with $m = 2$, the average range becomes

$$\text{range} = \tfrac{1}{5}(|56.85 - 43.82| + \ldots + |58.98 - 50.41|) = 12.84$$

and the standard deviation of the square of the errors is

$$S = \frac{12.84}{1.128} = 11.38$$

Hence,

$$E_{\text{low}} = -3\sqrt{\tfrac{1}{2}}(11.38) = -24.14$$

$$E_{\text{high}} = 3\sqrt{\tfrac{1}{2}}(11.38) = 24.14$$

Since neither E_α or E_β are outside these limits, then none of the parameter values are altered. The coefficients and forecasts are updated as before.

At $t = 4$, with $x_4 = 52$, the new forecast errors are measured and the corresponding squares of the errors are listed in the table. In this situation

$$S_2 = \tfrac{1}{3}(60.53 + 53.73 + 47.20) = 53.82$$

$$E_\alpha = \tfrac{1}{2}(35.67 + 38.80 - 53.82 - 50.42) = -9.92$$

$$E_\beta = \tfrac{1}{2}(35.67 + 50.42 - 53.82 - 38.80) = -2.18$$

$$\text{range} = \tfrac{1}{5}(|56.85 - 32.38| + \ldots + |58.98 - 41.86|) = 23.61$$

$$S = \frac{23.61}{1.693} = 13.95$$

$$E_{\text{low}} = -3\sqrt{\tfrac{1}{3}}(13.95) = -24.15$$

$$E_{\text{high}} = 3\sqrt{\tfrac{1}{3}}(13.95) = 24.15$$

Since neither E_α nor E_β lies outside the limits, the values of α and β are unchanged. Once again, the coefficients and forecasts for the five design points are updated.

In this way the system continues until α and/or β is altered. At that time, the system will start anew.

10-6 BLENDING TWO INDEPENDENT FORECASTS

Cohen [1] shows how two independent forecasts can be blended together to yield one combined forecast. The method is particularly suitable to items that are relatively new and for which not much demand history is available. The two independent forecasts are the following

$$\bar{x} = \frac{1}{T}(x_1 + x_2 + \ldots + x_T) = \text{average of the first } T \text{ demand entries}$$

$$\hat{x} = \text{forecast that is derived without using the demand history}$$

The forecast $\hat{x}$ is generally available to the forecaster before the first demand entry (x_1) becomes known. It may be derived from an estimate given by the sales force or the marketing department.

In addition to the two forecasts above, two standard deviation estimates are also needed in order to proceed. These are

$$S_x = \text{standard deviation of the demands for each time period}$$

$$\hat{\sigma} = \text{standard deviation of the forecast } \hat{x}$$

Here, S_x is an estimate of the standard deviation of x and may be generated without the use of the demand history. Possibly the estimate is based on experience with like items. Now $\hat{\sigma}$ is a measure of the reliability of the forecast $\hat{x}$. This may be based on the accuracy of prior such estimates.

As each new demand entry becomes available, an updated estimate of the standard deviation of $\bar{x}$ is formed. This is

$$S_{\bar{x}} = \frac{S_x}{\sqrt{T}}$$

Now with $\bar{x}$, $\hat{x}$, $S_{\bar{x}}$, and $\hat{\sigma}$ available at time $t = T$, the blended forecast becomes

$$\hat{x}_T(\tau) = w\bar{x} + (1 - w)\hat{x}$$

where

$$w = \frac{\hat{\sigma}^2}{\hat{\sigma}^2 + S_{\bar{x}}^2}$$

As each new time period passes on, the values of $\bar{x}$, $S_{\bar{x}}$, and w are updated. So also is the blended forecast.

Example 10.7

Consider a new item where the sales force estimates the monthly demand to be $\hat{x} = 30$ units. Through past experience with the sales-force estimates, the forecaster sets $\hat{\sigma} = 4$ as the standard deviation of $\hat{x}$. Also, through past experience of like items, the forecaster estimates the standard deviation of the monthly demands as $S_x = 8$.

Now suppose that at $t = 1$, the demand entry is $x_1 = 38$. This gives

$$\bar{x}_1 = 38 \quad \text{and} \quad S_{\bar{x}} = 8$$

Hence,

$$w = \frac{4^2}{4^2 + 8^2} = \frac{16}{16 + 64} = 0.20$$

$$\hat{x}_1(\tau) = 0.20(38) + 0.80(30) = 31.6$$

Assume that at $t = 2$, $x_2 = 30$. Hence, $\bar{x} = \frac{1}{2}(38 + 30) = 34$, $S_{\bar{x}} = 8/\sqrt{2} = 5.66$, and $w = 16/(16 + 32) = 0.333$. Also,

$$\hat{x}_2(\tau) = 0.333(34) + 0.667(30) = 31.3$$

In this way, the forecasts are updated as each new demand entry becomes available. The example continues in Table 10-7 for five time periods.

Table 10-7. WORKSHEET FOR EXAMPLE 10.7

t	x_t	$\hat{x}$	$\bar{x}$	$\hat{\sigma}$	$S_{\bar{x}}$	w	$\hat{x}_t(1)$
1	38	30	38.00	4	8.00	0.20	31.6
2	30	30	34.00	4	5.67	0.33	31.3
3	36	30	34.67	4	4.62	0.43	32.0
4	25	30	32.25	4	4.00	0.50	31.1
5	38	30	33.40	4	3.58	0.56	31.8

PROBLEMS

10-1. Calculate the smoothed average error (E_t), the smoothed absolute error (A_t), and the tracking signal (TS_t) for the first 10 time periods of Table 4-6 by use of the Trigg and Leach approach and with $\gamma = 0.2$.

10-2. Consider the following 10 demand entries for an item: 3, 5, 4, 8, 20, 16, 24, 18, 17, and 25. Suppose at the outset that the first forecast is $\hat{x}_1(\tau) = x_1 = 3$, and thereafter the forecasts will be generated using single smoothing with an adaptive response rate where $\gamma = 0.3$, and $\alpha = 0.1$ is the initial smoothing parameter. Find the smoothing parameter and corresponding forecasts for each time period.

10-3. Suppose that the first 10 demand entries for an item are the following: 5, 6, 8, 12, 15, 30, 33, 35, 34, and 38. The item is being forecast using the adaptive smoothing trend model with a modification via the Trigg and Leach method. At the outset, the coefficients estimated are $\hat{a}_1(0) = 1$ and $a_2(0) = 3$. Also, the smoothing parameters are $\gamma = 0.3$ and $\beta = 0.8$. Find the one-period-ahead forecasts for each time period.

10-4. Consider the following demands and corresponding one-period-ahead forecasts from two alternative forecasting methods (F_1 and F_2):

t	x_t	$\hat{x}_{1t}(1)$	$\hat{x}_{2t}(1)$
1	3	2	10
2	5	3	9
3	8	4	8
4	10	6	8
5	12	8	9
6	14	12	11
7	20	14	12
8	22	15	14
9	23	16	16
10	20	19	18

Now using the combination forecasting method, calculate the weights (w_t) that would result at each time period using the options listed in Section 10-2. That is, find the weights by using:

a. Option 1.
b. Option 2 with $\theta = 0.5$.
c. Option 3 with $\beta = 0.8$.
d. Option 4 with $\beta = 0.8$.
e. Option 5 with $\theta = 0.5$.

10-5. Using the results of Problem 10-4c, find the forecasts that result for the 10 time periods by use of the combination forecasting techniques.

10-6. Using the results of Problem 10-4d, find the forecasts that result for the 10 time periods by using the combination forecasting technique.

10-7. Using the results from Table 10-3, calculate the weights and forecasts that would result when option 2 is used with $\theta = 0.7$.

10-8. Using the results from Table 10-3, calculate the weights and forecasts that would result when option 5 is used with $\theta = 0.6$.

10-9. Assume that single smoothing is in use for an item where the smoothing parameter will be controlled by Chow's adaptive method of Section 10-3. Suppose that the forecaster chooses $\delta = 0.05$, $\alpha = 0.20$ as the initial smoothing parameter, and does not wish for α to run outside the limits of 0.10 and 0.30. At the outset, the first one-period-ahead forecast is taken as $\hat{x}_1 = x_1$. Now show a worksheet on how the forecasts are developed for the following 10 demand entries: 10, 8, 9, 11, 15, 18, 20, 24, 22, and 21.

10-10. Consider the results of Table 10-4, and assume that after $t = 18$, the forecaster chooses to increase δ from 0.05 to 0.10. Show the forecasts that result from this change for the remaining time periods. Use limits of 0.10 and 0.80.

10-11. Consider Example 10.5 and Table 10-5. Show the subsequent worksheet entries if the demand at $t = T + 1$ is 120.

10-12. Consider Example 10.6, which uses the factorial design method of adaptive control. Continue with the worksheet of Table 10-6 if the demand entry at $t = 5$ is $x_5 = 55$.

10-13. Continue with Example 10.7, which uses the blending method of two independent forecasts to find the forecasts when the subsequent four demand entries are the following:

$$x_6 = 37,\ x_8 = 40,\ x_9 = 34,\ x_{10} = 35$$

10-14. Assume that an item's season covers 25 weeks and over the past season, the average demand was 100 pieces and the standard deviation of the weekly demand was $S_x = 10$ pieces. Now using $\hat{x} = 100$, $\hat{\sigma} = 10/\sqrt{25} = 2$, the forecaster wishes to use the blending method to generate updated forecasts for the new season. Show how these progress when the demands for the first 10 weeks are the following: 80, 75, 90, 87, 82, 95, 74, 104, 88, and 90.

REFERENCES

[1] Trigg, D. W., and D. H. Leach, "Exponentially Smoothing with an Adaptive Response Rate." *Operational Research Quarterly*, Vol. 18, 1967, pp. 53–59.

[2] Bates, J. M., and C. W. J. Granger, "Combination of Forecast." *Operational Research Quarterly*, Vol. 20, No. 4, 1969, pp. 451–468.

[3] Chow, W. M., "Adaptive Control of the Exponential Smoothing Constant." *Journal of Industrial Engineering*, Vol. 16, No. 5, 1965, pp. 314–317.

[4] MONTGOMERY, D. C., "Adaptive Control of Exponential Smoothing Parameters by Evolutionary Operation." *AIIE Transactions*, Vol. 2, No. 3, 1970, pp. 268–269.

[5] ROBERTS, S. D., AND R. REED, "The Development of a Self-Adaptive Forecasting Technique," *AIIE Transactions*, Vol. 1, No. 4, 1969, pp. 314–322.

[6] NATRELLA, M. G., *Experimental Statistics*. Washington, D.C.: National Bureau of Standards Handbook 91, 1963, Chap. 2.

[7] COHEN, G. D., "Bayesian Adjustment of Sales Forecasts in Multi-item Inventory Control Systems." *Journal of Industrial Engineering*, Vol. 17, No. 9, 1966, pp. 474–479.

11

BOX-JENKINS MODELS

Box and Jenkins [1] have developed a forecasting technique that seeks in a systematic manner the forecasting model that is best suited to each item under investigation. The method can lead to better forecasts than the other models described in this book. This is because, at the outset, a statistical type analysis on the history of demands is conducted to find the forecast model that gives the best fit. The forecast model is selected from a collection of models that represent the Box–Jenkins family of models.

The technique is highly sophisticated in both its mathematical and computational aspects. It requires at least 50 observations of historical demand entries and some manual interpretation and decision making. For these reasons it is not practical to use the method for all items in the inventory. Instead, the technique is better suited to a smaller set of time series, such as those which play a more vital role in the overall planning and direction of the company.

Although the method is applicable to seasonal models, the discussion in this chapter is limited to the nonseasonal models. The reader may refer to [2] for a description of the seasonal models.

At the outset, the history of demands $(x_1, x_2, \ldots, x_T)$ is used to seek the forecast model for each item. This is where the large amount of computations and the manual intervention is needed. Once the model is selected, it is generally applicable for a number of time periods (1 to 2 years) before the analysis must be repeated. In the intermittent time periods, the forecasts are updated in a simple recursive manner as each new demand entry becomes available.

Essentially, three basic stages are carried out in the modeling process. These stages are called identification, fitting, and diagnostic checking. A short review of these are given below.

1. *Identification*—The objective of this first stage is to select the forecast model that seems most appropriate to the time series under study. The data are used to generate a series of sample autocorrelation functions. These are now compared to certain theoretical autocorrelation functions from known forecast models to seek the best match between the sample and theoretical results. With this, the forecast model is identified and selected. The principle of parsimony is applied in that the model with the smallest number of coefficients that is suitable for the item is the model that is selected.
2. *Fitting*—Upon selecting the model, the second stage is carried on whereby the coefficients are estimated. The estimates are found so that they yield the fit of the past demands which produces the minimum sum of squared residual errors.
3. *Diagnostic Checking*—Using the fitted results, the residual errors are examined to determine the adequacy of the fit. A good fit will yield residual errors that are randomly distributed with mean zero and a common variance. The check is made by way of the autocorrelation function of the residual errors.

Should the diagnostic check fail, the three stages above are repeated (in part) until a model is found which gives appropriate results.

Having selected the model and the corresponding coefficients, the model is used to forecast the future demands. The forecasts are updated at each time period as each new demand entry becomes available. In the event that the time series seems to be changing, the coefficients of the model may be reestimated or an entirely new model may be selected.

The chapter contains seven sections. The first introduces various basic concepts and definitions. The second, third, and fourth pertain to the identification, fitting, and diagnostic checking stages, respectively. The fifth section shows how the forecasts are generated, and the sixth is on confidence limits of the forecasts. The seventh section gives the mathematical basis for certain key results.

11-1 BASIC CONCEPTS

The purpose of this section is to introduce some of the basic concepts that are used in the Box–Jenkins method. These include the stationary and nonstationary time series, stationary models, autocorrelation, backshift operators, differencing, and general models.

Stationary and Nonstationary Time Series

A stationary time series is one where all the demands are in equilibrium with a common mean μ and a common variance σ^2. A nonstationary time series is a series where the above conditions do not apply. Example plots of the two types of time series are shown in Figure 11-1.

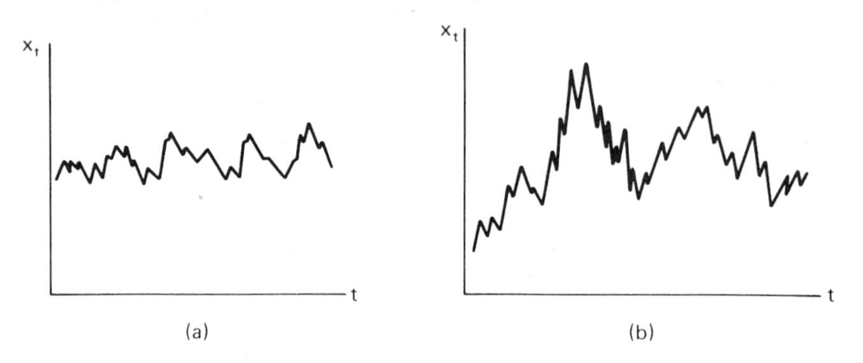

Figure 11-1. (a) Stationary and (b) nonstationary time series.

The time series is represented by x_t $(t = 1, 2, \ldots, T)$ and may or may not be a stationary series. For convenience in this chapter the notation z_t $(t = 1, 2, \ldots, T)$ will be used to identify a series that is stationary. For the case when x_t is also stationary, $z_t = x_t$ for all t. The mean of the stationary series will be denoted here by μ, whereby μ becomes the expected value of the z entries.

Also, for convenience, the series

$$w_t = z_t - \mu \qquad t = 1, 2, \ldots, T$$

is used. Again, w_t is a stationary series, but now the mean of the w entries is zero. It is this series that is used subsequently to describe the various types of forecasting models that are a part of the Box–Jenkins family of models.

Stationary Models

Two basic type of stationary models are encountered: the *autoregressive* and the *moving-average* models. Combinations of these models are also possible and are called the *mixed autoregressive/moving average* models. The various type of models are denoted by the notation (p, q), where p gives the number of coefficients in the autoregressive model and q is the corresponding number in the moving-average model.

AUTOREGRESSIVE MODELS $(p, 0)$. In the autoregressive model, the current entry w_t is related in a linear manner to its p most current entries and to an unknown noise a_t by the relation

$$w_t = \phi_1 w_{t-1} + \phi_2 w_{t-2} + \ldots + \phi_p w_{t-p} + a_t$$

The coefficient ϕ_i $(i = 1, 2, \ldots, p)$ gives the assigned weight to the ith prior entry. The noise (a_t) is a random occurrence with mean zero and variance σ^2.

Two commonly encountered autoregressive models are the (1, 0) and (2, 0) models. These become

$$(1, 0): \quad w_t = \phi_1 w_{t-1} + a_t$$

$$(2, 0): \quad w_t = \phi_1 w_{t-1} + \phi_2 w_{t-2} + a_t$$

In order for the stationary condition to be upheld, ϕ_1 in the (1, 0) model must lie within the range $(-1 < \phi_1 < 1)$. The corresponding bounds for the (2, 0) model are $(-1 < \phi_2 < 1)$, $(\phi_1 + \phi_2 < 1)$, and $(\phi_2 - \phi_1 < 1)$. The regions contained within these bounds are called the *admissible regions*, as shown in Figure 11-2.

Note in the (1, 0) model that since $w_t = z_t - \mu$, then

$$z_t = \mu(1 - \phi_1) + \phi_1 z_{t-1} + a_t$$

The corresponding relation for the (2, 0) model is

$$z_t = \mu(1 - \phi_1 - \phi_2) + \phi_1 z_{t-1} + \phi_2 z_{t-2} + a_t$$

MOVING-AVERAGE MODELS $(0, q)$. In the moving-average model, the current entry (w_t) is related to the q most current one-period-ahead forecast errors $(a_{t-1}, a_{t-2}, \ldots, a_{t-q})$ and the current noise a_t. Letting $\hat{z}_{t-j}$ represent the one-period-ahead forecast of z_{t-j}, then

$$a_{t-j} = z_{t-j} - \hat{z}_{t-j} \qquad \text{for } j = 1, 2, \ldots, q$$

The model becomes

$$w_t = -\theta_1 a_{t-1} - \theta_2 a_{t-2} - \ldots - \theta_q a_{t-q} + a_t$$

where $-\theta_j$ is the weight given to the jth prior forecast error.

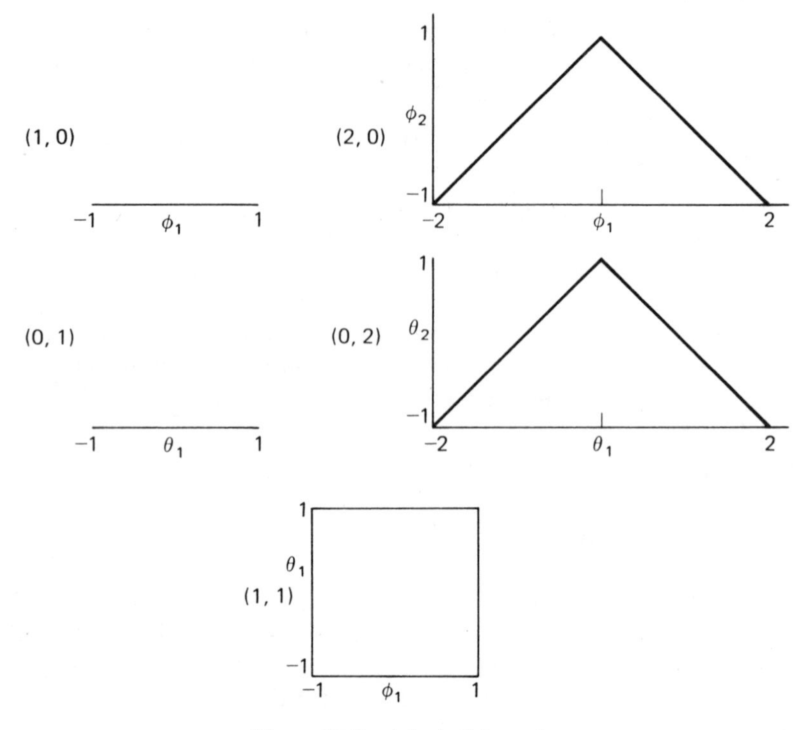

Figure 11-2. Admissible regions.

The (0, 1) and (0, 2) models are the most commonly occurring models of this type. These become

$$(0, 1):\quad w_t = -\theta_1 a_{t-1} + a_t$$

$$(0, 2):\quad w_t = -\theta_1 a_{t-1} - \theta_2 a_{t-2} + a_t$$

The admissible regions for these models are $(-1 < \theta_1 < 1)$ for (0, 1), and $(-1 < \theta_2 < 1)$, $(\theta_1 + \theta_2 < 1)$, and $(\theta_2 - \theta_1 < 1)$ for the (0, 2) model.

Since $w_t = z_t - \mu$, then (0, 1) gives the following relation with respect to z_t,

$$z_t = \mu - \theta_1 a_{t-1} + a_t$$

The corresponding relation for (0, 2) is

$$z_t = \mu - \theta_1 a_{t-1} - \theta_2 a_{t-2} + a_t$$

Mixed Autoregressive/Moving Average Model (p, q)**.** In the mixed autoregressive-moving average model, the entry w_t is related to the p most recent entries $(w_{t-1}, w_{t-2}, \ldots, w_{t-p})$, the q most recent forecast errors

$(a_{t-1}, a_{t-2}, \ldots, a_{t-q})$ and the current noise a_t. This gives

$$w_t = \phi_1 w_{t-1} + \ldots + \phi_p w_{t-p} - \theta_1 a_{t-1} - \ldots - \theta_q a_{t-q} + a_t$$

The most common model of this type is the (1, 1) model, which becomes

$$w_t = \phi_1 w_{t-1} - \theta_1 a_{t-1} + a_t$$

For this model the admissible region is defined by $(-1 < \phi_1 < 1)$ and $(-1 < \theta_1 < 1)$. Also, the relation with respect to z_t is readily found. This is

$$z_t = \mu(1 - \phi_1) + \phi_1 z_{t-1} - \theta_1 a_{t-1} + a_t$$

Autocorrelation Function

The autocorrelation gives a measure of the relation among the entries $w_1, w_2, \ldots, w_T$. The autocorrelation with a lag of k time periods is the correlation between w_t and w_{t-k}. The theoretical value is denoted by ρ_k, where $k = 0, 1, 2, \ldots, \rho_0 = 1$ and $(-1 \leq \rho_k \leq 1)$ for $k > 1$. The autocorrelation function represents the relationship of all the values of ρ_k over the range of all values of k.

A unique autocorrelation function exists for each of the stationary models described earlier. It is this function that is subsequently used to identify the model that is most representative to the time series under study.

The theoretical autocorrelation functions for the five models cited earlier are as follows:

$$(1, 0): \quad \rho_k = \phi_1^k \qquad k \geq 1$$

$$(2, 0): \quad \rho_1 = \frac{\phi_1}{1 - \phi_2}$$

$$\rho_k = \phi_1 \rho_{k-1} + \phi_2 \rho_{k-2} \qquad k \geq 2$$

$$(0, 1): \quad \rho_1 = \frac{-\theta_1}{1 + \theta_1^2}$$

$$\rho_k = 0 \qquad k \geq 2$$

$$(0, 2): \quad \rho_1 = \frac{-\theta_1(1 - \theta_2)}{1 + \theta_1^2 + \theta_2^2}$$

$$\rho_2 = \frac{-\theta_2}{1 + \theta_1^2 + \theta_2^2}$$

$$\rho_k = 0 \qquad k \geq 3$$

$$(1, 1): \quad \rho_1 = \frac{(1 - \theta_1\phi_1)(\phi_1 - \theta_1)}{1 + \theta_1^2 - 2\phi_1\theta_1}$$

$$\rho_k = \rho_{k-1}\phi_1 \qquad k \geq 2$$

Representative plots of the autocorrelation functions are shown in Figure 11-3. These plots are shown with various values for the coefficients to depict their elusive nature.

Backshift Operator

In the Box–Jenkins technique, it is often convenient to employ a mathematical procedure called the *backshift operator* (B). The backshift operator to the entry y_t gives the corresponding entry that occurred one period ahead, i.e.,

$$By_t = y_{t-1}$$

In the same manner,

$$B(By_t) = B^2 y_t = y_{t-2}$$
$$B(B^2 y_t) = B^3 y_t = y_{t-3}$$

and so forth.

Using the backshift operator, the five models cited earlier can be written in the following manner:

$$(1, 0):\quad (1 - \phi_1 B)w_t = a_t$$
$$(2, 0):\quad (1 - \phi_1 B - \phi_2 B^2)w_t = a_t$$
$$(0, 1):\quad w_t = (1 - \theta_1 B)a_t$$
$$(0, 2):\quad w_t = (1 - \theta_1 B - \theta_2 B^2)a_t$$
$$(1, 1):\quad (1 - \phi_1 B)w_t = (1 - \theta_1 B)a_t$$

Differencing

Although the concepts above are applicable for stationary time series, they may be extended for nonstationary series in a simple manner. This is accomplished by converting the nonstationary time series to a stationary time series with use of a differencing technique.

The difference operator is denoted by ∇ and, when applied to the demand entry x_t, it yields the difference

$$\nabla x_t = x_t - x_{t-1}$$

In the same manner,

$$\nabla(\nabla x_t) = \nabla x_t - \nabla x_{t-1} = x_t - 2x_{t-1} + x_{t-2}$$
$$\nabla(\nabla^2 x_t) = \nabla x_t - 2\nabla x_{t-1} + \nabla x_{t-2}$$
$$= x_t - 3x_{t-1} + 3x_{t-2} - x_{t-3}$$

and so on.

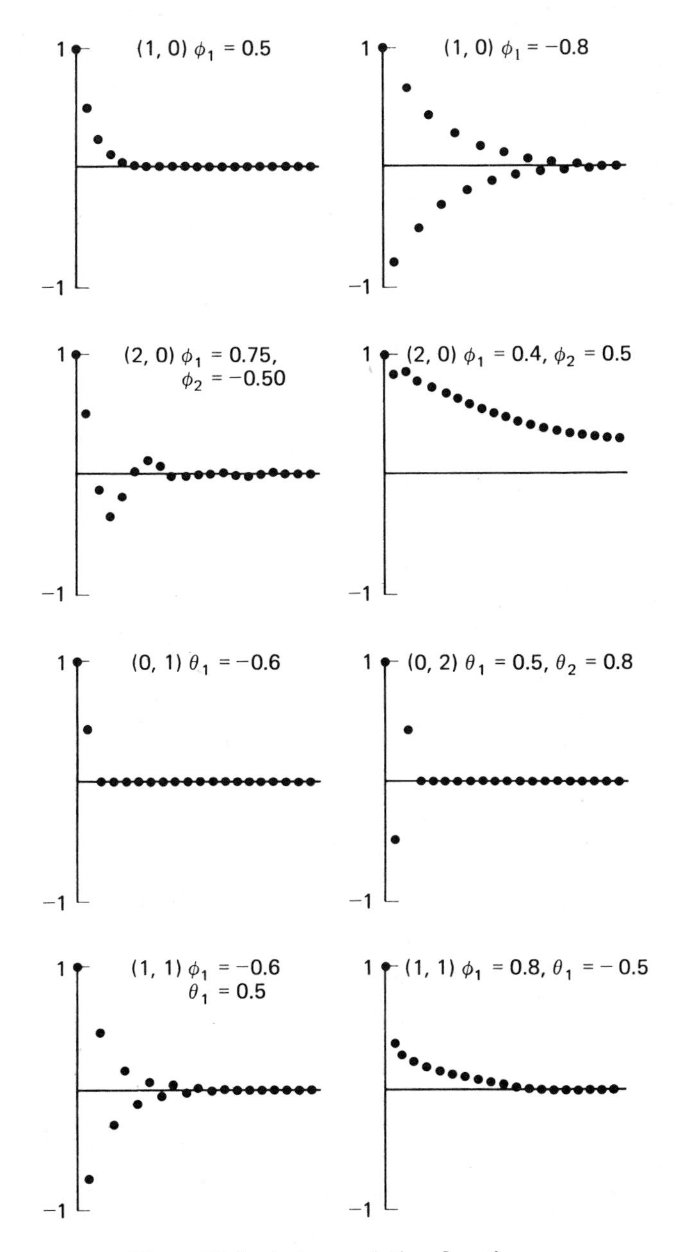

Figure 11-3. Autocorrelation functions.

The number of differences needed to convert the original series to a stationary time series is denoted as d. Hence, the stationary series (z_t) is related to the original series (x_t) by the relation

$$z_t = \nabla^d x_t$$

As before, $w_t = z_t - \mu$, where μ is the mean of z.

To illustrate the concepts in differencing, consider the entries of x_t listed in Table 11-1 along with the first and second differences. A plot of these results is shown in Figure 11-4.

Table 11-1. Example of First and Second Differences

t	x_t	∇x_t	$\nabla^2 x_t$
1	8	—	—
2	9	1	—
3	12	3	2
4	11	−1	−4
5	15	4	5
6	17	2	−2
7	20	3	1
8	25	5	2
9	31	6	1
10	34	3	−3
11	37	3	0
12	38	1	−2

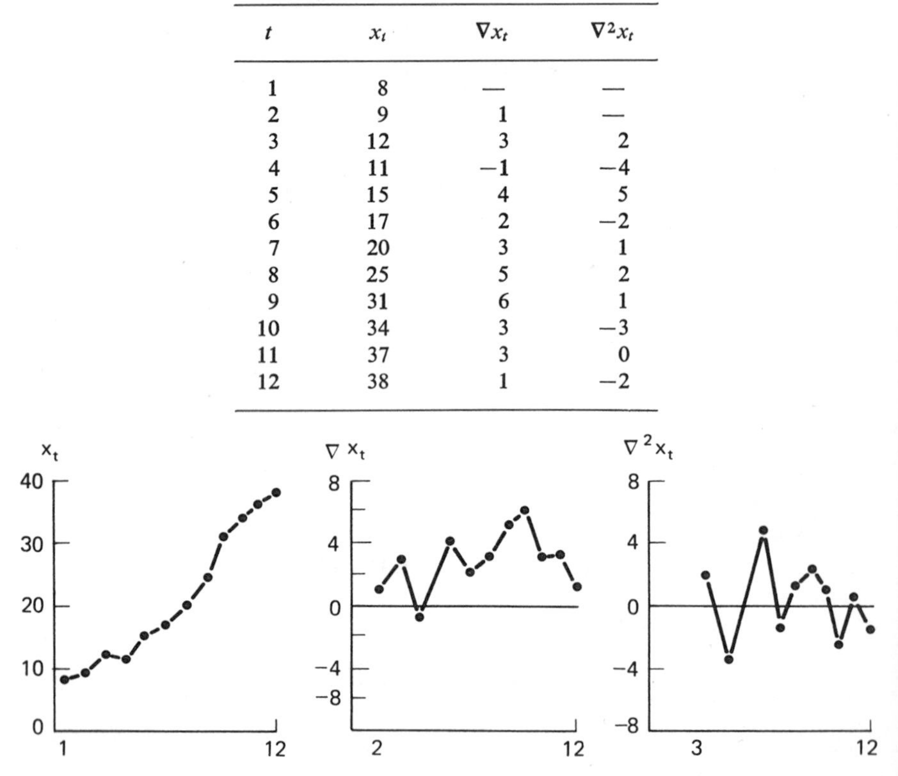

Figure 11-4. Plot of data and first and second differences.

General Models

At this time it is possible to redefine the models in more general terms, where the models apply to both stationary and nonstationary time series. The models are denoted by (p, d, q), where p and q have the same meaning as given earlier, and d represents the number of differences needed to obtain a stationary series.

In all situations, the following relations are applicable:

$$z_t = \nabla^d x_t$$

and

$$w_t = z_t - \mu$$

The models are the following:

$$\begin{aligned}
(1, d, 0)&: \quad w_t = \phi_1 w_{t-1} + a_t \\
(2, d, 0)&: \quad w_t = \phi_1 w_{t-1} + \phi_2 w_{t-2} + a_t \\
(0, d, 1)&: \quad w_t = -\theta_1 a_{t-1} + a_t \\
(0, d, 2)&: \quad w_t = -\theta_1 a_{t-1} - \theta_2 a_{t-2} + a_t \\
(1, d, 1)&: \quad w_t = \phi_1 w_{t-1} - \theta_1 a_{t-1} + a_t
\end{aligned}$$

11-2 IDENTIFICATION

For a particular time series under consideration $(x_1, x_2, \ldots, x_T)$, the forecaster's first task is to identify the model that best represents the series. This requires finding estimates of the parameters (p, d, q) from an analysis of the historical demands.

The principal tool used for this purpose is the autocorrelation function. For the dth difference, the function is denoted by

$$r_k(d) \qquad k = 0, 1, 2, \ldots, K$$

where $d = 0, 1, 2, \ldots$. The value of K depends on the number of demand entries (T) available in the history. A rule of thumb [3] is to use $K \leq \frac{1}{4}T$.

When $d = 0$, the autocorrelation with a lag of k is found by

$$r_k(0) = \frac{\sum_{t=k+1}^{T} (x_t - \bar{x})(x_{t-k} - \bar{x})}{\sum_{t=1}^{T} (x_t - \bar{x})^2}$$

where

$$\bar{x} = \frac{1}{T} \sum_{t=1}^{T} x_t$$

For $d = 1$, then $\nabla x_t = x_t - x_{t-1}$ and

$$r_k(1) = \frac{\sum_{t=k+2}^{T} (\nabla x_t - \bar{\nabla} x)(\nabla x_{t-k} - \bar{\nabla} x)}{\sum_{t=2}^{T} (\nabla x_t - \bar{\nabla} x)^2}$$

where

$$\bar{\nabla} x = \frac{1}{T-1} \sum_{t=2}^{T} \nabla x_t$$

In the same manner the corresponding entries for $d \geq 2$ are generated should they be needed.

The value of d that is selected is the smallest one which yields an autocorrelation function that is characteristic of a stationary time series. In this respect the autocorrelation function will taper down to zero fairly readily, whereas for a nonstationary series the function will tend to linger on with high values for many time periods.

Having selected d, the forecaster investigates $r_k(d)$, $k = 0, 1, 2, \ldots, K$, to seek which values of p and q are most appropriate. This is done by comparing a plot of the function $[r_k(d)]$ with known functions (ρ_k) for given values of p and q. The best match is chosen and the corresponding values of p and q are selected. Hence, the model (p, d, q) is identified.

Figure 11-5 shows five simulated stationary time series and their corresponding autocorrelation functions. Figure 11-6 shows examples of five simulated nonstationary time series and the autocorrelation functions with $d = 0$ and $d = 1$.

11-3 FITTING

Having identified the model, the next step is to seek estimates for the unknown coefficients of ϕ and θ that are contained in the model. The estimates are denoted as $(\hat{\phi}_1, \ldots, \hat{\phi}_p, \hat{\theta}_1, \ldots, \hat{\theta}_q)$ and are derived from either the least-squares or nonlinear-least-squares method.

The series used in this fitting stage is the dth difference of the original series

$$z_t = \nabla^d x_t$$

from which d is established in the identification stage.
Hence, the original series is

$$x_t \qquad t = 1, 2, \ldots, T$$

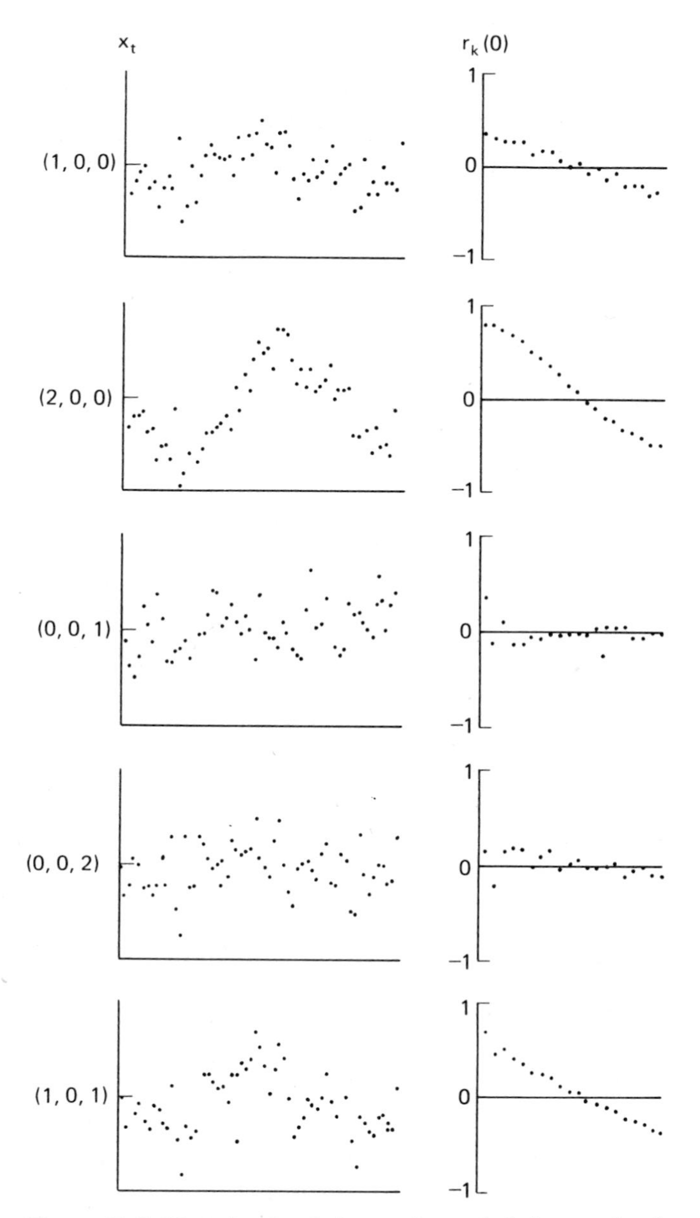

Figure 11-5. Five simulated time series and their associated sample autocorrelation functions.

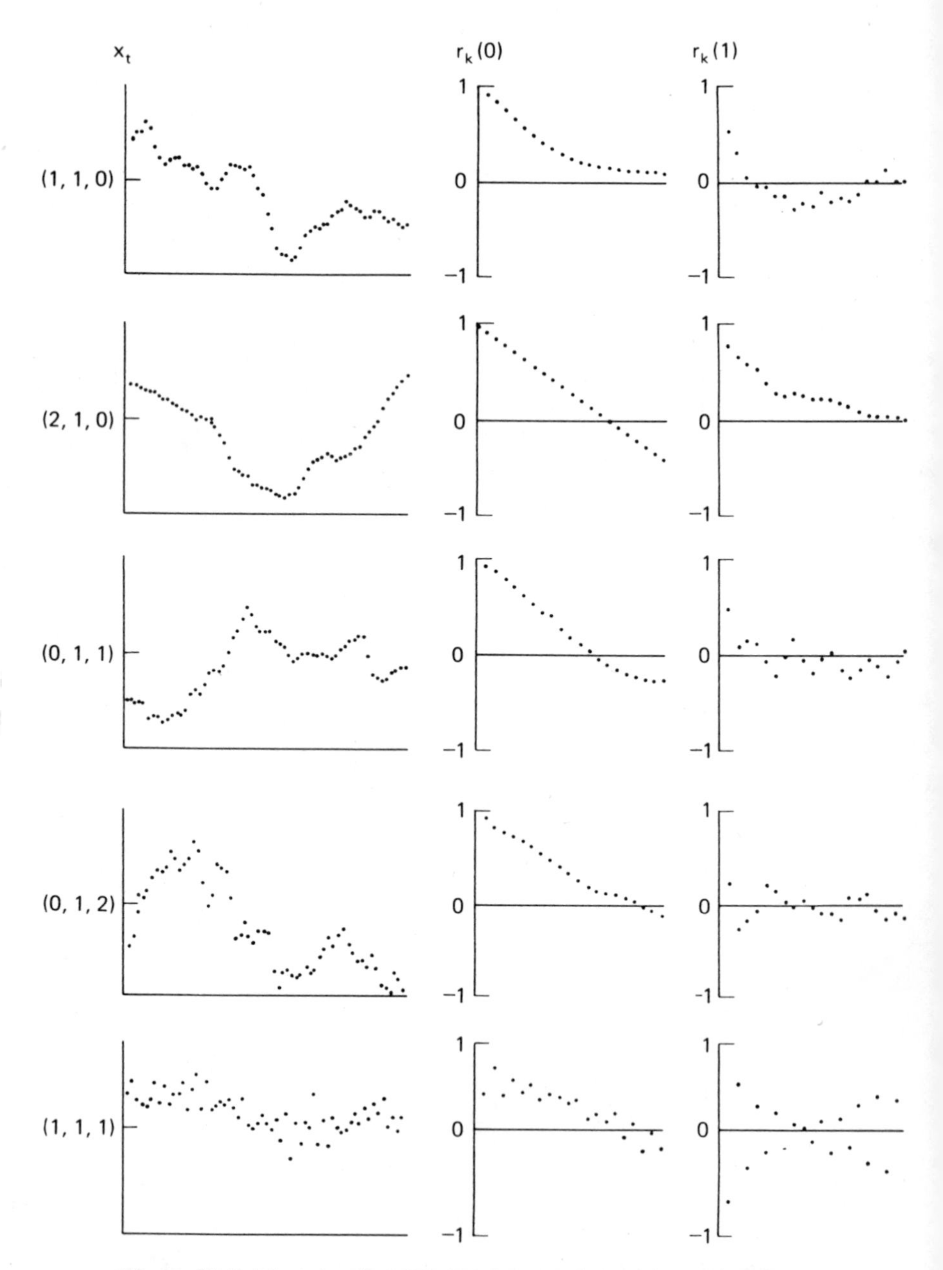

Figure 11-6. Five simulated nonstationary time series and their associated sample autocorrelation functions.

and the series with the dth difference is

$$z_t \qquad t = d + 1, \ldots, T$$

For convenience in later calculations the series is converted to

$$w_t = z_t - \bar{z} \qquad t = d + 1, \ldots, T$$

where

$$\bar{z} = \frac{\sum_{t=d+1}^{T} z_t}{T - d}$$

For the (1, d, 0) model, the unknown coefficient (ϕ_1) is estimated by

$$\hat{\phi}_1 = \frac{\sum w_t w_{t-1}}{\sum w_{t-1}^2}$$

where the summations run from $t = d + 2$ to T. In the (2, d, 0) model the coefficients ϕ_1 and ϕ_2 are estimated from

$$\hat{\phi}_2 = \frac{\sum w_t w_{t-1} \sum w_{t-1} w_{t-2} - \sum w_t w_{t-2} \sum (w_{t-1})^2}{(\sum w_{t-1} w_{t-2})^2 - \sum w_{t-1}^2 \sum w_{t-2}^2}$$

$$\hat{\phi}_1 = \frac{\sum w_t w_{t-1} - \hat{\phi}_2 \sum w_{t-2} w_{t-1}}{\sum w_{t-1}^2}$$

The results are obtained using the least-squares method.

When the (0, d, 1), (0, d, 2), or (1, d, 1) models are in use, the estimates of the unknown coefficients are obtained by using the nonlinear-least-squares method. This is a procedure that seeks solutions in an iterative manner.

The nonlinear-least-squares method is described in brief for the (0, d, 1) model, where an estimate of θ_1 is needed. At the outset, a value of θ_1 is selected, say $\hat{\theta}_1^0$. With this, the sum of squared residuals

$$S(\hat{\theta}_1^0) = \sum (w_t - \hat{w}_t)^2$$

is found where $\hat{w}_t$ is the fit of w_t using $\hat{\theta}_1^0$. Using the results above, another value $\hat{\theta}_1^1$ is selected. Now the sum

$$S(\hat{\theta}_1^1) = \sum (w_t - \hat{w}_t)^2$$

is found, where $\hat{w}_t$ is obtained from $\hat{\theta}_1^1$. This process continues in an optimum search manner until the coefficient is found that yields the minimum sum of squared residual errors. The value of θ_1 that gives this result is now denoted by $\hat{\theta}_1$ and is used subsequently in the forecasting phase.

11-4 DIAGNOSTIC CHECKING

The next phase in the Box–Jenkins model is to check on the adequacy of the fit. This check is carried out using the residual errors

$$e_t = z_t - \hat{z}_t$$

where $\hat{z}_t$ is the fitted value of z_t. When the model is fitted properly, the residual errors (e_t) will be independently distributed with a mean of zero.

This feature is tested by first generating the autocorrelation function of the residual errors. For a lag of k, the autocorrelation is

$$r_k = \frac{\sum (e_t - \bar{e})(e_{t-k} - \bar{e})}{\sum (e_t - \bar{e})^2}$$

where the summations run for the values of t where the residual error is measured. The average of the errors is $\bar{e}$ and should be very close to zero.

A chi-square test is now applied. The following sum is found:

$$\chi^2 = n \sum_{k=1}^{K} r_k^2$$

where

$$n = T - (p + q + d)$$

The preceding value of χ^2 (chi-square) is compared to a table value of chi-square with $m = K - (p + q)$ degrees of freedom. This table value is denoted by $\chi^2 m(\alpha)$, where α is the probability that a random occurrence of χ^2 will exceed the value $\chi^2 m(\alpha)$. Hence, choosing an α (say, $\alpha = 0.05$), the fit is accepted if $\chi^2 \leq \chi^2 m(0.05)$. Otherwise, the fit is deemed not proper and possibly some systematic error is contained in the data. In this event the forecaster must reinvestigate the data used in the identification stage to determine whether another model is more appropriate.

He/she may also detect a pattern in the autocorrelations of the residual errors. When a particular such value is high (say, r_{ko}), there is an indication that the forecast error with a lag of k_0 should be included in the model.

A way to detect when a particular residual error is high is by way of measuring the standard error of the autocorrelation. The standard error of the kth autocorrelation is

$$S_{r_k} = \sqrt{\frac{1}{K}\left(1 + 2\sum_{j=1}^{k-1} r_j^2\right)} \qquad k = 1, 2, \ldots, K$$

where K is the number of autocorrelations measured. Approximate 95% limits are found by $\pm 2S_{r_k}$. When r_k lies within these limits, the autocorrelation is adequate (not significantly different from zero); otherwise, the value is too high.

In Figure 11-7, examples are given showing how the entries x_t and

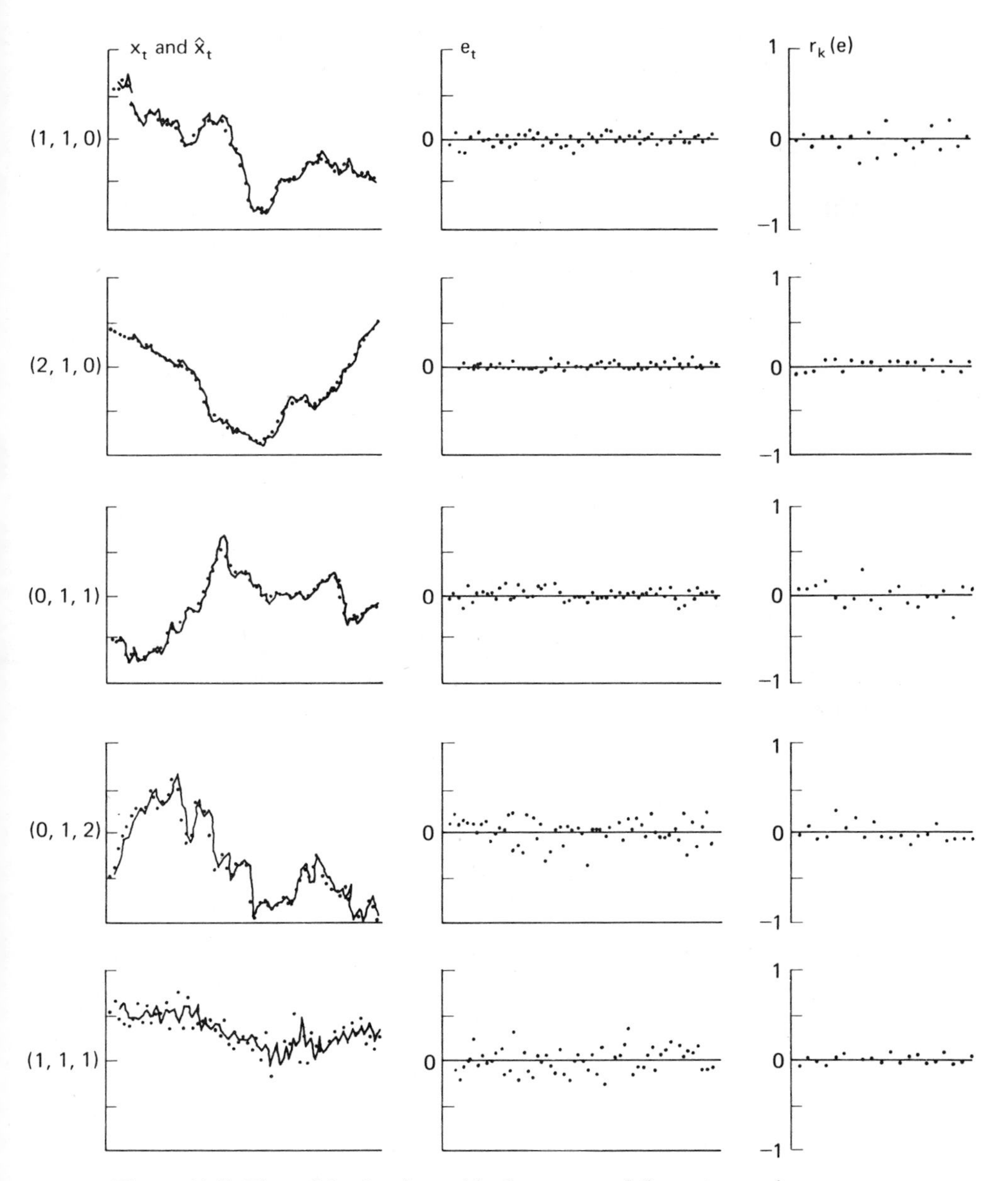

Figure 11-7. Plot of the fit, the residual errors, and the autocorrelation of the errors for the five simulated nonstationary time series.

their corresponding fitted values $\hat{x}_t$ are related for the time series originally plotted in Figure 11-5. Included also are plots of the associated residual errors and the autocorrelation function of the errors.

11-5 FORECASTING

Once the forecaster is satisfied that he/she has the appropriate model and coefficient estimates, he/she may begin to use the results to forecast the future entries of x_t. In this endeavor it is first necessary to convert the relations found using the w_t entries to a corresponding set of relations that use the original entries of the time series (x_t). The conversion depends on the number of differences (d) taken. Recall, for example, the following relations between w_t and x_t:

$$(d = 0): \quad w_t = x_t - \bar{x}$$

$$(d = 1): \quad w_t = x_t - x_{t-1}$$

$$(d = 2): \quad w_t = x_t - 2x_{t-1} + x_{t-2}$$

The forecasting equations for the models (1, d, 0), (2, d, 0), (0, d, 1), (0, d, 2), and (1, d, 1) with $d = 0$, 1, and 2 are listed in Table 11-2. For convenience the coefficients are listed as ϕ and θ instead of $\hat{\phi}$ and $\hat{\theta}$. In the tables the entries x_T, x_{T-1}, and x_{T-2} are the three most current demand entries, and a_T and a_{T-1} are the two most current one-period-ahead forecast errors. That is,

$$a_T = x_T - \hat{x}_{T-1}(1)$$

$$a_{T-1} = x_{T-1} - \hat{x}_{T-2}(1)$$

Also, $\bar{x}$ is the average of the prior values of x_t.

Table 11-2. BOX–JENKINS FORECAST EQUATIONS

Model	*Forecast Equations*
(100)	$\hat{x}_T(1) = \phi_1 x_T + (1 - \phi_1)\bar{x}$
	$\hat{x}_T(\tau) = \phi_1 \hat{x}_T(\tau - 1) + (1 - \phi_1)\bar{x} \quad \tau = 2, 3, \ldots$
(110)	$\hat{x}_T(1) = (1 + \phi_1)x_T - \phi_1 x_{T-1}$
	$\hat{x}_T(2) = (1 + \phi_1)\hat{x}_T(1) - \phi_1 x_T$
	$\hat{x}_T(\tau) = (1 + \phi_1)\hat{x}_T(\tau - 1) - \phi_1 \hat{x}_T(\tau - 2) \quad \tau = 3, 4, \ldots$
(120)	$\hat{x}_T(1) = (2 + \phi_1)x_T - (1 + 2\phi_1)x_{T-1} + \phi_1 x_{T-2}$
	$\hat{x}_T(2) = (2 + \phi_1)\hat{x}_T(1) - (1 + 2\phi_1)x_T + \phi_1 x_{T-1}$
	$\hat{x}_T(3) = (2 + \phi_1)\hat{x}_T(2) - (1 + 2\phi_1)\hat{x}_T(1) + \phi_1 x_T$
	$\hat{x}_T(\tau) = (2 + \phi_1)\hat{x}_T(\tau - 1) - (1 + 2\phi_1)\hat{x}_T(\tau - 2) + \phi_1 \hat{x}_T(\tau - 3)$
	$\tau = 4, 5, \ldots$

Table 11-2. *(continued)*

Model	*Forecast Equations*
(200)	$\hat{x}_T(1) = \phi_1 x_T + \phi_2 x_{T-1} + (1 - \phi_1 - \phi_2)\bar{x}$
	$\hat{x}_T(2) = \phi_1 \hat{x}_T(1) + \phi_2 x_T + (1 - \phi_1 - \phi_2)\bar{x}$
	$\hat{x}_T(\tau) = \phi_1 \hat{x}_T(\tau - 1) + \phi_2 \hat{x}_T(\tau - 2) + (1 - \phi_1 - \phi_2)\bar{x} \qquad \tau = 3, 4, \ldots$
(210)	$\hat{x}_T(1) = (1 + \phi_1)x_T + (\phi_2 - \phi_1)x_{T-1} - \phi_2 x_{T-2}$
	$\hat{x}_T(2) = (1 + \phi_1)\hat{x}_T(1) + (\phi_2 - \phi_1)x_T - \phi_2 x_{T-1}$
	$\hat{x}_T(3) = (1 + \phi_1)\hat{x}_T(2) + (\phi_2 - \phi_1)\hat{x}_T(1) - \phi_2 x_T$
	$\hat{x}_T(\tau) = (1 + \phi_1)\hat{x}_T(\tau - 1) + (\phi_2 - \phi_1)\hat{x}_T(\tau - 2) - \phi_2 \hat{x}_T(\tau - 3)$
	$\tau = 4, 5, \ldots$
(220)	$\hat{x}_T(1) = (2 + \phi_1)x_T + (\phi_2 - 2\phi_1 - 1)x_{T-1} + (\phi_1 - 2\phi_2)x_{T-2} + \phi_2 x_{T-3}$
	$\hat{x}_T(2) = (2 + \phi_1)\hat{x}_T(1) + (\phi_2 - 2\phi_1 - 1)x_T + (\phi_1 - 2\phi_2)x_{T-1} + \phi_2 x_{T-2}$
	$\hat{x}_T(3) = (2 + \phi_1)\hat{x}_T(2) + (\phi_2 - 2\phi_1 - 1)\hat{x}_T(1) + (\phi_1 - 2\phi_2)x_T + \phi_2 x_{T-1}$
	$\hat{x}_T(4) = (2 + \phi_1)\hat{x}_T(3) + (\phi_2 - 2\phi_1 - 1)\hat{x}_T(2) + (\phi_1 - 2\phi_2)\hat{x}_T(1) + \phi_2 x_T$
	$\hat{x}_T(\tau) = (2 + \phi_1)\hat{x}_T(\tau - 1) + (\phi_2 - 2\phi_1 - 1)\hat{x}_T(\tau - 2) + (\phi_1 - 2\phi_2)\hat{x}_T(\tau - 3)$
	$+ \phi_2 \hat{x}_T(\tau - 4) \qquad \tau = 5, 6, \ldots$
(001)	$\hat{x}_T(1) = -\theta_1 a_T + \bar{x}$
	$\hat{x}_T(\tau) = \bar{x} \qquad \tau = 2, 3, \ldots$
(011)	$\hat{x}_T(1) = -\theta_1 a_T + x_T$
	$\hat{x}_T(\tau) = \hat{x}_T(\tau - 1) \qquad \tau = 2, 3, \ldots$
(021)	$\hat{x}_T(1) = -\theta_1 a_T + 2x_T - x_{T-1}$
	$\hat{x}_T(2) = 2\hat{x}_T(1) - x_T$
	$\hat{x}_T(\tau) = 2\hat{x}_T(\tau - 1) - \hat{x}_T(\tau - 2) \qquad \tau = 3, 4, \ldots$
(002)	$\hat{x}_T(1) = -\theta_1 a_T - \theta_2 a_{T-1} + \bar{x}$
	$\hat{x}_T(2) = -\theta_2 a_T + \bar{x}$
	$\hat{x}_T(\tau) = \bar{x} \qquad \tau = 3, 4, \ldots$
(012)	$\hat{x}_T(1) = -\theta_1 a_T - \theta_2 a_{T-1} + x_T$
	$\hat{x}_T(2) = -\theta_2 a_T + \hat{x}_T(1)$
	$\hat{x}_T(\tau) = \hat{x}_T(\tau - 1) \qquad \tau = 3, 4, \ldots$
(022)	$\hat{x}_T(1) = -\theta_1 a_T - \theta_2 a_{T-1} + 2x_T - x_{T-1}$
	$\hat{x}_T(2) = -\theta_2 a_T + 2\hat{x}_T(1) - x_T$
	$\hat{x}_T(\tau) = 2\hat{x}_T(\tau - 1) - \hat{x}_T(\tau - 2) \qquad \tau = 3, 4, \ldots$
(101)	$\hat{x}_T(1) = \phi_1 x_T - \theta_1 a_T + (1 - \phi_1)\bar{x}$
	$\hat{x}_T(\tau) = \phi_1 \hat{x}_T(\tau - 1) + (1 - \phi_1)\bar{x} \qquad \tau = 2, 3, \ldots$
(111)	$\hat{x}_T(1) = (1 + \phi_1)x_T - \phi_1 x_{T-1} - \theta a_T$
	$\hat{x}_T(2) = (1 + \phi_1)\hat{x}_T(1) - \phi_1 x_T$
	$\hat{x}_T(\tau) = (1 + \phi_1)\hat{x}_T(\tau - 1) - \phi_1 \hat{x}_T(\tau - 2) \qquad \tau = 3, 4, \ldots$
(121)	$\hat{x}_T(1) = (2 + \phi_1)x_T - (1 + 2\phi_1)x_{T-1} + \phi_1 x_{T-2} - \theta_1 a_T$
	$\hat{x}_T(2) = (2 + \phi_1)\hat{x}_T(1) - (1 + 2\phi_1)x_T + \phi_1 x_{T-1}$
	$\hat{x}_T(3) = (2 + \phi_1)\hat{x}_T(2) - (1 + 2\phi_1)\hat{x}_T(1) + \phi_1 x_T$
	$\hat{x}_T(\tau) = (2 + \phi_1)\hat{x}_T(\tau - 1) - (1 + 2\phi_1)\hat{x}_T(\tau - 2) + \phi_1 \hat{x}_T(\tau - 3)$
	$\tau = 4, 5, \ldots$

To illustrate, suppose that the (2, 1, 0) model has been selected with $\phi_1 = 0.3$ and $\phi_2 = -0.6$. Assume further that $T = 50$, $x_{50} = 95$, $x_{49} = 103$, and $x_{48} = 98$. Using the forecasting equations listed in Table 11-2, the forecasts for the next four time periods are as follows:

$$\hat{x}_{50}(1) = 1.3(95) - 0.9(103) + 0.6(98) = 89.60$$

$$\hat{x}_{50}(2) = 1.3(89.6) - 0.9(95) + 0.6(103) = 92.78$$

$$\hat{x}_{50}(3) = 1.3(92.78) - 0.9(89.6) + 0.6(95) = 96.97$$

$$\hat{x}_{50}(4) = 1.3(96.97) - 0.9(92.78) + 0.6(89.6) = 96.32$$

At each subsequent time period, the forecasts are updated accordingly. For example, suppose that when $T = 51$, the entry $x_{51} = 92$ is observed. This gives the new forecasts,

$$\hat{x}_{51}(1) = 1.3(92) - 0.9(95) + 0.6(103) = 95.90$$

$$\hat{x}_{51}(2) = 1.3(95.9) - 0.9(92) + 0.6(95) = 98.87$$

and so on.

A second example is now cited using the model (0, 1, 2). Suppose in this situation that $\theta_1 = 0.6$ and $\theta_2 = -0.2$. Assume that the model identification and fit is carried out with $T = 50$ demand entries and that $x_{50} = 96$, $a_T = 0$, and $a_{T-1} = 0$. The forecasts at $T = 50$ are the following:

$$\hat{x}_{50}(1) = 96$$

$$\hat{x}_{50}(2) = 96$$

$$\hat{x}_{50}(\tau) = 96 \qquad \tau \geq 3$$

Now suppose that at $T = 51$, the entry $x_{51} = 104$ is observed. Hence, $a_{51} = 104 - 96 = 8$, and the updated forecasts become

$$\hat{x}_{51}(1) = -0.6(8) + 104 = 99.2$$

$$\hat{x}_{51}(2) = 0.2(8) + 99.2 = 100.8$$

$$\hat{x}_{51}(\tau) = 100.8 \qquad \tau \geq 3$$

At $T = 52$, suppose that the entry becomes $x_{52} = 112$. This gives $a_{52} = 112 - 99.2 = 12.8$ and

$$\hat{x}_{52}(1) = -0.6(12.8) + 0.2(8) + 112 = 105.92$$

$$\hat{x}_{52}(2) = 0.2(12.8) + 105.92 = 108.48$$

$$\hat{x}_{52}(\tau) = 108.48 \qquad \tau \geq 3$$

11-6 CONFIDENCE LIMITS

Confidence limits of the forecasts can now be derived. Since the reliability of the forecasts is smaller as the forecasts are projected further into the future, the corresponding confidence interval becomes larger.

The upper (U_τ) and lower (L_τ) limits for the τ-period-ahead forecast are found by

$$U_\tau = \hat{x}_T(\tau) + zV(\tau)S_a$$

$$L_\tau = \hat{x}_T(\tau) - zV(\tau)S_a$$

where z is selected from the normal distribution table to correspond with a confidence level desired. For a 95% confidence level, $z = 1.96$; for a 90% level, $z = 1.645$; and so on. S_a is the standard error of the forecast errors, and $V(\tau)$ is a function that depends on τ and the estimated coefficients in the model.

In the general case where the coefficients are $\phi_1, \ldots, \phi_p, \theta_1, \ldots, \theta_q$, then

$$V(\tau) = \left(1 + \sum_{j=1}^{\tau-1} \gamma_j^2\right)^{0.5}$$

with

$$\gamma_1 = \phi_1 - \theta_1$$

$$\gamma_2 = \phi_1\gamma_1 + \phi_2 - \theta_2$$

$$\gamma_3 = \phi_1\gamma_2 + \phi_2\gamma_1 - \theta_3$$

$$\gamma_4 = \phi_1\gamma_3 + \phi_2\gamma_2 + \phi_3\gamma_1 - \theta_4$$

$$\vdots$$

$$\gamma_j = \phi_1\gamma_{j-1} + \phi_2\gamma_{j-2} + \ldots + \phi_p\gamma_{j-p} - \theta_q$$

To illustrate, suppose that an item with model (2, 1, 0) is under review. Assume that $\phi_1 = 0.3$, $\phi_2 = -0.6$, $S_a = 10$, and a 95% confidence limit is needed for each of the first four future time periods. (Hence, $z = 1.96$ will be used.)

To find $V(1)$, $V(2)$, $V(3)$, and $V(4)$, the coefficients γ_1, γ_2, and γ_3 are needed. These become

$$\gamma_1 = 0.3$$

$$\gamma_2 = 0.3(0.3) - 0.6 = -0.51$$

$$\gamma_3 = 0.3(-0.51) - 0.6(0.3) = -0.333$$

Hence,

$$V(1) = 1$$
$$V(2) = (1 + 0.3^2)^{0.5} = 1.044$$
$$V(3) = (1 + 0.3^2 + 0.51^2)^{0.5} = 1.162$$
$$V(4) = (1 + 0.3^2 + 0.51^2 + 0.333^2)^{0.5} = 1.209$$

The upper and lower limits can now easily be evaluated. These are

$$L_1 = \hat{x}_T(1) - 1.96(1.000)10 = \hat{x}_T(1) - 19.60$$
$$U_1 = \hat{x}_T(1) + 1.96(1.000)10 = \hat{x}_T(1) + 19.60$$
$$L_2 = \hat{x}_T(2) - 1.96(1.044)10 = \hat{x}_T(2) - 20.46$$
$$U_2 = \hat{x}_T(2) + 1.96(1.044)10 = \hat{x}_T(2) + 20.46$$
$$L_3 = \hat{x}_T(3) - 1.96(1.162)10 = \hat{x}_T(3) - 22.78$$
$$U_3 = \hat{x}_T(3) + 1.96(1.162)10 = \hat{x}_T(3) + 22.78$$
$$L_4 = \hat{x}_T(4) - 1.96(1.209)10 = \hat{x}_T(4) - 23.70$$
$$U_4 = \hat{x}_T(4) + 1.96(1.209)10 = \hat{x}_T(4) + 23.70$$

11-7 MATHEMATICAL BASIS

Autocorrelation Function

The autocorrelation functions for the models shown earlier are derived here. Starting with the (1, 0) model,

$$w_t = \phi_1 w_{t-1} + a_t$$

it is noted that

$$w_t w_{t-1} = \phi_1 w_{t-1}^2 + a_t w_{t-1}$$

Taking expected values yields

$$C(1) = \phi_1 C(0) + 0$$

where $C(k)$ is the covariance between w_t and w_{t-k}. Note that $E(a_t w_{t-1}) = 0$, since a_t and w_{t-1} are independent. Now since the autocorrelation with a lag of k is

$$\rho_k = \frac{C(k)}{C(0)}$$

then, in the relation above,

$$\frac{C(1)}{C(0)} = \phi_1 \frac{C(0)}{C(0)}$$

gives

$$\rho_1 = \phi_1$$

In the general case,

$$w_t w_{t-k} = \phi_1 w_{t-1} w_{t-k} + a_t w_{t-k}$$

Following the same procedure as above yields

$$C(k) = \phi_1 C(k-1)$$

or

$$\rho_k = \phi_1 \rho_{k-1}$$

The autocorrelation function sought becomes

$$\rho_k = \phi_1^k$$

For the (2, 0) model,

$$w_t = \phi_1 w_{t-1} + \phi_2 w_{t-2} + a_t$$

The following three steps are taken to find ρ_1:

$$w_t w_{t-1} = \phi_1 w_{t-1}^2 + \phi_2 w_{t-2} w_{t-1} + a_t w_{t-1}$$
$$C(1) = \phi_1 C(0) + \phi_2 C(1) + 0$$
$$\rho_1 = \frac{\phi_1}{1 - \phi_2}$$

For ρ_k with $k = 2, 3, \ldots,$ the steps taken are the following:

$$w_t w_{t-k} = \phi_1 w_{t-1} w_{t-k} + \phi_2 w_{t-2} w_{t-k} + a_t w_{t-k}$$
$$C(k) = \phi_1 C(k-1) + \phi_2 C(k-2)$$
$$\rho_k = \phi_1 \rho_{k-1} + \phi_2 \rho_{k-2}$$

Recall the (0, 1) model, where

$$w_t = -\theta_1 a_{t-1} + a_t$$

In this situation it is first noted that the variance of w is related to the variance of the noise a by

$$\sigma_w^2 = (\theta_1^2 + 1)\sigma_a^2$$

Next, the relation

$$w_t w_{t-1} = (-\theta_1 a_{t-1} + a_t)(-\theta_1 a_{t-2} + a_{t-1})$$

yields the expected values

$$E(w_t w_{t-1}) = -\theta_1^2 E(a_{t-1}a_{t-2}) - \theta_1 E(a_{t-1}^2) - \theta_1 E(a_t a_{t-2}) + E(a_t a_{t-1})$$

Since $E(a_{t-1}^2) = \sigma_a^2$ and all other expectations listed on the right-hand side of the equation are zero, then

$$C(1) = -\theta_1 \sigma_a^2$$

Now since $C(0) = \sigma_w^2$, then

$$\frac{C(1)}{C(0)} = \frac{-\theta_1 \sigma_a^2}{(\theta_1^2 + 1)\sigma_a^2}$$

or

$$\rho_1 = \frac{-\theta_1}{\theta_1^2 + 1}$$

Continuing in this manner,

$$w_t w_{t-2} = (-\theta_1 a_{t-1} + a_t)(-\theta_1 a_{t-3} + a_{t-2})$$

and

$$E(w_t w_{t-2}) = 0$$

Hence, $\rho_2 = 0$. It is readily shown that $\rho_k = 0$ when $k > 1$.

In the (0, 2) model where

$$w_t = -\theta_1 a_{t-1} - \theta_2 a_{t-2} + a_t$$

the variance of w is

$$C(0) = \sigma_w^2 = (\theta_1^2 + \theta_2^2 + 1)\sigma_a^2$$

To find ρ_1, the following three steps are taken:

$$w_t w_{t-1} = (-\theta_1 a_{t-1} - \theta_2 a_{t-2} + a_t)(-\theta_1 a_{t-2} - \theta_2 a_{t-3} + a_{t-1})$$

$$C(1) = -\theta_1 \sigma_a^2 + \theta_1 \theta_2 \sigma_a^2$$

$$\rho_1 = \frac{-\theta_1 + \theta_1 \theta_2}{1 + \theta_1^2 + \theta_2^2}$$

For ρ_2 the corresponding steps are the following:

$$w_t w_{t-2} = (-\theta_1 a_{t-1} - \theta_2 a_{t-2} + a_t)(-\theta_1 a_{t-3} - \theta_2 a_{t-4} + a_{t-2})$$

$$C(2) = -\theta_2 \sigma_a^2$$

$$\rho_2 = \frac{-\theta_2}{1 + \theta_1^2 + \theta_2^2}$$

Continuing in this fashion, it is seen that $\rho_k = 0$ when $k > 2$.

Partial Autocorrelation Function

Another measure which is helpful in identifying autoregressive models is the partial autocorrelation function (g_k). This is particularly useful in determining the order of the autoregressive model. It gives a measure of the strength of the relationship between the periods of time. For example, if w_t is related to w_{t-1} and w_{t-2}, then g_1 and g_2 will have large values and g_k for $k \geq 3$ will be relatively small.

The partial autocorrelation is defined as

$$g_k = g_{kk} = \begin{cases} r_1 & k = 1 \\ \left[\dfrac{r_k - \sum_{j=1}^{k-1} g_{k-1,j} r_{k-j}}{1 - \sum_{j=1}^{k-1} g_{k-1,j} r_j}\right] & k = 2, 3, \ldots \end{cases}$$

where

$$g_{kj} = g_{k-1,j} - g_{kk} g_{k-1,k-j} \qquad j = 1, 2, \ldots, k-1$$

To illustrate the computations, consider a situation where $r_1 = 0.60$, $r_2 = 0.50$, $r_3 = 0.40$, $r_4 = 0.20$, and so on. Using the relations above,

$$g_1 = g_{11} = 0.60$$

$$g_2 = g_{22} = \frac{(0.50) - (0.60)(0.60)}{1 - (0.60)(0.60)} = 0.22$$

$$g_{21} = (0.60) - (0.22)(0.60) = 0.47$$

$$g_3 = g_{33} = \frac{(0.40) - (0.47)(0.50) - (0.22)(0.60)}{1 - (0.47)(0.60) - (0.22)(0.50)} = 0.05$$

$$g_{31} = (0.47) - (0.05)(0.22) = 0.46$$

$$g_{32} = (0.22) - (0.05)(0.47) = 0.20$$

$$g_4 = g_{44} = \frac{(0.20) - (0.46)(0.40) - (0.20)(0.50) - (0.05)(0.60)}{1 - (0.46)(0.60) - (0.20)(0.50) - (0.05)(0.40)} = -0.18$$

Hence, the first four partial autocorrelations are 0.60, 0.22, 0.05, and -0.18 and indicates a strong relation between w_t and w_{t-1}.

Fitting

At this time, the results given in the fitting stage are described. Recall that the coefficients are estimated by use of either the least-squares or nonlinear-least-squares methods.

The least-squares method is always used with the autoregressive models

$$\hat{w}_t = \sum_{j=1}^{p} \hat{\phi}_j w_{t-j}$$

The objective is to find the values of $\hat{\phi}_1, \ldots, \hat{\phi}_p$ which generate $\hat{w}_t$ and give the minimum sum of squared residual errors,

$$S(e) = \sum (w_t - \hat{w}_t)^2$$

Taking the partial derivatives $\partial S(e)/\partial \phi_i$ for $i = 1, 2, \ldots, p$ and setting each to zero yields

$$\begin{aligned}
\sum w_t w_{t-1} &= \hat{\phi}_1 \sum w_{t-1}^2 + \hat{\phi}_2 \sum w_{t-2} w_{t-1} + \ldots + \hat{\phi}_p \sum w_{t-p} w_{t-1} \\
\sum w_t w_{t-2} &= \hat{\phi}_1 \sum w_{t-1} w_{t-2} + \hat{\phi}_2 \sum w_{t-2}^2 + \ldots + \hat{\phi}_p \sum w_{t-p} w_{t-2} \\
&\cdot \\
&\cdot \\
&\cdot \\
\sum w_t w_{t-p} &= \hat{\phi}_1 \sum w_{t-1} w_{t-p} + \hat{\phi}_2 \sum w_{t-2} w_{t-p} + \ldots + \hat{\phi}_p \sum w_{t-p}^2
\end{aligned}$$

Hence, the results are p equations with p unknowns. The unknown coefficients can be derived by matrix methods.

When the moving-average model is in effect, then nonlinear-least-squares methods are needed to seek the estimates $\hat{\theta}_1, \hat{\theta}_2, \ldots, \hat{\theta}_q$. It is beyond the scope of this book to delve into the theoretical characteristics and manipulations of the nonlinear-least-squares methods. A good reference is given in [4].

At this time, however, the nonlinear-least-squares method will be described briefly for the (0, 1) model, where

$$w_t = -\theta_1 a_{t-1} + a_t$$

Upon choosing an initial value of $\hat{\theta}_1$, say θ_1^0, then by way of the Taylor expansion,

$$a_t \sim a_t^0 + \left.\frac{da_t}{d\hat{\theta}_1^0}\right|_{\hat{\theta}_1} (\theta_1 - \theta_1^0) = a_t^0 + y_t^0(\theta_1 - \theta_1^0)$$

with

$$y_t^0 = \left.\frac{da_t}{d\hat{\theta}_1}\right|_{\theta^0_1}$$

The sum of squares of the residual errors is

$$\sum a_t^2 \sim \sum (a_t^0 - y_t^0 \delta^0)^2$$

for $\delta^0 = \theta_1 - \theta_1^0$. The least-squares solution of δ^0 is found and becomes

$$\delta^0 = \frac{\sum a_t^0 y_t^0}{\sum (y_t^0)^2}$$

With this result the next estimate of θ_1 is

$$\theta_1^1 = \delta^0 + \theta_1^0$$

In this fashion the iterative procedure continues until $\delta \longrightarrow 0$.

Now, to find the values corresponding to y_t, the following relationships are observed. Note in this model that

$$w_t = (1 - \theta_1 B) a_t$$

$$a_t = w_t (1 - \theta_1 B)^{-1}$$

where B is the backshift operator. Now

$$\frac{da_t}{d\theta_1} = \frac{-B}{(1 - \theta_1 B)^2} w_t$$

$$= \frac{-(1 - \theta_1 B) a_{t-1}}{(1 - \theta_1 B)^2}$$

$$= \frac{-a_{t-1}}{1 - \theta_1 B}$$

Hence,

$$(1 - \theta_1 B) \frac{da_t}{d\theta_1} = -a_{t-1}$$

$$\frac{da_t}{d\theta_1} - \theta_1 \frac{da_{t-1}}{d\theta_1} = -a_{t-1}$$

Letting $y_t = da_t/d\theta_1$ and $y_{t-1} = da_{t-1}/d\theta_1$, then

$$y_t - \theta_1 y_{t-1} = -a_{t-1}$$

or

$$y_t = \theta_1 y_{t-1} - a_{t-1}$$

For example, consider the first six demand entries listed in Table 11-3. Assume here that $\theta_1 = -0.5$. The table shows that the first observed value of y_t^0 is measured when $t = 3$.

Table 11-3. WORKSHEET TO FIND y_t^o WHEN $\theta_1^0 = -0.5$ AND THE (0, 0, 1) MODEL IS IN USE

t	w_t	$\hat{w}_t = 0.5a_{t-1}^o$	a_t^o	$\hat{y}_t = -05y_{t-1} - a_{t-1}^o$
1	2	0	—	—
2	−3	0	−3.00	—
3	1	−1.50	2.50	3.00
4	2	1.25	0.75	−4.00
5	0	0.37	−0.37	1.25
6	3	0.18	2.82	0.25

Forecasting Equations

In seeking the forecast equations, the following type of manipulation is carried on. Suppose that the (1, 2, 1) model is in use. Here

$$w_t = \phi_1 w_{t-1} - \theta_1 a_{t-1} + a_t$$

Further,

$$\begin{aligned} w_t &= z_t - z_{t-1} \\ &= x_t - 2x_{t-1} + x_{t-2} \end{aligned}$$

Hence,

$$\begin{aligned} x_t - 2x_{t-1} + x_{t-2} &= \phi_1(x_{t-1} - 2x_{t-2} + x_{t-3}) - \theta_1 a_{t-1} + a_t \\ x_t &= (2 + \phi_1)x_{t-1} - (2\phi_1 + 1)x_{t-2} + \phi_1 x_{t-3} - \theta_1 a_{t-1} + a_t \end{aligned}$$

For $t = T + \tau$, this is

$$x_{T+\tau} = (2 + \phi_1)x_{T+\tau-1} - (2\phi_1 + 1)x_{T+\tau-2} + \phi_1 x_{T+\tau-3} - \theta_1 a_{T+\tau-1} + a_{T+\tau}$$

Now the following identities are used:

$$E(x_{T+\tau}) = \begin{cases} x_{T+\tau} & \text{if } \tau \leq 0 \\ \hat{x}_T(\tau) & \text{if } \tau \geq 1 \end{cases}$$

$$E(a_{T+\tau}) = \begin{cases} a_{T+\tau} & \text{if } \tau \leq 0 \\ 0 & \text{if } \tau \geq 1 \end{cases}$$

Upon taking expected values of $x_{T+\tau}$, the results yield the following relations:

$$\hat{x}_T(1) = (2 + \phi_1)x_T - (2\phi_1 + 1)x_{T-1} + \phi_1 x_{T-2} - \theta_1 a_T$$

$$\hat{x}_T(2) = (2 + \phi_1)\hat{x}_T(1) - (2\phi_1 + 1)x_T + \phi_1 x_{T-1}$$

$$\hat{x}_T(3) = (2 + \phi_1)\hat{x}_T(2) - (2\phi_1 + 1)\hat{x}_T(1) + \phi_1 x_T$$

$$\hat{x}_T(\tau) = (2 + \phi_1)\hat{x}_T(\tau - 1) - (2\phi_1 + 1)\hat{x}_T(\tau - 2) + \phi_1 \hat{x}_T(\tau - 3) \quad \text{for } \tau = 4, 5, 6, \ldots$$

Stationarity

An interesting duality exists between the autoregressive and moving average models. Note that the autoregressive model (1, 0)

$$w_t = \phi_1 w_{t-1} + a_t$$

becomes

$$\begin{aligned} w_t &= a_t + \phi_1 a_{t-1} + \phi_1^2 w_{t-2} \\ &= a_t + \phi_1 a_{t-1} + \phi_1^2 a_{t-2} + \phi_1^3 w_{t-3} \end{aligned}$$

and eventually

$$w_t = \sum_{j=0}^{\infty} \phi_1^j \, a_{t-j}$$

In this manner, it is seen that stationarity is ensured when ϕ_1 lies between -1 and $+1$. If it should lie outside of these limits, then w_t would be dominated by remote events. On the other hand, with ϕ_1 within the limits, then w_t is dominated by the most recent noise a_t. Also notice that this (1, 0) model becomes a moving average model of the type (0, ∞) with $\theta_j = -\phi_1^j$.

In the same fashion, the moving average model (0, 1)

$$w_t = -\theta_1 a_{t-1} + a_t$$

becomes

$$a_t = \sum_{j=0}^{\infty} \theta_1^j w_{t-j}$$

which is the autoregressive model (∞, 0) with $\phi_j = \theta_1^j$. Again it is seen that θ_1 must lie between -1 and $+1$ if a_t is not to be dominated by remote events. If this condition is satisfied, the moving average model is said to be *invertible*.

The above dualities apply also for the autoregressive model $(p, 0)$ and the moving average model $(0, q)$. Again, the $(p, 0)$ model can be inverted to a $(0, \infty)$ model, and the $(0, q)$ model can be inverted to a $(\infty, 0)$ model.

PROBLEMS

11-1. Show in the (1, 0) model that

$$w_t = \phi_1^3 w_{t-3} + \phi_1^2 a_{t-2} + \phi_1 a_{t-1} + a_t$$

Find a relation of w_t that uses all the past forecast errors.

11-2. Show in the (0, 1) model that

$$w_t = -\theta_1 w_{t-1} - \theta_1^2 w_{t-2} - \theta_1^3 w_{t-3} - \theta_1^4 a_{t-4} + a_t$$

Find a relation of w_t that uses all the past entries of w.

11-3. Use the results of Problem 11-1 to show why ϕ_1 must lie within the range $(-1, 1)$ in order to have a stationary time series.

11-4. Use the results of Problem 11-2 to show why θ_1 must lie within the range $(-1, 1)$ in order to have a stationary time series.

11-5. Find ρ_k for $k = 0, 1, 2, 3, 4, 5$, when:
a. $\phi_1 = 0.8$ in the (1, 0) model.
b. $\phi_1 = -0.8$ in the (1, 0) model.
c. $\phi_1 = 0.5$ and $\phi_2 = -0.4$ in the (2, 0) model.
d. $\phi_1 = -0.6$ and $\phi_2 = 0.2$ in the (2, 0) model.
e. $\theta_1 = 0.8$ in the (0, 1) model.
f. $\theta_1 = -0.5$ in the (0, 1) model.
g. $\theta_1 = -0.2$ and $\theta_2 = 0.7$ in the (0, 2) model.
h. $\theta_1 = 0.4$ and $\theta_2 = 0.5$ in the (0, 2) model.
i. $\phi_1 = 0.8$ and $\theta_1 = -0.6$ in the (1, 1) model.
j. $\phi_1 = -0.7$ and $\theta_1 = 0.5$ in the (1, 1) model.

11-6. State which of the following coefficients are within the admissible range:
a. $\phi_1 = -0.9$ in the (1, 0) model.
b. $\phi_1 = -0.6$ and $\phi_2 = 0.5$ in the (2, 0) model.
c. $\phi_1 = 0.6$ and $\phi_2 = 0.5$ in the (2, 0) model.
d. $\theta_1 = 0.5$ in the (0, 1) model.
e. $\theta_1 = -1.1$ in the (0, 1) model.
f. $\theta_1 = 0.8$ and $\theta_2 = 0.3$ in the (0, 2) model.
g. $\phi_1 = 0.9$ and $\theta_1 = 0.8$ in the (1, 1) model.

11-7. Find Bx_t, B^2x_t, B^3x_t when the first 10 observations of x_t are the following: 3, 5, 8, 2, 6, 9, 10, 4, 5, and 7.

11-8. Find ∇x_t, $\nabla^2 x_t$, $\nabla^3 x_t$, and $\nabla^4 x_t$ when the first 10 observations of x_t are the following: 1, 8, 28, 62, 130, 220, 380, 513, 730, and 1012.

11-9. Write an expression in terms of x_{t-j} for the following:

a. $\nabla B x_t$
b. $\nabla^2 B x_t$
c. $\nabla B^2 x_t$
d. $\nabla^3 B^2 x_t$
e. $B \nabla x_t$
f. $B^3 \nabla^2 x_t$

11-10. Find the autocorrelations with $k = 1$ and $k = 2$ for the following observation of x_t: 2, 3, 5, 8, 9, 7, 8, 7, 9, 10, and 8.

a. When $d = 0$
b. When $d = 1$

11-11. Consider the following demand entries (x_t): 3, 5, 9, 12, 10, 11, 15, 13, 18, 19, 20, and 17.

a. List the corresponding entries of z_t when $d = 1$.
b. Use the results of part a to list the entries of w_t.
c. Use the results of part b to find an estimate of ϕ_1 by use of the least-squares method and assuming that the (1, 1, 0) model applies.

11-12. Suppose that 10 entries of w_t are the following: 2, 1, −1, −3, −2, 0, 3, 1, 4, and 3.

a. Find the sum of square of errors ($S(\theta_1)$) if $\theta_1 = 0.5$ and the (0, 1) model is in use.
b. Compare the result in part a with that when $\theta_1 = 0.3$.

11-13. Consider the following residual errors: −1, −3, −4, 2, 0, 1, 2, 1, −1, and 3. Find the autocorrelations of the residual errors when $k = 1$ and 2.

11-14. Find the forecasts for $\tau = 1, 2, 3, 4$ in the (1, 2, 0) model when $\phi_1 = 0.5$, $x_T = 100$, $x_{T-1} = 90$, and $x_{T-2} = 95$.

11-15. Find the forecasts for $\tau = 1, 2, 3, 4, 5$ in the (1, 2, 1) model when $\phi_1 = 0.6$, $\theta_1 = 0.4$, $a_T = -10$, $x_T = 80$, $x_{T-1} = 91$, and $x_{T-2} = 86$.

11-16. Find the forecasts for $\tau = 1, 2, 3, 4$ in the (0, 0, 2) model when $\bar{x} = 100$, $a_T = 10$, $a_{T-1} = -6$, $\theta_1 = 0.6$, and $\theta_2 = -0.4$.

11-17. Find the 95% confidence intervals of the forecasts in Problem 11-14 when the standard deviation of the forecast errors in $S_a = 8$.

11-18. Find the 95% confidence intervals of the forecasts in Problem 11-15 when the standard deviation of the forecast errors is $S_a = 12$.

11-19. Find the 95% confidence intervals of the forecasts in Problem 11-16 when the standard deviation of the forecast errors is $S_a = 10$.

11-20. Find the partial autocorrelations for $k = 1, 2, 3, 4$, and 5, when $r_1 = 0.80$, $r_2 = 0.60$, $r_3 = 0.50$, $r_4 = 0.35$, $r_5 = 0.25$, $r_6 = 0.10$, $r_7 = 0.08$, and $r_8 =$ 0.10.

REFERENCES

[1] Box, G. E. P., and G. M. Jenkins, *Time Series Analysis, Forecasting and Control.* San Francisco: Holden-Day, Inc., 1976.

[2] Nelson, C. R., *Applied Time Series Analysis for Managerial Forecasting.* San Francisco: Holden-Day, Inc., 1973.

[3] Mabert, V. A. "An Introduction to Short-Term Forecasting Using the Box–Jenkins Methodology." *AIIE* (PP&G—75-1), Norcross, Ga., 1975.

[4] Marquart, D. W., "An Algorithm for Least-Squares Estimation of Nonlinear Parameters." *Society of Industrial and Applied Mathematics*, Vol. 11, No. 2, June 1963.

REFERENCES FOR FURTHER STUDY

Box, G. E. P., and G. M. Jenkins, "Some Recent Advances in Forecasting and Control." *Applied Statistics*, Vol. 17, 1968, pp. 91–109.

Naylor, T. H., and T. G. Seaks, "Box–Jenkins Methods: An Alternative to Econometric Models." *International Statistical Review*, Vol. 40, No. 2, 1972, pp. 123–137.

12

SPECIAL TECHNIQUES IN FORECASTING

This chapter presents various techniques in forecasting that have been used for special purposes in the management of the total inventory. The methods include the forecasting and control of lumpy items, forecasting of items where periodic demands are not posted, and techniques in the accumulated sum (CUSUM) method. The chapter also describes methods of forecasting all time requirements for declining items, and how to dampen the forecasts of fast dropping demands so that the forecasts die out in a gradual manner. Finally, the chapter concludes with a description on how growth rates can be measured from the forecasts and how these rates may be used to detect shifts in a group or line of items.

No doubt a large variety of methods are in use throughout the world to meet the special needs for individual companies and items. The purpose here is to highlight some of the problems that are important to inventory management and yet are not generally in the mainstream of a forecaster's activities. The reader may wish to see Michael [1], who describes a method of forecasting with heuristic programming, for example. His method copes with two types of problems: those that are too large for traditional models and those that are too loosely structured or ill-structured to be expressed in the mathematical terms necessary for the traditional algorithmic model.

12-1 LUMPY DEMANDS

Items are called *lumpy* when their demand history is highly irregular. This may occur when the demand entries have frequent occurrences around a given level and occasional but random occurrences that are irregularly high. For example, consider the 12 demand entries 0, 0, 30, 0, 0, 0, 90, 0, 0, 0, 0, and 0. The average demand of 10 pieces per time period is really not representative of any of the entries themselves. It is difficult to forecast the demands of lumpy items, since they do not fall into any of the standard demand patterns. Perhaps a horizontal model based upon smoothing or moving averages will give the best results.

Brown [2] defines a lumpy item as one where the ratio of the standard deviation over the average is greater than 1. In the example above, the standard deviation is $S = 26.6$ and the average is $\bar{x} = 10$. Hence, the ratio $S/\bar{x} = 2.66$ certainly satisfies Brown's criterion.

Method 1

Wilcox [3] states that all the techniques he has investigated for forecasting lumpy items were unsatisfactory for inventory control purposes because they either would overstock the items or they would not meet the desired level of customer service. Wilcox has developed a method of forecasting and controlling inventory for lumpy items that is easy to apply and yields the required level of service. A discussion of the method is given below.

Consider an item with the following history of demand entries ($x_1, x_2, \ldots, x_T$) and where the procurement lead time for the item is L time periods. The most recent 12 lead-time interval demands (X_t) are saved. For example, if $L = 3$ and $T = 14$, then

$$
\begin{aligned}
X_1 &= x_1 + x_2 + x_3 & X_7 &= x_7 + x_8 + x_9 \\
X_2 &= x_2 + x_3 + x_4 & X_8 &= x_8 + x_9 + x_{10} \\
X_3 &= x_3 + x_4 + x_5 & X_9 &= x_9 + x_{10} + x_{11} \\
X_4 &= x_4 + x_5 + x_6 & X_{10} &= x_{10} + x_{11} + x_{12} \\
X_5 &= x_5 + x_6 + x_7 & X_{11} &= x_{11} + x_{12} + x_{13} \\
X_6 &= x_6 + x_7 + x_8 & X_{12} &= x_{12} + x_{13} + x_{14}
\end{aligned}
$$

If the 14 demand entries were 0, 0, 30, 0, 0, 0, 90, 0, 0, 0, 0, 0, 20, and 0, then

$$
\begin{aligned}
X_1 &= 30 & X_4 &= 0 \\
X_2 &= 30 & X_5 &= 90 \\
X_3 &= 30 & X_6 &= 90
\end{aligned}
$$

$$X_7 = 90 \qquad X_{10} = 0$$

$$X_8 = 0 \qquad X_{11} = 20$$

$$X_9 = 0 \qquad X_{12} = 20$$

The lead-time sums are now sorted from low to high, yielding 0, 0, 0, 0, 20, 20, 30, 30, 30, 90, 90, and 90. This gives a scale of the lowest to highest lead-time demand entries based on a history of 12 lead-time periods.

The next step is to determine how far into the scale to go in order to meet the desired level of service. This is done by multiplying the specified service by the number of table entries. For example, if service is specified at 80% and 12 table entries are available, $0.80 \times 12 = 9.6$. Hence, the order point is taken as the entry in position 9 plus 0.6 of the change from position 9 to position 10. In the example,

$$\text{order point} = 30 + 0.6(90 - 30) = 66$$

Method 2

An item that is lumpy may also be controlled by the more conventional methods of inventory management, where an order point is calculated with the following relation:

$$\text{order point} = \hat{X}_L + k\hat{\sigma}_L$$

In the above, $\hat{X}_L$ represents the forecast of the lead-time demand, k is a safety factor ($k \geq 0$), and $\hat{\sigma}_L$ is an estimate of the standard deviation of the lead-time demand. Whenever the on-hand plus on-order inventory level falls below the order point, an order quantity (Q) is ordered from the supplier, where Q is found using economical considerations.

Because of the nature of lumpy items, the order point may best not be updated with the passing of each time period, but perhaps on a once-a-year basis to allow more stability. On the other hand, the estimates $\hat{X}_L$ and $\hat{\sigma}_L$ can be updated more regularly to avoid the need to retain a large quantity of past demand entries. At year end, their current estimates may then be used to update the order point.

There are several ways to update $\hat{X}_L$ and $\hat{\sigma}_L$. Suppose, at the outset, that estimates of the average lead-time demand and the square of the lead-time demand are available. For convenience these are labeled $\hat{X}_{L0}$ and $\hat{X}^2_{L0}$, respectively. Now when the first L time periods elapse (L = lead time), the total demand over the lead time is accumulated and here labeled as X. Using single smoothing with parameter α, the estimates of $\hat{X}_L$ and $\hat{\sigma}_L$ are udpated. These are by

$$\hat{X}_L = \alpha X + (1 - \alpha)\hat{X}_{L0}$$
$$\hat{X}_L^2 = \alpha X^2 + (1 - \alpha)(\hat{X}_{L0}^2)$$
$$\hat{\sigma}_L = [(\hat{X}_L^2) - (\hat{X}_L)^2]^{0.5}$$

Now, for convenience, the new estimates of $\hat{X}_L$ and $\hat{X}_L^2$ are labeled with the subscript zero for the next updating period. Hence, $\hat{X}_{L0} = \hat{X}_L$ and $(\hat{X}_{L0}^2) = (\hat{X}_L^2)$.

To illustrate, suppose that the initial estimates of the lead-time average demand and standard deviation for an item are $\hat{X}_L = 20$ and $\hat{\sigma}_L = 30$, respectively. Hence, $\hat{X}_{L0} = \hat{X}_L = 20$ and $\hat{X}_{L0}^2 = \hat{X}_L^2 + \hat{\sigma}_L^2 = 1300$. Suppose also that the lead time is $L = 3$ and the first three time-period demand entries are (0, 0, 30), whereby $X = 30$. Using single smoothing with $\alpha = 0.2$, the updated estimates become

$$\hat{X}_L = 0.2(30) + 0.8(20) = 22$$
$$\hat{X}_L^2 = 0.2(30^2) + 0.8(1300) = 1{,}220$$
$$\hat{\sigma}_L = [1220 - 22^2]^{0.5} = 27.1$$

Now if the demands for the next three time periods are (0, 0, 0), then $X = 0$ and

$$\hat{X}_L = 0.2(0) + 0.8(22) = 17.6$$
$$\hat{X}_L^2 = 0.2(0^2) + 0.8(1220) = 976$$
$$\hat{\sigma}_L = [976 - 17.6^2]^{0.5} = 25.8$$

Continuing in this fashion, if the following entries are (90, 0, 0) then $\hat{X}_L = 32.1$ and $\hat{\sigma}_L = 37.0$. Also, if the next three time period demands become (0, 0, 0), then $\hat{X}_L = 25.7$ and $\hat{\sigma}_L = 35.5$.

At this time the order point is calculated as

$$\text{order point} = 25.7 + k35.5$$

Chapter 15 gives a full discussion of how the safety factor can be established. For our purposes here, assume that $k = 1$ is in use, whereby the order point becomes

$$\text{order point} = 25.7 + 35.5 = 61.2$$

12-2 FORECASTING WITHOUT POSTING DEMANDS

Brown [4] shows how forecasts for an item can be obtained when the demands are not posted at regular time intervals. Many items necessarily fall into this category, because they are stored in boxes or barrels and are used as required without the need of an order or requisition, e.g., certain fluids, screws, bolts, and nuts. Generally, these are inexpensive items and the collective value is a small percentage of the total inventory.

The following steps are taken in order to arrive at a forecast for these items. Let t_1 represent the most current time period where a replenishment was obtained. The inventory at this time (including the procurement) is denoted as I_1. Now suppose that at some future time period where no intermittent replenishments have occurred, say t_2, the inventory is I_2. The forecast of demands per time period becomes

$$\hat{x}_{t_2}(\tau) = \frac{I_1 - I_2}{t_2 - t_1}$$

This forecast may be improved by combining the preceding forecast entry with corresponding forecast entries of the past. A horizontal model using smoothing should be adequate in most situations.

Example 12.1

Consider a bolt that is stored in a 5-gallon barrel and is used as needed to fasten certain containers. When a need for the bolt arises, the workman procures the bolts without filling out forms or requisitions. Suppose that the most current replenishment of the bolt occurred in the first week of May and the inventory at that time was estimated[1] at 10,000 units. Assume further that 7 weeks later, the inventory is low and is estimated to be 1600 units. Using the method described above, the forecast of demand per week is

$$\hat{x}_T(\tau) = \frac{10{,}000 - 1600}{7} = \frac{8400}{7} = 1200$$

12-3 FORECASTING WHEN SOME FUTURE DEMANDS ARE KNOWN IN ADVANCE

Adjustments in the forecasts of an item may be necessary when some of the future demands for the item are known in advance. For example, the salesperson may notify the forecaster in May that he/she has a pending order for 20 units from a particular customer, but the customer cannot activate the order until the coming July. This delay may be caused because of the customer's fiscal year budget or because the units cannot be used until July. The demand is not recorded in May, since it is not a true demand at that time; it should rightfully be recorded in July. In this section a method is presented which is used to adjust forecasts that are already available. It is assumed that

[1]The estimate may be based on the weight of the barrel or the distance from the bottom of the barrel to where the bolts lie at the current time.

the already available forecasts have been generated without any use of the information known on prior demands.

Consider a situation where at the current time period T, the forecaster learns that a demand of size x_0 is scheduled for $t = T + \tau$. The current forecast for time $T + \tau$ is $\hat{x}_T(\tau)$, and the standard deviation of the forecast error for this time period is assumed known and denoted by $\hat{\sigma}$. The forecast is generated using the historical demand entries only and an adjustment is now needed in lieu of the advance knowledge of demand.

The actual demand for $t = T + \tau$ is denoted as x and has an estimated mean of $\hat{\mu} = \hat{x}_T(\tau)$ and a standard deviation $\hat{\sigma}$. The entry x_o is a portion of x that is known at time T, not an addition to x. For purposes of this discussion, the demand x is assumed to be normally distributed. With the information above the adjusted forecast, $\tilde{x}_T(\tau)$, becomes

$$\tilde{x}_T(\tau) = x_o + \hat{\sigma}E(z > k)$$

where z = standardized normal variate

$$k = \frac{x_o - \hat{\mu}}{\hat{\sigma}}$$

$E(z > k)$ = partial expectation of z above k

The table entries for $E(z > k)$ are found in Appendix B. Note that the adjustment above concerns only the forecast for $t = T + \tau$; the forecasts for all other time periods are unaffected.

An adjustment is also needed for the standard deviation $\hat{\sigma}$. This becomes

$$\tilde{\sigma} = \hat{\sigma}\sigma_{(z>k)}$$

where $\sigma_{(z>k)}$ is the partial standard deviation of $z > k$ and entries of $\sigma_{(z>k)}$ are listed in Appendix B.

Example 12.2

Suppose that at time T, knowledge of a future demand of $x_o = 75$ pieces for $t = T + 4$ becomes available. Assume that the current forecast and standard deviation for this period are $\hat{x}_T(4) = 100$ and $\hat{\sigma} = 25$, respectively. Assuming further that the true demand is normally distributed, find the adjusted forecast for $t = T + 4$.

Since

$$k = \frac{75 - 100}{25} = -1$$

then, using Appendix B,

$$E(z > -1) = 1.08$$

and

$$\sigma_{(z>-1)} = 0.867$$

Hence, the adjusted forecast is

$$\tilde{x}_T(4) = 75 + 25(1.08) = 102$$

and the standard deviation becomes

$$\tilde{\sigma} = 25(0.867) = 21.7$$

MATHEMATICAL BASIS. When x is normal with mean $\hat{\mu}$ and standard deviation $\hat{\sigma}$, the expected value of x greater than x_o is

$$\begin{aligned} E(x > x_o) &= \int_{x_o}^{\infty} (x - x_o)\phi(x)\,dx \\ &= \hat{\sigma} \int_{k}^{\infty} (z - k)\phi(z)\,dz \\ &= \hat{\sigma} E(z > k) \end{aligned}$$

where

$$k = \frac{x_o - \hat{\mu}}{\hat{\sigma}}$$

and z is a standardized normal variate. Hence, the adjusted mean is

$$\tilde{\mu} = x_o + \hat{\sigma} E(z > k)$$

The adjusted variance $\tilde{\sigma}^2 = V(x > x_o)$ is

$$\begin{aligned} V(x > x_o) &= E(x > x_o)^2 - [E(x > x_o)]^2 \\ &= \hat{\sigma}^2 V(z > k) \\ &= \hat{\sigma}^2 \sigma^2_{(z>k)} \end{aligned}$$

where

$$\begin{aligned} E(x > x_o)^2 &= \int_{x_o}^{\infty} (x - x_o)^2 \phi(x)\,dx \\ &= \hat{\sigma}^2 \int_{k}^{\infty} (z - k)^2 \phi(z)\,dz \\ &= \hat{\sigma}^2 E(z > k)^2 \end{aligned}$$

In Appendix B the equations for $E(z > k)$, $E(z > k)^2$, and $\sigma_{(z>k)}$ are given.

12-4 THE CUMULATIVE SUM TECHNIQUE

The cumulative sum technique (CUSUM) has been used by many researchers and practitioners of forecasting and quality control in a variety of ways. This section describes two methods that have been applied with success in forecasting. These are the CUSUM Chart and CUSUM Series techniques.

The CUSUM Chart

The first form of CUSUM that is presented here comes from Wild [5] and takes the form of a chart with a plot of the cumulative difference between each demand entry and a given fixed constant x_o. The constant is an initial estimate of the average value of the demands. The purpose of the chart is to allow the forecaster to identify when changes in the average of the demands take place. The chart is interpreted solely on the basis of its slope. A horizontal graph reveals stability in the data, while a positive slope senses an increase in the average of the demand. Likewise, a negative slope indicates that a decrease in the average of the demands has occurred.

When the fixed constant x_o is less than the average value of the demands, the CUSUM chart will rise steadily, and vice versa. When this happens the forecaster adjusts x_0 accordingly and the chart is started anew.

Example 12.3

Consider the demand entries listed in Table 12-1 and the plot of demands in Figure 12-1. At first glance, there seems to be no change in the average level of demands. Show how the CUSUM method applies to these data using $x_o = 10$.

Table 12-1. DEMANDS FOR EXAMPLE 12.3

1–6	7–12	13–18	19–24
6	8	8	10
13	13	14	8
14	15	6	16
8	12	13	9
11	7	7	12
4	10	15	13

Figure 12-1. Demands in Example 12.3.

The calculations are carried out in Table 12-2 and the CUSUM plot is shown in Figure 12-2. Now, there is a clear indication that the average value of the demands has taken place in the latter half of the time horizon and that the level $x_o = 10$ is no longer applicable.

Table 12-2. CUSUM CALCULATION FOR EXAMPLE 12.3

t	x_t	$x_t - 10$	$\sum_{i=1}^{t} (x_i - 10)$
1	6	−4	−4
2	13	3	−1
3	14	4	3
4	8	−2	1
5	11	1	2
6	4	−6	−4
7	8	−2	−6
8	13	3	−3
9	15	5	2
10	12	2	4
11	7	−3	1
12	10	0	1
13	8	−2	−1
14	14	4	3
15	6	−4	−1
16	13	3	2
17	7	−3	−1
18	15	5	4
19	10	0	4
20	8	−2	2
21	16	6	8
22	9	−1	7
23	12	2	9
24	13	3	12

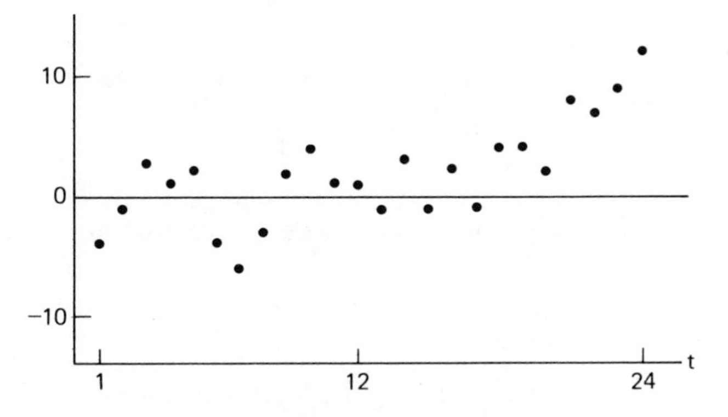

Figure 12-2. CUSUM plot for Example 12.3.

CUSUM in Series

Another CUSUM method is given by Harrison and Davies [6] with the objective of monitoring a forecasting system to signal an alarm should the forecasts seem to be going astray. In this method a series of accumulative summations of forecast errors are kept on a chart or stored in a computer. When one of these sums exceeds a predefined limit, the alarm is sounded and a change in the parameter(s) of the forecast model is then made. The CUSUM chart is then reestablished, and the cycle begins again.

Let e_T represent the current one-period-ahead forecast error,

$$e_T = x_T - \hat{x}_{T-1}(1)$$

Saving the prior N such errors, the forecaster compiles the following N sums at time period T:

$$\begin{aligned} S_1 &= e_T \\ S_2 &= e_T + e_{T-1} \\ &\cdot \\ &\cdot \\ &\cdot \\ S_N &= e_T + e_{T-1} + \dots + e_{T-N+1} \end{aligned}$$

The calculations can be simplified by noting that the sums at time T are related to those at $t = T - 1$. This is by

$$\begin{aligned} S_2 &= S_1^* + e_T \\ S_3 &= S_2^* + e_T \\ &\cdot \\ &\cdot \\ &\cdot \\ S_N &= S_{N-1}^* + e_T \end{aligned}$$

where S_i^* $i = 1, 2, \dots, N - 1$ designates the sums at time $t = T - 1$.

The signal to imply the change in the demand pattern is set off when any of

$$|S_i| > L_i \qquad i = 1, 2, \dots, N$$

where $L_1, L_2, \dots, L_N$ are limits set by the forecaster. Further, if $|S_n| > L_n$, it is assumed that the change in demand occurred back n periods. Based on this information, a change in the forecasting parameter(s) is required.

The limits L_i ($i = 1, 2, \dots, N$) may be fixed by statistical methods depending on the forecasting model in use. Generally,[2] the limits are related by

$$L_n = \sqrt{n}\,L_1$$

So if $L_1 = 10$, then $L_2 = \sqrt{2}(10) = 14.1$, $L_3 = \sqrt{3}(10) = 17.3$, and so on.

Example 12.4

Suppose that the single smoothing model is to be used for an item. At the outset $\alpha = 0.1$ will be used, but should a signal occur, then $\alpha = 0.2$ will be used as a replacement, and vice versa. Also, the smoothed average will be adjusted by the average of the largest S_n that is out of limits when an alarm is set off. Assume further that the forecaster chooses to keep $N = 5$ sums with limits:

$$L_1 = 30, \quad L_2 = 42, \quad L_3 = 52, \quad L_4 = 60, \quad L_5 = 67$$

The example is followed in Table 12-3.

Table 12-3. CUSUM WORKSHEET FOR EXAMPLE 12.4

t	x_t	e_t	S_1	S_2	S_3	S_4	S_5	α	$\hat{x}_t(1)$
1	12	—	0	0	0	0	0	0.1	12.0
2	15	3.0	3.0	3.0	3.0	3.0	3.0	0.1	12.3
3	10	−2.3	−2.3	0.7	0.7	0.7	0.7	0.1	12.1
4	16	3.9	3.9	1.6	4.6	4.6	4.6	0.1	12.5
5	15	2.5	2.5	6.4	4.1	7.1	7.1	0.1	12.8
6	14	1.2	1.2	3.7	7.6	5.3	8.3	0.1	12.9
7	20	7.1	7.1	8.3	10.8	14.7	12.4	0.1	13.6
8	40	26.4	26.4	33.5	34.7	37.2	41.1	0.1	16.2
9	31	14.8	14.8	41.2	48.3	49.5	52.0	0.1	17.7
10	34	16.3	16.3	31.1	57.5†	64.6†	65.8	0.2	33.8‡
11	33	−0.8	−0.8	−0.8	−0.8	−0.8	−0.8	0.2	33.6
12	40	6.4	6.4	5.6	5.6	5.6	5.6	0.2	34.9

†Alarm.
‡The forecast is adjusted using the largest sum with an alarm. Since this is $S_4 = 64.6$, then

$$\hat{x}_{10}(1) = \hat{x}_9(1) + \tfrac{1}{4}S_4 = 17.7 + \tfrac{1}{4}(64.6) = 33.8$$

Note that the average demands rise to a higher level at time $t = 7$. With this CUSUM method the sums S_3 and S_4 are set off at $t = 10$. This indicates to the forecaster that the demand pattern has changed four periods back (i.e., at $t = 7$). In this situation the smoothing parameter is changed to $\alpha = 0.2$, the smoothed average is adjusted as shown in the table, and the sums are reset to zero.

[2]When $S_n = e_T + e_{T-1} + \ldots + e_{T-n+1}$ and the errors are independent with standard deviation σ, then $\sigma_{s_n}^2 = n\sigma^2$ and $\sigma_{s_n} = \sqrt{n}\,\sigma$.

12-5 ALL-TIME REQUIREMENTS

Typically, when a new item is introduced into the inventory, the demand gradually rises from time period to time period until it reaches an equilibrium level. Eventually, the need for the item drops and the demand steadily dwindles to zero. The latter event occurs because the item has been replaced or is no longer in style. For a service part the parent machine(s) may be out of production, and the need for the part drops as the population of the machine becomes less and less.

An important role in inventory management is to estimate requirements for all future time periods. This is to allow management to assess its current inventory and order one last quantity for the item—this order is often called the "all-time buy." Should the item be produced within its own manufacturing facilities, the manufacturing management is notified and the die (if any) is scrapped.

No doubt many methods of estimating all-time requirements have been used by individual companies for their special variety of items. In this section three such techniques are described: the linear decline, exponential decay, and geometric methods. The former two are described by Brown [7].

Linear Decline

The *linear decline method* assumes that the demand of an item follows a trend demand pattern which is declining. The trend is projected from the current time period until it reaches zero and the sum of all demands over this period is recorded. This sum is the estimate of the all-time requirements for the item.

The τ-period-ahead forecast as of the current time period ($t = T$) is

$$\hat{x}_T(\tau) = a + b\tau$$

where, for a declining trend, the slope is negative ($b < 0$). The number of time periods (τ_0) until a zero demand is reached becomes

$$\tau_0 = -\frac{a}{b}$$

Hence, the forecast of all-time requirements ($\hat{X}_T$) is

$$\hat{X}_T = \hat{x}_T(1) + \hat{x}_T(2) + \ldots + \hat{x}_T(\tau_o)$$

or

$$\hat{X}_T = \sum_{\tau=1}^{\tau_o} (a + b\tau) = \tau_o a + \frac{\tau_o(\tau_o + 1)}{2} b$$

Example 12.5

Find the all-time requirements for an item that has the following forecast equation:

$$\hat{x}_T(\tau) = 100 - 5\tau$$

Since

$$\tau_o = -\left(\frac{100}{-5}\right) = 20$$

then

$$\hat{X}_T = 20(100) - 5\left(\frac{20 \times 21}{2}\right) = 950$$

Example 12.6

If $\hat{x}_T(\tau) = 100 - 6\tau$ for an item, find the all-time requirements.

Here

$$\tau_o = \frac{100}{6} = 16.67$$

and an approximate is

$$\hat{X}_T = 16.67(100) - 6\frac{(16.67)(17.67)}{2} = 783$$

Exponential Decay

Another method of estimating the all-time requirements is by way of the *exponential decay method* and is more applicable for items that are service parts. This method is based on the declining rate of the parent machine associated with the item. Through population studies, the forecaster establishes that the population size is dropping by a fixed percentage (θ) at each time period. Hence, if the population size at the end of time period t is P_t, then at $t + 1$ it is

$$P_{t+1} = (1 - \theta)P_t$$

and for $t + 2$,

$$P_{t+2} = (1 - \theta)P_{t+1} = (1 - \theta)^2 P_t$$

In general, for time period $t + \tau$, the population size becomes

$$P_{t+\tau} = (1 - \theta)^\tau P_t$$

Now for the item, it is assumed that the demand rate per machine in the population will remain steady. However, since the machine population is dropping, the demand of the item is correspondingly falling. To use this method, the forecaster must have estimates of θ and an estimate of the level for the item at time T ($\hat{x}_T$). Hence, the forecast for $t = T + 1$ is

$$\hat{x}_T(1) = (1 - \theta)\hat{x}_T$$

and the forecast for $t = T + \tau$ becomes

$$\hat{x}_T(\tau) = (1 - \theta)^\tau \hat{x}_T$$

With this relation, the estimate of all-time requirements is found from

$$\begin{aligned}\hat{X}_T &= \sum_{\tau=1}^{\infty} \hat{x}_T(\tau) \\ &= \sum_{\tau=1}^{\infty} (1 - \theta)^\tau \hat{x}_T \\ &= \hat{x}_T\left(\frac{1 - \theta}{\theta}\right)\end{aligned}$$

Example 12.7

Consider an item whose level at time T is estimated as $\hat{x}_T = 100$. Suppose that the item is a service part for a machine which is declining by 5% each month. Find an estimate of the all-time requirements.

Since $\theta = 0.05$, then

$$\hat{X}_T = 100\left(\frac{0.95}{0.05}\right) = 1900$$

Geometric Method

The *geometric method* is like the exponential decay method except that the demands of the past are used to estimate θ and the current level. These estimates are here denoted as $\hat{\theta}$ and $\hat{x}_T$, respectively. With these available, the forecast for all future time periods is

$$\begin{aligned}\hat{X}_T &= \sum_{\tau=1}^{\infty} \hat{x}_T(\tau) \\ &= \sum_{\tau=1}^{\infty} \hat{x}_T(1 - \hat{\theta})^\tau \\ &= \hat{x}_T \frac{1 - \hat{\theta}}{\hat{\theta}}\end{aligned}$$

Assume that at time T, the past demands are $x_1, x_2, \ldots, x_T$. For this time horizon, a fit of the type

$$\hat{x}_t = \hat{x}_1(1 - \theta)^{t-1} \qquad t = 1, \ldots, T$$

is found by the least-squares method.

The estimate of θ is

$$\hat{\theta} = 1 - e^{\hat{B}}$$

where

$$\hat{B} = \frac{\sum (t-1) \sum \ln x_t - T \sum (t-1) \ln x_t}{[\sum (t-1)]^2 - T \sum (t-1)^2}$$

and the estimate of $\hat{x}_1$ is

$$\hat{x}_1 = e^{\hat{A}}$$

where

$$\hat{A} = \frac{\sum \ln x_t - \hat{B} \sum (t-1)}{T}$$

and the summations range from $t = 1$ to T. Note also that ln is the natural logarithm. The estimate of $\hat{x}_T$ becomes

$$\hat{x}_T = \hat{x}_1(1 - \hat{\theta})^{T-1}$$

Example 12.8

Assume that an item has the following six annual demand entries: 50, 41, 33, 25, 20, and 16. Using the geometric method, find estimates of $\hat{x}_6$, $\hat{\theta}$, and the all-time requirements.

For this purpose, the following table is used as a worksheet:

t	x_t	$\ln x_t$	$(t-1) \ln x_t$
1	50	3.91	0.00
2	41	3.71	3.71
3	33	3.50	7.00
4	25	3.22	9.66
5	20	3.00	12.00
6	16	2.77	13.85
Sums		20.11	46.22

Now, because $T = 6$, $\sum (t-1) = 15$, and $\sum (t-1)^2 = 55$, then

$$\hat{B} = \frac{15(20.11) - 6(46.22)}{15^2 - 6(55)} = -0.23$$

$$\hat{A} = \frac{20.11 - (-0.23)(15)}{6} = 3.93$$

$$\hat{\theta} = 1 - e^{-0.23} = 0.21$$

$$\hat{x}_1 = e^{3.93} = 50.74$$

$$\hat{x}_6 = 50.74(0.79)^5 = 15.61$$

So now the all-time requirements is estimated as

$$\hat{X}_6 = 15.61\left(\frac{0.79}{0.21}\right) = 58.72$$

Mathematical Basis. The equation

$$x_t = \hat{x}_1(1 - \hat{\theta})^{t-1}$$

becomes

$$\ln x_t = \ln \hat{x}_1 + (t - 1) \ln (1 - \hat{\theta})$$

From this relation, the sum of squared errors is

$$S(e) = \sum_{t=1}^{T} [\ln x_t - \ln \hat{x}_1 - (t - 1) \ln (1 - \hat{\theta})]^2$$

$$= \sum_{t=1}^{T} [\ln x_T - \hat{A} - (t - 1)\hat{B}]^2$$

where

$$\hat{A} = \ln \hat{x}_1$$

$$\hat{B} = \ln (1 - \hat{\theta})$$

Now the least-square estimates $\hat{A}$ and $\hat{B}$ are found. The relations are as shown earlier. Hence,

$$\hat{x}_1 = e^{\hat{A}}$$

$$\hat{\theta} = 1 - e^{\hat{B}}$$

12-6 DAMPENING OF FAST-DROPPING FORECASTS

Consider an item where a trend forecast model is being applied and where the trend is negative and is swiftly approaching zero. Even though the trend is derived using the history of demands in a sound manner, the forecaster may wish to change the drop (from a straight line leading to zero) with a more gradual decline. This in essence may well be how most items die out based on the history of other items in the inventory. Such a technique is called *dampening* and is often used to adjust fast-declining forecasts.

Of first concern is selection of a decision rule that identifies which items are candidates for dampening. Certainly, the majority of items need not have their forecasts adjusted in any way. One possible decision rule is to apply dampening to all items where the forecasts will otherwise reach zero within the coming year.

Once an item is identified as one where dampening is to be applied, the forecasts for that item are altered. Many schemes are possible for this purpose, depending on the nature of items in the inventory. A technique is described here which is associated with the all-time requirement forecasts of the previous section.

Using the results from Section 12-5, let $\hat{X}_{LT}$ represent the all-time requirements when the linear decline method is applied, and let $\hat{X}_{ET}$ represent the corresponding forecasts for the exponential decay model. Hence,

$$\hat{X}_{LT} = a\tau_o + \frac{\tau_o(\tau_o + 1)}{2} b$$

$$\hat{X}_{ET} = \hat{x}_T \frac{1 - \theta}{\theta}$$

where $\tau_o = -a/b$ and θ is the percent decrease at each time period.

For the item with fast-dropping forecasts, a and b are known and

$$\hat{x}_T(\tau) = a + b\tau$$

Hence, the level at time T is

$$\hat{x}_T(0) = \hat{x}_T = a$$

and the quantity $\hat{X}_{LT}$ can be found as shown above. Now these results can be used to find estimates of θ for the exponential decay method. This estimate is

$$\hat{\theta} = \frac{\hat{x}_T}{\hat{x}_T + \hat{X}_{LT}}$$

With the estimate $\hat{\theta}$ available, the forecaster can now generate forecasts using the decay method. These are

$$\hat{x}_T(\tau) = a(1 - \hat{\theta})^\tau$$

At first thought, one may reason that the exponential decay forecasts should be used in place of the trend forecasts. This is because the exponential forecasts are, in fact, declining at a gradual rate. This switch in forecasts may be too hasty, however, since the exponential decay forecasts will drop faster in the short run than the forecasts generated with the trend model.

Because of the reason above, a more conservative approach seems appropriate. This is to choose the maximum of the two forecasts as the one to use. In this way

$$\hat{x}_T(\tau) = \max\,(a + b\tau, a(1 - \hat{\theta})^\tau) \qquad \tau = 1, 2, \ldots$$

The forecasts will follow a linear decline for the short time horizon and then will gradually decay to zero in the longer time horizon. It should be noted, however, that the total forecast for all future time periods is greater with this scheme than it would otherwise be.

Example 12.9

Consider an item where at the current time period, the forecasts are estimated by

$$\hat{x}_T(\tau) = 100 - 20\tau$$

Alter the forecasts by use of the dampening procedure.

Using the results above, the level at time T is $\hat{x}_T = 100$, and the all-time requirements become

$$\begin{aligned}\hat{X}_{LT} &= \hat{x}_T(1) + \hat{x}_T(2) + \hat{x}_T(3) + \hat{x}_T(4) + \hat{x}_T(5)\\ &= 80 + 60 + 40 + 20 + 0 = 200\end{aligned}$$

Hence,

$$\hat{\theta} = \frac{100}{100 + 200} = \frac{1}{3}$$

and for the exponential decay method,

$$\hat{x}_T(\tau) = 100(2/3)^\tau$$

So the dampened forecasts are the following:

$$\begin{aligned}\hat{x}_T(1) &= \max\,(80, 66.7) = 80\\ \hat{x}_T(2) &= \max\,(60, 44.4) = 60\\ \hat{x}_T(3) &= \max\,(40, 29.6) = 40\\ \hat{x}_T(4) &= \max\,(20, 19.7) = 20\\ \hat{x}_T(5) &= \max\,(0, 13.2) = 13\\ \hat{x}_T(6) &= \max\,(0, 8.7) = 9\\ \hat{x}_T(7) &= \max\,(0, 6.0) = 6\\ \hat{x}_T(8) &= \max\,(0, 4.0) = 4\\ \hat{x}_T(9) &= \max\,(0, 3.1) = 3\\ \hat{x}_T(10) &= \max\,(0, 2.1) = 2\\ \hat{x}_T(11) &= \max\,(0, 1.4) = 1\end{aligned}$$

and so on.

12-7 GROWTH RATES

For all but the horizontal-type demand patterns, it could be meaningful to measure the rate of change in levels from the current time period T to a projected 1-year-ahead time period. For a company with hundreds or thousands of items, it may be too cumbersome to attempt to study these rates on each item individually. However, compiling the rates over a homogeneous group of items may yield some informative results on directional shifts in the demands of items by groups or lines.

The growth rate for an item can be measured in a variety of ways, depending on its most useful form to the forecaster. The following 1-year-ahead rate is easy to apply and gives meaningful results.

$$R = \frac{\hat{\mu}_{T+M}}{\hat{\mu}_T} - 1$$

Here, $\hat{\mu}_T$ is an estimate of the level as of the current time period, M is the number of time periods in a year, and $\hat{\mu}_{T+M}$ is the estimated level at $t = T + M$.

Example 12.10

Using $M = 12$, find the growth rates for items A, B, C, D, and E. Below is a list of the forecast relations for each.

Item A: $\hat{x}_T(\tau) = 100 + 2\tau$

Item B: $\hat{x}_T(\tau) = 60 - 1\tau$

Item C: $\hat{x}_T(\tau) = 80 + 2\tau - 0.1\tau^2$

Item D: $\hat{x}_T(\tau) = (400 + 5\tau)\rho_{T+\tau}$ where $\rho_T = \rho_{T+12} = 1.2$

Item E: $\hat{x}_T(\tau) = 70 + 2\tau + \delta_{T+\tau}$ where $\delta_T = \delta_{T+12} = 20$

The growth rates for the items become

Item A: $R = \frac{124}{100} - 1 = 0.24$

Item B: $R = \frac{48}{60} - 1 = -0.20$

Item C: $R = \frac{89.6}{80} - 1 = 0.12$

Item D: $R = \frac{552}{480} - 1 = 0.15$

Item E: $R = \frac{114}{90} - 1 = 0.27$

PROBLEMS

12-1. Consider an item with a lead time of 4 weeks and with the following history of weekly demands: 0, 2, 0, 1, 3, 0, 4, 0, 33, 0, 0, 2, 3, 50, 0, 0, and 4.

a. Find the mean and standard deviation of demands and compare with Brown's criterion for a lumpy item.

b. Using Wilcox's method, find an order point that would correspond with a 75% level of service.

12-2. Suppose that an item is replenished at intermittent time periods. Assume that at the fourth week of the year a replenishment was made and brought the inventory level to 600 pieces and that no replenishments were received since then. Now suppose that at the twentieth week of the year, the inventory level is down to 50 pieces. What is the forecast of the weekly demand for the item?

12-3. Continuing with Problem 12-2, suppose that a replenishment of 700 pieces is received during the twenty-first week. Now if during the forty-second week of the year (with no intermittent replenishments) the inventory is down to 100 pieces, find an updated forecast of the weekly demand. Suggest a way to combine the total information to arrive at a forecast of the weekly demands.

12-4. Suppose that the forecasts of U.S. demands for an item as of time T are the following for the τth future months:

τ	1	2	3	4
$x_{T(\tau)}$	40	50	60	70

Assume also that the standard deviation of the forecasts is $\sigma = 20$ for each of the four time periods.

a. If it is known currently that an order for 30 pieces from New York will be received in the third future month ($\tau = 3$), find the updated forecast and the corresponding standard deviation of the forecast. (This order is assumed to be a part of the regular demand for $\tau = 3$ and not an addition to it.)

b. If for time period $T + 2$ ($\tau = 2$), the forecaster knows that a special order of size 20 will be received from Europe, how does this information affect the forecasts? (This order is assumed not a part of the regular demand but an addition to it.)

12-5. Redo Example 12.4 (CUSUM in series) using $\alpha = 0.2$ at the outset and $\alpha = 0.3$ as a replacement when a signal occurs, and vice versa.

12-6. Suppose that the forecast of τth future requirements of an item is $\hat{x}_T(\tau) = 100 - 2\tau$. Find the forecast of the all-time requirements.

12-7. Assume that parts A, B, and C are only used on a parent machine that is not in production and whose population is dropping by 20% per year. If the estimated levels of A, B, and C are, respectively, $\hat{x}_A = 100$, $\hat{x}_B = 20$, and $\hat{x}_C = 200$, find the all-time-requirement estimates for each of the parts.

12-8. For the item in Problem 12-6, estimate the percent decrease per time period (θ) by use of the results in Section 12-6. Now use the exponential decay method to estimate the all-time requirements.

12-9. Suppose that the forecast for a part is of the form

$$\hat{x}_T(\tau) = (50 - 5\tau)r_\tau$$

where $M = 6$ and the seasonal ratios are $r_1 = 1.5$, $r_2 = 1.3$, $r_3 = 0.9$, $r_4 = 0.8$, $r_5 = 0.7$, and $r_6 = 0.8$. Find the forecast of all-time requirements.

12-10. Find the all-time requirements for an item with the following forecast equation:

$$\hat{x}_T(\tau) = 50 + 1\tau - 0.2\tau^2$$

12-11. Suppose that an item has the following forecast equation as of the current time period T:

$$\hat{x}_T(\tau) = 200 - 20\tau$$

a. Plot the forecasts for the future time periods using the linear decline method.
b. Estimate θ via Section 2-6 and plot the forecasts using the exponential decay method.
c. Use the dampening procedure and plot the forecasts for the future time periods.
d. Compare the all-time requirements for the three methods shown.

12-12. Use the dampening method to adjust the forecasts for the item in Problem 12-9. In estimating θ, use only the linear portion of the forecast, then apply the seasonal ratios accordingly.

12-13. Find the one-year-ahead growth rates for the items listed below. Use $M = 12$ in all cases.

Item	*Forecast Equation*
A	$\hat{x}_T(\tau) = 60 + 2\tau$
B	$\hat{x}_T(\tau) = 44 + 0.5\tau$
C	$\hat{x}_T(\tau) = 100$
D	$\hat{x}_T(\tau) = 100 - 2\tau + 0.1\tau^2$
E	$\hat{x}_T(\tau) = 200(0.9)^\tau$
F	$\hat{x}_T(\tau) = (20 + 2\tau)r_{T+\tau}$ with $r_T = r_{T+12} = 0.8$
G	$\hat{x}_T(\tau) = 60 - \tau + d_{T+\tau}$ with $d_T = d_{T+12} = 5$

12-14. Estimate the all-time requirements using the geometric method when the prior five most recent yearly demands are the following: $x_1 = 100$, $x_2 = 73$, $x_3 = 47$, $x_4 = 38$, and $x_5 = 20$. Note that the most recent entry is x_5.

REFERENCES

[1] MICHAEL, G. C., "A Computer Simulation Model for Forecasting Catalog Sales." *Journal of Marketing Research*, May 1971, pp. 224–229.

[2] BROWN, R. G., *Materials Management Systems*. New York: John Wiley & Sons, Inc., 1977, pp. 245–246.

[3] WILCOX, J. E., "How to Forecast Lumpy Items." *American Production and Inventory Management*, First Quarter, 1970, pp. 51–54.

[4] BROWN, R. G., *Smoothing, Forecasting and Prediction of Discrete Time Series*. Englewood Cliffs, N.J.: Prentice-Hall, Inc., 1962, pp. 256–259.

[5] WILD, R., *The Techniques of Production Management*. New York: Holt, Rinehart and Winston, Inc., 1971, pp. 471–473.

[6] HARRISON, P. J., AND O. L. DAVIES, "The Use of Cumulative Sum (CUSUM) Techniques for the Control of Routine Forecasts of Product Demand." *Operations Research*, Vol. 12, No. 2, 1964, pp. 325–333.

[7] BROWN, R. G., *Smoothing, Forecasting and Prediction of Discrete Time Series*. Englewood Cliffs, N.J.: Prentice-Hall, Inc., 1962, pp. 259–261.

REFERENCE FOR FURTHER STUDY

WADE, J. B., "Determining Reorder Points When Demand Is Lumpy." *Management Science*, Vol. 24, No. 6, Feb. 1978, pp. 623–632.

13

MULTIDIMENSIONAL FORECASTING MODELS

This chapter describes various forecasting techniques that are used in situations which call for more than one dimension. In this context the multiple dimensions may pertain to a multiple of items belonging to a product group or line, or to a part that is stocked in a multiple number of locations. Still further, the product may have a multiple of sizes and colors associated with it. The goal of this chapter is to show how forecasts can be generated simultaneously for each element of the whole.

The chapter contains eight sections. The first describes a ranking model where an estimate of the probability distribution of the demands for an item is found. Next is a presentation of vector smoothing which is a method of updating estimates on the probability distribution of demands. Sections 13-3 and 13-4 show how vector smoothing is used to generate forecasts for each item in a line of items, and for each region when a part is stocked in various regions. Sections 13-5 and 13-6 show how blending methods are applied to forecast each item of a line when the items are classified into one or more categories. Sections 13-7 and 13-8 show how the percent done and percent of aggregate methods are used to forecast individual items belonging to a line.

13-1 RANKING MODELS

Brown [1] shows how ranking models are used when the forecaster is interested in estimating the shape of the distribution of demands. Using the most recent T demand entries, the forecaster can rank the entries and thereupon estimate various probability relations on the distribution of the demands. The model should only be used with items whose demand pattern is horizontal or is slowly changing in time.

Using the T demand entries $x_1, x_2, \ldots, x_T$, the forecaster ranks these from low to high, i.e.,

$$x_{(1)} \leq x_{(2)} \leq \ldots \leq x_{(T)}$$

where $x_{(1)}$ is the smallest demand, $x_{(2)}$ is the next smallest, and so on. In general, $x_{(k)}$ will represent the kth smallest demand. The probability that a future demand will be less than $x_{(k)}$ is estimated by

$$P_k = P(x < x_{(k)}) = \frac{k}{T+1}$$

The standard deviation of these estimates can also be found. For P_k, the standard deviation becomes

$$\sigma_k = \sqrt{\frac{k(T+1-k)}{(T+1)^2(T+2)}}$$

Example 13.1

Suppose that $T = 39$, and the demand entries are the following: 0, 7, 4, 8, 17, 25, 6, 0, 4, 3, 9, 6, 10, 31, 4, 3, 18, 5, 16, 9, 14, 21, 0, 4, 8, 7, 6, 17, 13, 4, 20, 8, 15, 16, 1, 3, 8, 7, and 4. The entries are ranked as shown in Table 13-1, and the corresponding probabilities (P_k) and standard deviations (σ_k) are listed. Since it is not possible for a demand to fall below zero, the probability $P_o = P(x < o)$ is not meaningful and is not calculated. For those demand entries with more than one value of k, the low value of k is used to estimate P_k and σ_k.

Using the results of the table, the forecaster can observe, for example, that the probability of a demand less than $x = 20$ is 0.900. The 95% confidence limit of this probability is approximately

$$0.90 \pm 2(0.047) = (0.806, 0.994)$$

It is now possible to find the demand x_o where the probability of x less than x_o is P_o, i.e.,

$$P(x < x_o) = P_o$$

Table 13-1. RANKING AND PROBABILITIES FOR EXAMPLE 13.1

k	$x_{(k)}$	$P_k = P(x < x_{(k)})$	σ_k
1–3	0	—	
4	1	0.100	0.047
5–7	3	0.125	0.052
8–13	4	0.200	0.062
14	5	0.350	0.074
15–17	6	0.375	0.076
18–20	7	0.450	0.078
21–24	8	0.525	0.078
25–26	9	0.625	0.076
27	10	0.675	0.073
28	13	0.700	0.072
29	14	0.725	0.070
30	15	0.750	0.068
31–32	16	0.775	0.065
33–34	17	0.825	0.059
35	18	0.875	0.052
36	20	0.900	0.047
37	21	0.925	0.041
38	25	0.950	0.034
39	31	0.975	0.024

This is obtained by interpolating with $x_{(k)}$, $x_{(k+1)}$, P_k, and P_{k+1}, where $P_k \leq P_o \leq P_{k+1}$. Using

$$\frac{P_o - P_k}{P_{k+1} - P_k} = \frac{x_o - x_{(k)}}{x_{(k+1)} - x_{(k)}}$$

then

$$x_o = x_{(k)} + (x_{(k+1)} - x_{(k)})\frac{P_o - P_k}{P_{k+1} - P_k}$$

Example 13.2

Using Table 13-1, find x_o where $P(x < x_o) = 0.94$. Since $P_o = 0.94$ lies between $P_{37} = 0.925$ and $P_{38} = 0.950$,

$$x_0 = 21 + (25 - 21)\frac{0.94 - 0.925}{0.950 - 0.925} = 23.4$$

and

$$P(x < 23.4) = 0.94$$

Another situation arises when the forecaster wishes to find the probability P_o corresponding to a particular demand x_o. Here, he wishes to find P_o, where

$$P(x < x_o) = P_o$$

Using the relation cited above and solving for P_o,

$$P_o = P_k + (P_{k+1} - P_k)\frac{x_o - x_{(k)}}{x_{(k+1)} - x_{(k)}}$$

Example 13.3

Using Table 13-1, find P_o when $x_o = 26$. Since x_o lies between $x = 25$ and 31,

$$P_o = 0.950 + (0.975 - 0.950)\frac{26 - 25}{31 - 25} = 0.954$$

or

$$P(x < 26) = 0.954$$

MATHEMATICAL BASIS. Given the ranked demands $(x_{(1)} \leq \ldots \leq x_{(k)} \leq \ldots \leq x_{(T)})$, then the probability density associated with

$$P_k = P(x < x_{(k)})$$

is

$$f(P_k) = \frac{T!}{(k-1)!(T-k)!} P_k^{k-1}(1 - P_k)^{T-k}$$

This density is obtained since, P_k^{k-1} is the probability of finding $(k - 1)$ entries below $x_{(k)}$, and $(1 - P_k)^{T-k}$ is the probability of $(T - k)$ entries above $x_{(k)}$. Also, the number of ways to arrange the T entries is

$$\frac{T!}{(k-1)!(T-k)!}$$

Now using

$$E(P_k) = \int_0^1 P_k f(P_k)\, dP_k$$

$$E(P_k^2) = \int_0^1 P_k^2 f(P_k)\, dP_k$$

and

$$\sigma_k^2 = E(P_k^2) - E(P_k)^2$$

then the expectancies $E(P_k)$ and $E(P_k^2)$ are easily found. Note first that the beta function gives

$$B(a, b) = \int_0^1 x^{a-1}(1 - x)^{b-1}\, dx = \frac{(a-1)!(b-1)!}{(a+b-1)!}$$

Hence,

$$E(P_k) = \frac{k}{T+1}$$

$$E(P_k^2) = \frac{(k+1)k}{(T+2)(T+1)}$$

and

$$\sigma_k^2 = \frac{k(T+1-k)}{(T+1)^2(T+2)}$$

13-2 VECTOR SMOOTHING

Brown [2] shows how smoothing techniques can be extended to generate current estimates of the probability distribution of demands. The range of demands over a time period is grouped into K intervals and the probability of a demand falling in an interval is estimated. As each new demand entry becomes available, the probabilities are updated using smoothing techniques. This method is called *vector smoothing* and is applicable when the demand pattern is horizontal or is shifting slowly over time.

Consider the K intervals

$$x \leq x_{(1)}, x_{(1)} < x \leq x_{(2)}, \ldots, x_{(K-2)} < x \leq x_{(K-1)}, x_{(K-1)} < x$$

Let P_k represent the probability associated with each interval, i.e.,

$$P_k = P(x_{(k-1)} < x \leq x_{(k)}) \qquad \text{for } k = 2, 3, \ldots, K-1$$

and

$$P_1 = P(x \leq x_{(1)})$$

$$P_K = P(x_{(K-1)} < x)$$

The sum of the probabilities above is equal to 1, yielding

$$\sum_{k=1}^{K} P_k = 1$$

Suppose that at $t = T - 1$, the probability estimates above are available and are denoted by

$$P_{k,T-1} \qquad \text{for } k = 1, 2, \ldots, K$$

Now at $t = T$, the demand x_T is available and falls in one of the preselected intervals, say $(x_{(k_o-1)} < x \leq x_{(k_o)})$. At this time, let $q_{k_o,T} = 1$ and $q_{kT} = 0$ for all $k \neq k_o$. The updated estimates become

$$P_{kT} = \alpha q_{kT} + (1 - \alpha)P_{k,T-1} \qquad k = 1, 2, \ldots, K$$

where α is a smoothing parameter within (0, 1) and chosen by the forecaster. Since

$$\sum_{k=1}^{K} P_{k,T-1} = 1$$

and

$$\sum_{k=1}^{K} q_{kT} = 1$$

then

$$\sum_{k=1}^{K} P_{kT} = 1$$

When the forecaster suspects the distribution of demands is stable, a low value of α is chosen (perhaps $\alpha = 0.1$). Should he feel that the distribution is slowly shifting, a higher value of α (say $\alpha = 0.3$) may be more appropriate.

To initialize the system, either of two techniques can be used. The first applies when no prior demands are known. The forecaster estimates the probability of demands falling in each interval so that they sum to 1. A value of α is selected and as new demand entries become available, the probabilities are updated, as shown earlier.

The second method is applicable when some prior demands are known. At $t = T$, the entries $x_1, x_2, \ldots, x_T$ are used to give starting values of the probabilities. Letting m_k designate the number of entries falling in the kth interval, then

$$P_{kT} = \frac{m_k}{T} \qquad k = 1, 2, \ldots, K$$

gives the initial probabilities that are needed. From here on, the probabilities are updated using vector smoothing.

Example 13.4

Consider the six intervals $x \leq 5$, $5 < x \leq 10$, $10 < x \leq 15$, $15 < x \leq 20$, $20 < x \leq 30$, and $30 < x$. Suppose that initial estimates of the probabilities associated with each interval are needed at $T = 39$ and the demands are those given in Example 13.1. Find the initial estimates.

The results are as follows:

k	*Interval*	m_k	$P_{k,39}$
1	0–5	14	0.359
2	6–10	13	0.333
3	11–15	3	0.077
4	16–20	6	0.154
5	21–30	2	0.051
6	31–	1	0.026
Sums		39	1.000

Example 13.5

Assume that $\alpha = 0.10$ is selected as the smoothing parameter and the probabilities found in Example 13.4 are used to start the system. Find the updated probabilities at $t = 40$, when $x_{40} = 12$.

Since 12 lies within (11, 15), $k_o = 3$ and $q_1 = 0, q_2 = 0, q_3 = 1, q_4 = 0$, $q_5 = 0$, and $q_6 = 0$. So now

$$P_{1,40} = 0.1(0) + 0.9(0.359) = 0.323$$
$$P_{2,40} = 0.1(0) + 0.9(0.333) = 0.300$$
$$P_{3,40} = 0.1(1) + 0.9(0.077) = 0.169$$
$$P_{4,40} = 0.1(0) + 0.9(0.154) = 0.139$$
$$P_{5,40} = 0.1(0) + 0.9(0.051) = 0.046$$
$$P_{6,40} = 0.1(0) + 0.9(0.026) = 0.023$$

The example continues for subsequent time periods in Table 13-2.

Table 13-2. VECTOR SMOOTHING RESULTS WITH EXAMPLE 13.5

t	x_t	(0–5) P_{1t}	(6–10) P_{2t}	(11–15) P_{3t}	(16–20) P_{4t}	(21–30) P_{5t}	(31–) P_{6t}
39		0.359	0.333	0.077	0.154	0.051	0.026
40	12	0.323	0.300	0.169	0.139	0.046	0.023
41	3	0.391	0.270	0.152	0.125	0.041	0.021
42	8	0.352	0.343	0.132	0.113	0.037	0.019
43	17	0.317	0.309	0.119	0.202	0.033	0.017
44	3	0.385	0.278	0.107	0.182	0.030	0.015
45	6	0.347	0.350	0.096	0.164	0.027	0.014
46	25	0.312	0.315	0.086	0.148	0.124	0.013
47	4	0.381	0.284	0.077	0.133	0.112	0.012
48	15	0.343	0.256	0.169	0.120	0.100	0.011
49	9	0.309	0.330	0.152	0.108	0.090	0.010
50	2	0.378	0.297	0.137	0.097	0.081	0.009

13-3 MULTIITEM FORECASTS USING VECTOR SMOOTHING

Vector smoothing can be used to generate forecasts for individual items making up a group or a line. In the retail market, the group may represent a product where individual items differ by size and/or color, such as styles of shoes, sweaters, coats, and so forth. In the industrial market, the group may represent a product with the items being various models of this product, e.g., the different models of an engine or a radio.

The model is applicable when an updated forecast for the total of the group is available to the forecaster at each time period. Vector smoothing is used to estimate the probability mix for each item in the group, and these results are extended to allocate the total forecast to each item.

Suppose that the group consists of K items and let P_{kt} represent the probability estimate at time t that a demand will occur for item k. As before, the sum of the probabilities must be 1, giving

$$\sum_{k=1}^{K} P_{kt} = 1$$

At $t = T - 1$, the probabilities that are available to the forecaster are $P_{k,T-1}$ $(k = 1, 2, \ldots, K)$. Now at time $t = T$, the mix of demands for the items becomes known. These are

$$x_{kt} = \text{demand of item } k \text{ for time period } T \qquad k = 1, 2, \ldots, K$$

The mix of demands during $t = T$ is converted to fractions by

$$q_{kT} = \frac{x_{kT}}{x_T} \qquad k = 1, 2, \ldots, K$$

where

$$x_T = x_{1T} + x_{2T} + \ldots + x_{KT}$$

The probability mix is now updated using

$$P_{kT} = \alpha q_{kT} + (1 - \alpha) P_{k,T-1} \qquad k = 1, 2, \ldots, K$$

where α is the smoothing parameter.

Using the most current forecast for the group $\hat{x}_T(\tau)$, the forecaster can now allocate the forecast to individual items. This is

$$\hat{x}_{kT}(\tau) = P_{kT} \hat{x}_T(\tau) \qquad k = 1, 2, \ldots, K$$

which is the forecast for item k as of time T for the the τth future time period. In this way the sum of the item forecasts is always the same as the forecast for the group, i.e.,

$$\hat{x}_T(\tau) = \sum_{k=1}^{K} \hat{x}_{kT}(\tau)$$

The forecaster may use one value of the smoothing parameter for all time periods, or he/she may vary α at each time period depending on the magnitude of the current demand (x_T). In the latter case, a smaller setting seems appropriate when x_T is small relative to its average value. One possible scheme is to set

$$\alpha = \min\left(\frac{x_T}{x}, 0.20\right)$$

where x is the group's total demand over the previous year. In this way α could range anywhere from 0 to 0.20. When the demand is low, a small weight is given to the current mix, and when the demand is high, a higher weight not exceeding 0.20 is used.

To initialize the system, the forecaster again has two options. When no prior demands are available for the mix of items, he may first estimate the market share (or probability mix) for the individual items and then proceed using vector smoothing. When prior demands are known, these demands are used as the basis of generating the initial probabilities that are needed.

Example 13.6

Consider a situation where the forecaster wishes to apply vector smoothing to a line with three items. The goal is to obtain one period ahead forecasts for each of the items. At the outset he estimates $P_1 = 0.2$, $P_2 = 0.3$, and $P_3 = 0.5$ as the probability mix. The smoothing constant will be found at each time period from the relation

$$\alpha = \min\left(\frac{x_T}{500}, 0.20\right)$$

The worksheet for this example is given in Table 13-3.

At $t = 1$, the demands for the items are $x_{11} = 3$, $x_{21} = 2$, and $x_{31} = 4$. Since $x_1 = 3 + 2 + 4 = 9$, then

$$\alpha = \min\left(\frac{9}{500}, 0.20\right) = 0.018$$

Hence, the probabilities are updated by

$$P_{11} = 0.018(0.333) + 0.982(0.200) = 0.202$$

$$P_{21} = 0.018(0.222) + 0.982(0.300) = 0.299$$

$$P_{31} = 0.018(0.444) + 0.982(0.500) = 0.499$$

Table 13-3. WORKSHEET FOR EXAMPLE 13.6 USING VECTOR SMOOTHING TO FORECAST THE DEMANDS FOR THE ITEMS IN A GROUP

t	x_{1t}	x_{2t}	x_{3t}	x_t	α	P_{1t}	P_{2t}	P_{3t}	$\hat{x}_t(1)$	$\hat{x}_{1t}(1)$	$\hat{x}_{2t}(1)$	$\hat{x}_{3t}(1)$
0						0.200	0.300	0.500				
1	3	2	4	9	0.018	0.202	0.299	0.499	15	3.04	4.48	7.48
2	1	3	4	8	0.016	0.201	0.300	0.499	29	5.84	8.69	14.47
3	2	4	7	13	0.026	0.200	0.300	0.500	24	4.79	7.20	12.01
4	5	10	20	35	0.070	0.196	0.299	0.505	37	7.25	11.06	18.69
5	10	10	28	48	0.096	0.197	0.290	0.513	47	9.25	13.65	24.09
6	9	16	30	55	0.110	0.193	0.290	0.516	84	16.25	24.39	43.36
7	15	28	30	73	0.146	0.195	0.304	0.501	90	17.57	27.36	45.07
8	9	18	51	78	0.156	0.183	0.293	0.525	66	12.06	19.31	34.63
9	18	30	35	83	0.166	0.188	0.304	0.508	31	5.84	9.42	15.74
10	6	10	18	34	0.068	0.188	0.303	0.509	10	1.87	3.03	5.09
11	0	8	4	12	0.024	0.183	0.312	0.505	3	0.55	0.94	1.51
12	1	0	3	4	0.008	0.184	0.310	0.506	13	2.39	4.02	6.59

Since the one-period-ahead forecast[1] for the group is $\hat{x}_1(1) = 15$, the corresponding forecasts for the items are

$$\hat{x}_{11}(1) = 0.202(15) = 3.04$$

$$\hat{x}_{21}(1) = 0.299(15) = 4.48$$

$$\hat{x}_{31}(1) = 0.499(15) = 7.48$$

The example continues in Table 13-3.

13-4 MULTIREGIONAL FORECASTS USING VECTOR SMOOTHING

Another application of vector smoothing is to generate forecasts for an item by regions. The need arises when the item is stocked in various regions of an area and forecasts of future demands for the item in the total area are available. Regional forecasts are required in order to stock the item properly in each location. This situation occurs frequently in the retail and industrial world. A store with several retail outlets in a metropolitan area requires forecasts for every item in each outlet. This is to have an economical supply available at each outlet to meet the customers' demands. A service part may be stocked in several parts depots throughout the country and forecasts by

[1]The forecasts are generated from the multiplicative trend seasonal model and are found in Table 9-4.

depot are used to allocate the supply. With good depot forecasts, customer service is maximized and inventory and transportation costs are minimized.

The method described in Section 13-3 for multiitem forecasts is used in parallel for multiregional forecasts. A summary is given below for an item stocked in K regions and where $k = 1, 2, \ldots, K$.

Data:

(1) $P_{k,T-1}$ = share for region k at time $T-1$ $\left(note: \sum_{k=1}^{K} P_{k,T-1} = 1\right)$

(2) x_{kT} = demand at region k for time T

(3) $x_T = \sum_{k=1}^{K} x_{kT}$ is the total demand of the item at time T

(4) $\hat{x}_T(\tau)$ = forecast of the item as of time T for the τth future time period

Calculations:

(1) $$q_{kT} = \frac{x_{kT}}{x_T} \qquad k = 1, \ldots, K$$

(2) $$P_{kT} = \alpha q_{kT} + (1-\alpha)P_{k,T-1} \qquad k = 1, \ldots, K$$

where α may be a fixed constant or may vary depending on the size of x_T. For example, the forecaster may choose

$$\alpha = \min\left(\frac{x_T}{X_T}, 0.20\right)$$

where X_T is the total demand for the item over the past year.

(3) $$\hat{x}_{kT}(\tau) = P_{kT}\hat{x}_T(\tau) \qquad k = 1, \ldots, K$$

Hence, the forecasts needed ($\hat{x}_{kt}(\tau)$) are generated for each region k.

Example 13.7

Table 13-4 shows results for a part stored in four regions. At the outset ($t = 0$) the regional shares are estimated as $P_{10} = 0.4$, $P_{20} = 0.2$, $P_{30} = 0.3$, and $P_{40} = 0.1$. In this situation the forecaster decides to allow the smoothing parameter to vary at each time period with the following scheme,

$$\alpha = \frac{x_t}{12\hat{x}_t}$$

where

$$\hat{x}_t = 0.1x_t + 0.9\hat{x}_{t-1}$$

Table 13-4. WORKSHEET FOR EXAMPLE 13.7 USING VECTOR SMOOTHING TO FORECAST DEMANDS BY REGIONS

t	x_{1t}	x_{2t}	x_{3t}	x_{4t}	x_t	$\hat{x}_t$	α	P_{1t}	P_{2t}	P_{3t}	P_{4t}	$\hat{x}_t(1)$	$\hat{x}_{1t}(1)$	$\hat{x}_{2t}(1)$	$\hat{x}_{3t}(1)$	$\hat{x}_{4t}(1)$
0								0.400	0.200	0.300	0.100					
1	41	15	22	10	88	88.0	0.08	0.405	0.198	0.296	0.101	91	36.9	18.0	26.9	9.2
2	36	13	24	6	79	87.1	0.08	0.409	0.195	0.297	0.099	95	38.9	18.5	28.2	9.4
3	44	25	34	12	115	89.9	0.11	0.406	0.197	0.297	0.100	91	36.9	17.9	27.0	9.1
4	44	14	29	7	94	90.3	0.09	0.412	0.193	0.298	0.097	78	32.1	15.1	23.2	7.6
5	13	7	15	6	41	85.4	0.04	0.401	0.188	0.292	0.119	58	23.3	10.9	16.9	6.9
6	29	18	40	13	100	86.8	0.10	0.390	0.188	0.303	0.120	48	18.7	9.0	14.5	5.7
7	9	10	16	2	37	81.8	0.04	0.384	0.191	0.308	0.117	41	15.7	7.8	12.6	4.8
8	20	8	10	8	46	78.3	0.05	0.387	0.190	0.304	0.120	46	17.8	8.7	14.0	5.5
9	3	1	1	0	5	70.9	0.01	0.389	0.190	0.302	0.119	57	22.2	10.8	17.2	6.8
10	16	12	21	7	56	69.4	0.07	0.381	0.192	0.308	0.119	81	30.9	15.6	24.9	9.6
11	40	30	46	18	134	75.9	0.15	0.369	0.197	0.313	0.121	114	42.1	22.5	35.7	13.8
12	45	18	38	9	110	79.3	0.12	0.374	0.193	0.317	0.117	137	51.2	26.4	43.4	15.9

When $t = 1$, $\hat{x}_1$ is set to $x_1 = 88$ and $\alpha = 88/(12 \times 88) = 0.08$. The regional shares are

$$P_{11} = 0.08\left(\frac{41}{88}\right) + 0.92(0.400) = 0.405$$

$$P_{21} = 0.08\left(\frac{15}{88}\right) + 0.92(0.200) = 0.198$$

$$P_{31} = 0.08\left(\frac{22}{88}\right) + 0.92(0.300) = 0.296$$

$$P_{41} = 0.08\left(\frac{10}{88}\right) + 0.92(0.100) = 0.101$$

Since the one-period-ahead forecast is $\hat{x}_1(1) = 91$,[2] the regional forecasts are

$$\hat{x}_{11}(1) = 91(0.405) = 36.9$$
$$\hat{x}_{21}(1) = 91(0.198) = 18.0$$
$$\hat{x}_{31}(1) = 91(0.296) = 26.9$$
$$\hat{x}_{41}(1) = 91(0.101) = 9.2$$

13-5 MULTIITEM FORECASTS USING BLENDING METHODS

Blending techniques (Section 10-6) can be extended to generate forecasts for each of k items belonging to a group. This method was developed by Cohen [3] and is based on blending one forecast for the total of the group with average demands for each item.

Let $\hat{x}$ represent the forecast for the group and $\hat{\sigma}$ the corresponding standard error of this forecast. Also, let $\bar{x}_k$ designate the average demand for item k, and S_k the associated standard error of $\bar{x}_k$. The sum of all the averages gives an average for the group, i.e.,

$$\bar{x} = \sum_{k=1}^{K} \bar{x}_k$$

Assuming independence among the items, the variance of $\bar{x}$ is

$$S_{\bar{x}}^2 = \sum_{k=1}^{K} S_k^2$$

With the data above available, it is now possible to find the forecasts for each item. For item k, the forecast is

$$\hat{x}_k = \bar{x}_k + w_k(\hat{x} - \bar{x})$$

[2]The forecasts are taken from Table 8-4.

where the weight (w_k) is

$$w_k = \frac{S_k^2}{S_{\bar{x}}^2 + \hat{\sigma}^2}$$

Example 13.8

Suppose that a group consists of $K = 3$ items and at the current time period the following information is known to the forecaster. The one-period-ahead forecast for the group and the associated standard error are $\hat{x} = 100$ and $\hat{\sigma} = 25$, respectively. The average demand and standard deviation for each item are

$$\bar{x}_1 = 60 \quad \text{and} \quad S_1 = 9$$
$$\bar{x}_2 = 20 \quad \text{and} \quad S_2 = 10$$
$$\bar{x}_3 = 10 \quad \text{and} \quad S_3 = 6$$

With the information above, the weights and forecast for each item can now be found. First,

$$\bar{x} = 60 + 20 + 10 = 90$$
$$S_{\bar{x}}^2 = 81 + 100 + 36 = 217$$

Hence, the weights become

$$w_1 = \frac{81}{625 + 217} = 0.0962$$

$$w_2 = \frac{100}{625 + 217} = 0.1188$$

$$w_3 = \frac{36}{625 + 217} = 0.0428$$

and the forecasts are

$$\hat{x}_{1T}(1) = 60 + 0.0962(100 - 90) = 60.96$$
$$\hat{x}_{2T}(1) = 20 + 0.1188(100 - 90) = 21.19$$
$$\hat{x}_{3T}(1) = 10 + 0.0428(100 - 90) = 10.43$$

Mathematical Basis. Let μ represent the true demand for all items in the group where

$$E(\hat{x}) = \mu$$
$$E(\bar{x}) = \mu$$

The forecast for item k is

$$\hat{x}_{kT}(\tau) = \bar{x}_k + w_k(\hat{x} - \bar{x})$$

which can be rewritten as

$$\hat{x}_{kT}(\tau) = (1 - w_k)\bar{x}_k + w_k(\hat{x} - \bar{x} + \bar{x}_k)$$

Now the variance of this forecast becomes

$$S^2_{\hat{x}k} = (1 - w_k)^2 S^2_{\bar{k}} + w^2_k(\hat{\sigma}^2 + S^2_{\bar{x}} - S^2_{\bar{k}})$$

In order to find the weight (w_k) which yields the minimum variance, the following partial derivative is set to zero, i.e.,

$$\frac{\partial S^2_{\hat{x}k}}{\partial w_k} = -2(1 - w_k)S^2_{\bar{k}} + 2w_k(\hat{\sigma}^2 + S^2_{\bar{x}} - S^2_{\bar{k}}) = 0$$

Hence,

$$w_k = \frac{S^2_{\bar{k}}}{\hat{\sigma}^2 + S^2_{\bar{x}}}$$

13-6 MULTICLASSIFIED ITEM FORECASTS USING BLENDING METHODS

Cohen [3] further extends the blending method to generate item-level forecasts when the items have more than one classification. For example, a group may consist of I items and each item is stocked in J locations. Forecasts may be available for each of the I items regardless of location, and forecasts may be available for the total demand in a location. Of need, however, is to generate forecasts for each item at every location.

Another example is in the retail area where a style of shoe is available in various colors and sizes. Forecasts of the demand for each color may be available, as well as forecasts by sizes (regardless of color). Here, however, the forecast for each combination of size and color is required.

The method is described assuming a two-way classification. Suppose that the first classification is of size I, and the second is of size J. Let M_{ij} represent an item with classification i and j, where $i = 1, 2, \ldots, I$ and $j = 1, 2, \ldots, J$.

The average demand for M_{ij} is denoted by $\bar{x}_{ij}$, and the standard deviation of this average is S_{ij}. The sums of all averages in categories i and j are

$$\bar{X}_{i\cdot} = \sum_{j=1}^{J} \bar{x}_{ij}$$

$$\bar{X}_{\cdot j} = \sum_{i=1}^{I} \bar{x}_{ij}$$

respectively. The corresponding variances of these estimates become

$$S^2_{i\cdot} = \sum_{j=1}^{J} S^2_{ij}$$

$$S^2_{\cdot j} = \sum_{i=1}^{I} S^2_{ij}$$

Now suppose that forecasts for categories i and j are available and are

$$\hat{x}_{i\cdot} \quad \text{for } i = 1, 2, \ldots, I$$
$$\hat{x}_{\cdot j} \quad \text{for } j = 1, 2, \ldots, J$$

The corresponding standard errors of these forecasts are also assumed available and are

$$\hat{\sigma}_{i\cdot} \quad \text{for } i = 1, 2, \ldots, I$$
$$\hat{\sigma}_{\cdot j} \quad \text{for } j = 1, 2, \ldots, J$$

Using the information above, the blended forecasts for each item can be found. For item M_{ij}, the forecast is

$$\hat{x}_{ij} = \bar{x}_{ij} + w_{ij}(\hat{x}_{i\cdot} - \bar{x}_{i\cdot}) + v_{ij}(\hat{x}_{\cdot j} - \bar{x}_{\cdot j})$$

where

$$w_{ij} = \frac{1/A_{ij}}{1/S_{ij}^2 + 1/A_{ij} + 1/B_{ij}}$$

$$v_{ij} = \frac{1/B_{ij}}{1/S_{ij}^2 + 1/A_{ij} + 1/B_{ij}}$$

$$A_{ij} = \hat{\sigma}_{i\cdot}^2 + (S_{i\cdot}^2 - S_{ij}^2)$$

$$B_{ij} = \hat{\sigma}_{\cdot j}^2 + (S_{\cdot j}^2 - S_{ij}^2)$$

Example 13.9

Suppose that an item consists of two categories (size and color) with two sizes ($i = 1, 2$), and two colors ($j = 1, 2$). A summary of the assumed data and results for this example is shown in Table 13-5. The data available at the outset are the average demands ($\bar{x}_{ij}$), standard error of the averages (S_{ij}), forecasts ($\hat{x}_{i\cdot}$, $\hat{x}_{\cdot j}$) and the standard error of the forecasts ($\hat{\sigma}_{i\cdot}$, $\hat{\sigma}_{\cdot j}$).

With the above, the weights are found. For M_{11}, the weights are

$$A_{11} = 400 + (41 - 25) = 416$$

$$B_{11} = 144 + (34 - 25) = 153$$

$$w_{11} = \frac{\frac{1}{416}}{\frac{1}{25} + \frac{1}{416} + \frac{1}{153}} = 0.049$$

$$v_{11} = \frac{\frac{1}{153}}{\frac{1}{25} + \frac{1}{416} + \frac{1}{153}} = 0.133$$

and the forecast is

$$\hat{x}_{11} = 100 + 0.049(220 - 180) + 0.133(120 - 140) = 99.3$$

In the same manner, the weights and forecasts for the other three items are found.

Table 13-5. DATA, WEIGHTS, AND FORECASTS FOR EXAMPLE 13-9

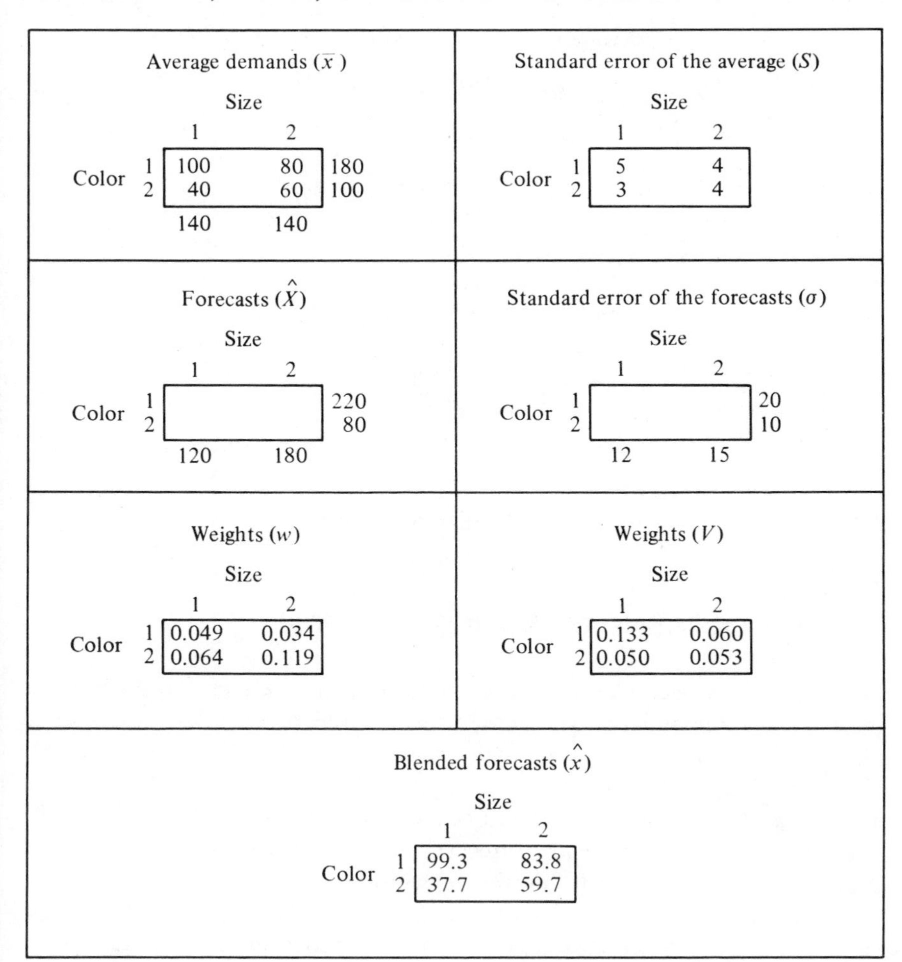

Average demands ($\bar{x}$)

Color \ Size	1	2	
1	100	80	180
2	40	60	100
	140	140	

Standard error of the average (S)

Color \ Size	1	2
1	5	4
2	3	4

Forecasts ($\hat{X}$)

Color \ Size	1	2	
1			220
2			80
	120	180	

Standard error of the forecasts (σ)

Color \ Size	1	2	
1			20
2			10
	12	15	

Weights (w)

Color \ Size	1	2
1	0.049	0.034
2	0.064	0.119

Weights (V)

Color \ Size	1	2
1	0.133	0.060
2	0.050	0.053

Blended forecasts ($\hat{x}$)

Color \ Size	1	2
1	99.3	83.8
2	37.7	59.7

MATHEMATICAL BASIS. The forecast for M_{ij} can be written as

$$\hat{x}_{ij} = (1 - w_{ij} - v_{ij})\bar{x}_{ij} + w_{ij}(\hat{x}_{i\cdot} - \bar{x}_{i\cdot} + \bar{x}_{ij}) + v_{ij}(\hat{x}_{\cdot j} - \bar{x}_{\cdot j} + \bar{x}_{ij})$$

whereby the variance is

$$S^2_{\hat{x}_{ij}} = (1 - w_{ij} - v_{ij})^2 S^2_{ij} + w^2_{ij}(\hat{\sigma}^2_{i\cdot} + S^2_{i\cdot} - S^2_{ij}) + v^2_{ij}(\hat{\sigma}^2_{\cdot j} + S^2_{\cdot j} - S^2_{ij})$$

The weights w_{ij} and v_{ij} are obtained from

$$\frac{\partial S\hat{x}_{ij}}{\partial w_{ij}} = 0 \quad \text{and} \quad \frac{\partial S\hat{x}_{ij}}{\partial v_{ij}} = 0$$

The results give

$$w_{ij}(\hat{\sigma}_{i\cdot}^2 + S_{i\cdot}^2) + v_{ij}S_{ij}^2 = S_{ij}^2$$

$$w_{ij}S_{ij}^2 + v_{ij}(\hat{\sigma}_{\cdot j}^2 + S_{\cdot j}^2) = S_{ij}^2$$

Hence,

$$v_{ij} = w_{ij}\left(\frac{\hat{\sigma}_{i\cdot}^2 + S_{i\cdot}^2 - S_{ij}^2}{\hat{\sigma}_{\cdot j}^2 + S_{\cdot j}^2 - S_{ij}^2}\right) = w_{ij}\frac{A_{ij}}{B_{ij}}$$

and

$$w_{ij} = \frac{S_{ij}^2}{\hat{\sigma}_{i\cdot}^2 + S_{i\cdot}^2 + S_{ij}^2(A_{ij}/B_{ij})}$$

$$= \frac{1/A_{ij}}{1/A_{ij} + 1/B_{ij} + 1/S_{ij}^2}$$

In the same manner,

$$v_{ij} = \frac{1/B_{ij}}{1/A_{ij} + 1/B_{ij} + 1/S_{ij}^2}$$

13-7 PERCENT-DONE ESTIMATING METHODS

Hartung [4] and Hertz and Schaffir [5] present methods of forecasting the total season demand for an item by use of the percent-done estimating method. The context that gives rise to this method is the style goods inventory problem, where an item is maintained in inventory for the length of a season. The season is of a fixed length and covers a portion of a year. Periodically over the season, total season forecasts are generated for each item and decisions are made on what items to replenish and how much.

For catalog items, the season length coincides with the life of the catalog. The items are clustered into homogeneous lines for purposes of estimating the percent done. A line is a group of more-or-less similar merchandise, e.g., men's long-sleeve sport shirts, dolls, plastic dinnerware, and so on. Most likely a line will be carried from one year to the next, but the line will not include the same items.

Two steps are required to carry out the technique. The first is to estimate the percent done (percent of total sales to date) of the line for each time period t. These estimates are obtained using the sum of demands for all items in the line from the prior year(s). The second step uses the current-year cumulative demands (to date) for each item in the line and the corresponding percent done estimate (for the line) to generate the item's forecast of the total season demand.

Two methods are given to calculate the percent-done estimates. The first is called the *weighted expected percent done* and is denoted here by P_{1t} $(t = 1, 2, \ldots, w)$. To find these estimates, the items in the line for the prior year are identified. Let K represent the number of items, and let X_{kt} designate the cumulative demand for item k up to and including time period t. The percent-done estimate for time t is

$$P_{1t} = \frac{\sum_{k=1}^{K} X_{kt}}{\sum_{k=1}^{K} X_{kw}} \qquad t = 1, 2, \ldots, w$$

Note that X_{kw} gives the total demand for item k for the prior year.

A second method of finding the percent done is called the *expected percent done* and is represented by P_{2t} $(t = 1, 2, \ldots, w)$. This is calculated by

$$P_{2t} = \frac{1}{\frac{1}{K} \sum_{k=1}^{K} \frac{X_{kw}}{X_{kt}}} \qquad t = 1, 2, \ldots, w$$

The forecaster selects one of the methods above and for convenience, P_t is denoted as the percent done as of time t regardless of the method chosen.

Now during the current season, the items belonging to the line are again identified. Let $t = T$ represent the most current week where cumulative sales for each item are available. For item k this is X_{kT}. The forecast of the total season demands for item k is

$$\hat{X}_{kT} = \frac{X_{kT}}{P_T}$$

where P_T is the percent-done estimate as of T for the line.

Example 13.10

Consider a line where the season covers $w = 8$ weeks. Assume for the previous season that $K = 4$ and the sales by item are those listed in Table 13-6. The table shows the sales for all items in the line and the estimates of P_{1t} and P_{2t}.

At $t = 1$,

$$P_{11} = \frac{43}{671} = 6.4\%$$

and

$$P_{21} = \frac{1}{\frac{1}{4}\left(\frac{182}{8} + \frac{155}{10} + \frac{201}{17} + \frac{133}{8}\right)} = 6.0\%$$

Table 13-6. CALCULATING THE PERCENT DONE (EXAMPLE 13.10)

	Cumulative Sales to Week t							
Items (k)	1	2	3	4	5	6	7	8
1	8	22	55	95	133	161	182	182
2	10	13	41	82	118	132	150	155
3	17	51	79	108	133	158	181	201
4	8	21	36	64	88	113	129	133
Sums	43	107	211	349	472	564	642	671
P_{1t}	6.4	15.9	31.4	52.0	70.3	84.1	95.7	100.0
P_{2t}	6.0	13.1	30.0	51.6	70.1	84.1	95.8	100.0

In the same manner, the corresponding percent-done estimates are found for all 8 weeks of the season.

Example 13.11

Carrying forward with Example 13.10, suppose that during the current season the line now includes $K = 3$ items. Suppose that at $T = 1$, the demands for the three items are $X_{11} = 12$, $X_{21} = 10$, and $X_{31} = 21$.

If P_{1t} (weighted percent done) is used, the forecasts of total season demands become

$$\hat{X}_{11} = \frac{12}{0.064} = 187.5$$

$$\hat{X}_{21} = \frac{10}{0.064} = 156.2$$

$$\hat{X}_{31} = \frac{21}{0.064} = 328.1$$

Should P_{2t} (expected percent done) be the method chosen, then

$$\hat{X}_{11} = \frac{12}{0.060} = 200.0$$

$$\hat{X}_{21} = \frac{10}{0.060} = 166.7$$

$$\hat{X}_{31} = \frac{21}{0.060} = 350.0$$

Continuing in this manner, the forecasts are updated as the season progresses. The forecasts are shown in Table 13-7 for the hypothetical demands listed.

Table 13-7. Forecasts Using Percent Done Method (Example 13.11)

	Week t							
Items (k)	1	2	3	4	5	6	7	8
	Cumulative Sales							
1	12	35	64	85	108	137	152	160
2	10	23	43	85	133	159	173	173
3	21	52	83	112	137	165	181	193
	Total Season Forecasts Using P_{1t}							
1	187	220	204	163	154	163	159	160
2	156	145	136	163	189	189	181	173
3	328	327	264	215	195	196	189	193
	Total Season Forecasts Using P_{2t}							
1	200	267	213	165	154	163	159	160
2	167	176	143	165	190	189	181	173
3	350	397	277	217	196	196	189	193

13-8 PERCENT OF AGGREGATE DEMANDS METHOD

Hausman and Sides [6] describe another method of forecasting for seasonal style good problems. They consider a line of items that are sold through a catalog and have a selling season of W weeks. The demand rate for the items is not uniform throughout the season. The demands are lowest at the start and end of the selling season and reach a peak somewhere in between. To achieve more uniformity, the weeks are consolidated into a smaller number of unequal-length time intervals. In their example, the 18-week season is regrouped into nine time periods of unequal length.

The goal of the method is to periodically update forecasts for each of K items in a line. As the season progresses, the item's percent of the total sales to date is measured. These percents are used with a total season forecast for the (entire) line to obtain forecasts for each item.

The procedure is straightforward and easy to use. Let K represent the number of items in the line and X_{kt} the cumulative sales for item k up to and including time period t. The percent of sales at time t for item k is

$$P_{kt} = \frac{X_{kt}}{\sum_{k=1}^{K} X_{kt}}$$

Now let $\hat{X}$ represent the total season forecast for all items in the line. With these data the forecasts are now generated. The total season forecast for item k as of time t is P_{kt} of $\hat{X}$. That is,

$$\hat{X}_{kt} = \hat{X}P_{kt}$$

The forecast of $\hat{X}$ is obtained external to this analysis and may remain fixed or vary as the season progresses. Needed, however, is that $\hat{X}$ should be greater or equal to the total sales to date for the sum of all items in the line.

It is also possible to forecast the remainder of season sales for each item. For item k this is

$$R_{kt} = \hat{X}_{kt} - X_{kt}$$

The standard error of this forecast can easily be found. This is

$$\sigma_{R_{kt}} = \left[(R_t P_{kt})\frac{R_t + X_t}{X_t}\right]^{1/2}$$

where

$$X_t = \sum_{k=1}^{K} X_{kt}$$

and

$$R_t = \hat{X} - X_t$$

Note however that this standard error is valid only when the total season demand is truly $\hat{X}$.

Example 13.12

Suppose that a line with $W = 8$ has $K = 3$ items and the sales at $t = 1$ are $X_{11} = 12$, $X_{21} = 10$, and $X_{31} = 21$. Also, assume that the total season forecast for the line is $\hat{X} = 560$. Hence, the percent of aggregate sales to date for the items is

$$P_{11} = \frac{12}{43} = 0.279$$

$$P_{21} = \frac{10}{43} = 0.233$$

$$P_{31} = \frac{21}{43} = 0.488$$

These results yield the following forecasts:

$$\hat{X}_{11} = 560(0.279) = 156$$
$$\hat{X}_{21} = 560(0.233) = 130$$
$$\hat{X}_{31} = 560(0.488) = 274$$

Now the remainder of season forecasts is

$$R_{11} = 156 - 12 = 144$$
$$R_{21} = 130 - 10 = 120$$
$$R_{31} = 274 - 21 = 253$$

The standard errors of the remainder forecasts become

$$\sigma_{R_{11}} = \left[517(0.279)\left(\frac{517 + 43}{43}\right)\right]^{1/2} = 43$$
$$\sigma_{R_{21}} = \left[517(0.233)\left(\frac{517 + 43}{43}\right)\right]^{1/2} = 40$$
$$\sigma_{R_{21}} = \left[517(0.488)\left(\frac{517 + 43}{43}\right)\right]^{1/2} = 57$$

The example continues in Table 13-8. In this example the forecast $\hat{X} = 560$ remains fixed for all time periods.

Table 13-8. Forecasts Using Percent of Aggregate Demands Method from Example 13.12

	Week t							
Items (k)	1	2	3	4	5	6	7	8
			Cumulative Sales					
1	12	35	64	85	108	137	152	160
2	10	23	43	85	133	159	173	170
3	21	52	83	112	137	165	181	193
Sums	43	110	190	282	378	461	506	523
			Total Season Forecasts					
1	156	178	189	169	160	166	168	–
2	130	117	127	169	197	193	191	–
3	274	265	244	222	203	201	201	–
Sums	560	560	560	560	560	560	560	–
			Remainder of Season Forecasts					
1	144	143	125	84	52	29	16	–
2	120	94	84	84	64	34	18	–
3	253	213	162	110	66	36	20	–
Sums	517	450	371	278	182	99	54	–
			Standard Error of Remainder Forecasts					
1	43	27	19	13	9	6	4	–
2	40	22	16	13	10	6	5	–
3	57	33	22	15	10	7	5	–

MATHEMATICAL BASIS. The probability distribution of R_{kt} is assumed Poisson, where

$$P(R_{kt} \mid \lambda_k R_t) = \frac{e^{-\lambda_k R_t}(\lambda_k R_t)^{R_{kt}}}{R_{kt}!}$$

and

$$\lambda_k \text{ is the probability of a sale for item } k$$

Now when λ_k is gamma-distributed,

$$f(\lambda_k \mid X_{kt}, X_t) = \frac{e^{-\lambda_k X_t}(\lambda_k X_t)^{X_{kt}-1} X_t}{(X_{kt} - 1)!}$$

and

$$P(R_{kt} \mid X_{kt} X_t R_t) = \int_0^\infty P(R_{kt} \mid \lambda_k R_t) f(\lambda_k \mid X_{kt}, X_t)\, d\lambda_k$$

$$= \frac{(R_{kt} + X_{kt} - 1)!}{R_{kt}!(X_{kt} - 1)!} \left(\frac{R_t}{R_t + X_t}\right)^{R_{kt}} \left(\frac{X_t}{R_t + X_t}\right)^{X_{kt}}$$

Since the result is the negative binomial distribution, then

$$E(R_{kt}) = R_t \left(\frac{X_{kt}}{X_t}\right)$$

$$V(R_{kt}) = R_t \left(\frac{X_{kt}}{X_t}\right)\left(\frac{R_t + X_t}{X_t}\right)$$

PROBLEMS

13-1. Consider the 24 demand entries in Table 4-6.

a. Rank the entries from low to high and estimate the probability and standard deviation corresponding to each entry.
b. Find x where the approximate probability of a demand less than x is closest to 0.90. Give the approximate 95% confidence interval for this probability.
c. Find the approximate 50% confidence limits of the demand.
d. What is the approximate probability that the demand will be greater than or equal to 100? What is the corresponding 95% confidence interval?

13-2. Using the results of Problem 13-1, find:

a. The approximate probability of a demand less than 110.
b. The approximate probability of a demand above 114.
c. The approximate probability of the demand to fall between 80 and 120.
d. The approximate value of x where the probability of the demand less than x is 0.95.

13-3. Using Table 4-6, consider the five intervals: $x \leq 80$, $80 < x \leq 90$, $90 < x \leq 100$, $100 < x \leq 120$, and $120 < x$.

a. As of $t = 24$, find estimates of the probability for each interval.

b. Suppose (at $t = 25$) that $x_{25} = 98$ and vector smoothing with $\alpha = 0.20$ is in use. Find the updated probabilities corresponding to each of the five intervals.

c. Continue this process assuming that the subsequent demand entries are $x_{26} = 126$, $x_{27} = 85$, and $x_{28} = 103$.

13-4. Consider a car model with the following options pertaining to the engine and radio:

Engine Options	*Radio Options*
(E_1) V6 engine	(R_1) No radio
(E_2) V8 engine	(R_2) AM
	(R_3) AM–FM

Suppose that probability estimates of the demand for each option are the following:

$$P(E_1) = 0.30 \quad \text{and} \quad P(E_2) = 0.70$$

$$P(R_1) = 0.20, \quad P(R_2) = 0.50, \quad \text{and} \quad P(R_3) = 0.30$$

and the radio and engine options are independent.

a. Find the one-quarter-ahead forecast of cars with options (E_1 and R_1) when the one-quarter-ahead forecast for all cars of the model is 10,000 units.

b. Do part a for all the option combinations.

c. Suppose that during the first quarter, the following sales were reported by combination of options:

Combination	*Sales*
E_1R_1	525
E_1R_2	1850
E_1R_3	1190
E_2R_1	1270
E_2R_2	4225
E_2R_3	2640

Using vector smoothing with $\alpha = 0.20$, find updated estimates of the probabilities for each of the options on engines alone, and on the radios alone.

d. Using the results of part c, find the updated forecasts for each combination of options. Use 12,000 as the total forecast.

13-5. Suppose that a line consists of three items A, B, and C and vector smoothing will be used to assist in generating forecasts for each item. At the outset ($t = 0$), the probabilities associated with the demand for each item are estimated as $P_{A0} = 0.2$, $P_{B0} = 0.5$, and $P_{C0} = 0.3$. The τ-period-ahead forecast for the line is 100 for $\tau = 1$ and 120 for $\tau = 2$.

a. Find the corresponding forecasts for each item as of $t = 0$.

b. If during the first period ($t = 1$), the demands for each item are $x_{A1} = 38$, $x_{B1} = 67$, and $x_{C1} = 25$, find the update probability estimates corresponding to each item when the smoothing parameter is updated by $\alpha = 0.25$.

c. Using the results of part b, give the forecasts for each item with $\tau = 1$ and 2 when the forecasts for the line are updated and now are 130 for $\tau = 1$ and 150 for $\tau = 2$.

13-6. A line has four items A, B, C, and D, where the average demand per period and the corresponding standard deviations for each item are:

$$\bar{x}_A = 20 \qquad S_A = 5$$
$$\bar{x}_B = 40 \qquad S_B = 10$$
$$\bar{x}_C = 10 \qquad S_C = 4$$
$$\bar{x}_D = 50 \qquad S_D = 12$$

Now the one-period-ahead forecast for the line is 140 and the associated standard deviation of this forecast is 30. Using the blending method, find the one-period-ahead forecasts for each of the items.

13-7. Suppose that a line consists of three models (A, B, C) and is sold in two locations (a, b). The average demands and standard deviation corresponding to each model and location are the following:

	Average	*Standard Deviation*
Aa	10	3
Ab	20	5
Ba	15	6
Bb	20	7
Ca	35	8
Cb	20	4

The one-period-ahead forecasts and associated standard deviations for each model and location are the following:

	Forecasts	*Standard Deviations*
A	40	10
B	30	8
C	60	20
a	50	15
b	80	25

a. Using the blending method, find the one-period-ahead forecasts for each model and location combination.
b. Using the results of part a, what are the one-period-ahead forecasts for each model regardless of location and for each location regardless of the model?

13-8. Suppose that an item is stocked in four locations *A*, *B*, *C*, and *D* and has a selling season of 10 weeks. Assume that the sales for the last season by location and week are as follows:

	Week									
Location	1	2	3	4	5	6	7	8	9	10
A	3	8	10	9	15	20	25	15	8	2
B	6	10	15	20	22	40	30	20	15	7
C	1	2	3	5	4	6	7	4	1	0
D	10	20	30	25	38	38	50	40	21	10

a. Find the weighted expected-percent-done estimates for the item by weeks.
b. Find the expected-percent-done estimates for the item by weeks.

13-9. Consider the results of Problem 13-8. Assume that during the current season, the item is now stocked in three locations *A*, *B*, and *D*.
a. If the first week's sales by location are 4, 10, and 15 for *A*, *B*, and *D*, respectively, find the total season forecasts by location using the expected-percent-done method.
b. Continue with part a, assuming the sales by location progress as shown below:

	Week									
Location	1	2	3	4	5	6	7	8	9	10
A	4	10	13	15	20	28	25	20	10	8
B	10	22	30	35	40	50	54	42	24	12
D	15	28	40	50	61	70	67	48	25	16

13-10. Consider the item in Problem 13-9, which is stocked in locations *A*, *B*, and *D*. For each location find the total season forecasts, the remainder of season forecasts, and the standard error of the remainder forecasts by time period when the total season forecasts vary and are those shown below.

Week	*Total Season Forecast*	*Week*	*Total Season Forecast*
1	1000	6	960
2	1000	7	930
3	980	8	930
4	980	9	900
5	960	10	900

REFERENCES

[1] BROWN, R. G., *Smoothing, Forecasting and Prediction of Discrete Time Series.* Englewood Cliffs, N.J.: Prentice-Hall, Inc., 1962, pp. 242–247, 253–256.

[2] BROWN, R. G., *Smoothing, Forecasting and Prediction of Discrete Time Series.* Englewood Cliffs, N.J.: Prentice-Hall, Inc., 1962, pp. 199–206.

[3] COHEN, G. D., "Bayesian Adjustment of Sales Forecasts in Multi-item Inventory Control Systems." *Journal of Industrial Engineering*, Vol. 17, No. 9, 1966, pp. 474–479.

[4] HARTUNG, P., "A Simple Style Goods Inventory Model." *Management Science*, Vol. 19, No. 2, August 1973, pp. 1452–1458.

[5] HERTZ, D. B., AND K. H. SCHAFFIR, "A Forecasting Model for Management of Seasonal Style Goods Inventories." *Operations Research*, Vol. 8, No. 2, 1960, pp. 45–52.

[6] HAUSMAN, W. H., AND R. S. G. SIDES, "Mail Order Demands for Style Goods: Theory and Data Analysis." *Management Science*, Vol. 20, No. 2, October 1973, pp. 191–202.

14

FORECAST ERRORS AND TRACKING SIGNALS

No matter what forecasting model is in use, forecasting errors will occur and should be taken into consideration in the overall planning of the inventory system. The forecast error is the difference between the forecast and the demand and is a highly useful observation in controlling both the forecasts and the inventory. Several measures of the error may be calculated and used for various purposes. One primary use is to gauge the accuracy of the forecasting system. Another is to track the flow of errors in order to monitor and control the system—seeking the best forecasting model and parameter combination. An extremely important use is to apply a measure of the error in setting the appropriate level of safety stock for each item. This is needed to provide an acceptable level of customer service. This latter area relating safety stocks and customer service is described in Chapter 15.

Two of the most common methods of measuring the forecast error are cited here. These are the standard deviation and the mean absolute deviation. A useful relation between these measures is given whereby one may be estimated from the other. In addition, a method is described which shows how a relation between the forecast errors and the average demands may be generated for the items in the inventory.

Two tracking signal methods are described. The tracking signal is used to monitor the forecasting system that is in use. It has the ability to warn the forecaster when the forecasts for an item are out of control. When such a signal is given, the forecaster may choose corrective action by changing the forecast model or the forecasting parameter or both. Such a warning will generally not be given as long as the forecasts are unbiased. An *unbiased* forecasting model is one where the forecasts are neither consistently low nor high with respect to the actual demands. Otherwise, the forecasts are *biased.*

14-1 THE MEAN ABSOLUTE DEVIATION AND STANDARD DEVIATION OF THE FORECAST ERROR

The one-period-ahead forecast error at time t is calculated by

$$e_t = x_t - \hat{x}_{t-1}(1)$$

where x_t is the demand at time t and $\hat{x}_{t-1}(1)$ is the one-period-ahead forecast that is generated at $t - 1$. For various forecasting and inventory control purposes, a measure of the magnitude of this error over the long run is needed. The two most common measures are the mean absolute deviation (MAD) and the standard deviation (σ_e). Generally, in an ongoing forecasting system, only one of these two measures needs to be updated regularly, since the one measure may be estimated from the other.

The estimates of MAD and σ_e are found by giving appropriate weights to the forecast errors of the past. A common way to assign the weights to past errors is to use the corresponding weights appropriated to the demand entries as they are used in the forecasts.

Certain forecast models, such as the moving-average and regression models, give equal weight to each of the N most current forecast errors. In this way the estimates of MAD and σ_e are

$$\text{MAD} = \frac{1}{N} \sum_{j=0}^{N-1} |e_{T-j}|$$ [1]

and

$$\sigma_e = \sqrt{\frac{1}{N-1} \sum_{j=0}^{N-1} e_{T-j}^2}$$

In the smoothing-type models with one smoothing or discounting parameter, the forecasts are generated by giving higher weights to the more current demand entries. In the same way, the estimates of MAD and σ_e are calculated by assigning correspondingly higher weights to the more current

[1] $|e_{T-j}|$ represents the absolute value of e_{T-j}.

forecast errors. In these situations the estimates at time T are

$$\text{MAD}_T = \alpha|e_T| + (1 - \alpha)\cdot\text{MAD}_{T-1}$$

$$\sigma_{e_T} = \sqrt{\alpha e_T^2 + (1 - \alpha)\sigma_{e_{T-1}}^2}$$

where MAD_{T-1} and $\sigma_{e_{T-1}}$ are the corresponding estimates at $T - 1$. For the single, double, or triple smoothing models, the smoothing parameter (α) above is the same as that used in forecasting. For adaptive smoothing or discounted regression models, the smoothing parameter is

$$\alpha = 1 - \beta$$

where β is the discounting parameter used in the forecast model.

The preceding scheme is also applied to those forecasting models which require two or more smoothing parameters. The smoothing parameter used to calculate MAD or σ_e is generally a fixed constant that is set by the forecaster for this purpose.

With σ_e and σ_{e_T} derived as shown above, σ_e^2 and $\sigma_{e_T}^2$ are used as estimates of the variance of the forecast error. It should be noted that these are true estimates of the variance when the forecasts are unbiased; otherwise, they are estimates of the mean square error.[2] In any respect, the measure reflects how far the actual demand deviates from the forecast and is the proper estimate to use in a forecasting system.

In most forecasting systems, the forecaster will maintain an updated estimate of MAD or σ_e but generally not both. This is because one can be estimated from the other by the relation

$$\text{MAD} = 0.8\sigma_e$$

or

$$\sigma_e = 1.25\text{MAD}$$

These relations are very good when the forecasting errors are normally distributed, and are reasonably close otherwise (see Brown [1]). It should be noted, however, that σ_e is best estimated directly rather than indirectly through MAD as shown above. Hence, if the forecaster really needs the estimate of σ_e for his/her use, he/she should not bother calculating MAD (as is often done), but should directly estimate σ_e as shown earlier.

---

[2]The mean square error and the variance are defined by

$$\text{MSE} = E(e^2) \quad \text{and} \quad \sigma^2 = E(e - \mu_e)^2$$

Where $E(\ )$ represents the expected value of the quantity within the parenthesis and μ_e is the mean of the error. When the forecasts are unbiased, then $\mu_e = 0$ and $\text{MSE} = \sigma^2$.

Example 14.1

Find estimates of MAD and σ_e for an item where the forecasts are generated by giving equal weight to the 10 most current demand entries. Suppose that the 10 most current forecast errors are: $-3, 4, 7, -2, -1, -6, 4, 3, -2$, and -1. The estimates become

$$\text{MAD} = \tfrac{1}{10}(|-3| + |4| + \ldots + |-2| + |-1|) = 3.3$$

and

$$\sigma_e = \sqrt{\tfrac{1}{9}[(-3)^2 + (4)^2 + \ldots + (-2)^2 + (-1)^2]} = 4.0$$

Note in this example that if MAD were used to estimate σ_e, then

$$\sigma_e = 1.25(3.3) = 4.1$$

Example 14.2

Consider a situation where MAD and σ_e are estimated using smoothing techniques with $\alpha = 0.1$. Now if $\text{MAD}_{T-1} = 3.5$, $\sigma_{e_{T-1}} = 4.4$, and $e_T = -6$, find the updated estimates.

These are

$$\text{MAD}_T = 0.1|-6| + 0.9(3.5) = 3.75$$

$$\sigma_{e_T} = \sqrt{0.1(-6)^2 + 0.9(4.4)^2} = 4.58$$

14-2 STANDARD DEVIATION OF THE FORECAST ERROR AS RELATED TO THE AVERAGE DEMAND

It is oftentimes helpful to have an easy method of estimating the value of σ_e for any item in the inventory. Holt et al. [2] show that a close relation between the standard deviation of demands (σ_x) and the average demands ($\bar{x}$) does exist in the form

$$\sigma_x = a\bar{x}^b$$

From a sample of 25 parts, their estimate yields

$$\sigma_x = 0.8\bar{x}^{0.75}$$

It is possible to use this approach to seek a relationship between σ_e and $\bar{x}$ of the form

$$\sigma_e = a\bar{x}^b$$

This relation can be exploited by the forecaster for all the items, or for a broad class of items in the inventory.

Such a relation may be beneficial to the forecaster in gaining a better grasp of the total safety stock requirements in the inventory. Also for new items, estimates of the safety stock and order point can be more properly established for the item using inventory control methods.

To illustrate the approach, 20 items from a real inventory were sampled and the average demands ($\bar{x}$) and corresponding standard deviations of forecast errors (σ_e) were recorded. The results are listed in Table 14-1 and are taken from a forecasting system that was run continuously for 24 months while using either single or double smoothing. A plot of the entries and a fit is given in Figure 14-1. The fit yields the relation

$$\sigma_e = 1.220\bar{x}^{0.748}$$

and is found by the least-squares method.[3]

The preceding relation should not be used universally and is only the result of a demonstration on how to proceed with this technique. In each inventory an analysis of the above type could be carried out to find the particular relation that best applies for the combination of items and forecasting system in use. Certainly, a larger sample is needed than the 20 used here.

Brown [3] has performed a study on 4674 items and reports that

$$\text{MAD} = \bar{x}^{0.79}$$

Table 14-1. Sample of Average Demand and Standard Deviation of Forecast Error for 20 Items

Item	$\bar{x}$	σ_e	*Item*	$\bar{x}$	σ_e
1	8.0	5.9	11	543.1	171.5
2	1.8	1.6	12	1164.2	277.1
3	100.9	23.6	13	261.7	66.8
4	626.6	162.1	14	12.4	9.9
5	242.2	73.2	15	25.3	13.9
6	8.3	5.3	16	342.6	115.9
7	4.1	7.6	17	18.4	10.9
8	21.4	18.0	18	1036.5	218.7
9	1593.2	324.5	19	7.2	3.9
10	23.6	6.5	20	1421.3	284.1

[3]The fit $\sigma_e = a\bar{x}^b$ is found by taking $Y = A + BX$, where $Y = \ln \sigma_e$, $A = \ln a$, $B = b$, and $X = \ln \bar{x}$. Upon finding the least-squares estimates for A and B, then $a = e^A$ and $b = B$.

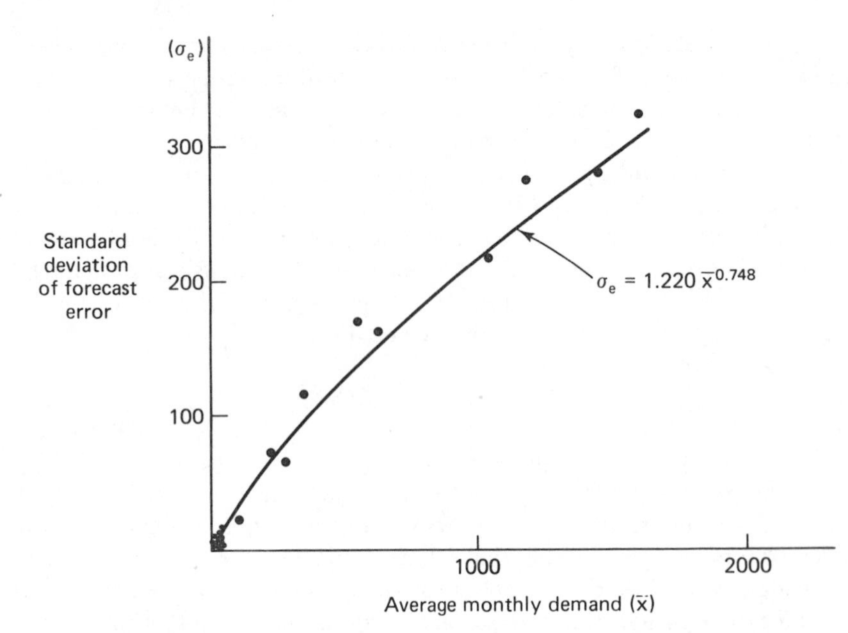

Figure 14-1. Sample results and fit of average demand and standard deviation of forecast error.

If it is assumed that $\sigma_e = 1.25$ MAD, his findings translate to

$$\sigma_e = 1.25\bar{x}^{0.79}$$

which is in close agreement to the sample of 20 items.

Example 14.3

Suppose that two new items are introduced to the inventory and estimates of the average monthly demands are $\bar{x}_1 = 30$ and $\bar{x}_2 = 300$, respectively. Find estimates of the standard deviation of the forecast error assuming that the relation $\sigma_e = 1.22\bar{x}^{0.748}$ is applicable.

The results give

$$\sigma_{e_1} = 1.22(30)^{0.748} = 15.53$$

$$\sigma_{e_2} = 1.22(300)^{0.748} = 86.95$$

14-3 STANDARD DEVIATION OF FORECAST ERROR OVER THE LEAD TIME

Throughout this book it is assumed that demand data become available at regular time intervals, whereupon the forecasts are updated and projected over the time periods of the future. At the same time, estimates of the standard deviation of the 1-month-ahead forecast errors are also maintained regularly. Generally, however, the forecast and standard deviation that is of most use to the forecaster covers the lead-time period for the item. The *lead time* represents the length of time beginning when a replenishment order is sent to the supplier and ending when the goods are received, stocked, and available for use.

The forecast for the lead-time period (L) is obtained from

$$\hat{X}_T(L) = \sum_{\tau=1}^{L} \hat{x}_T(\tau)$$

when L is an integer. For L a fraction,

$$\hat{X}_T(L) = \sum_{\tau=1}^{H} \hat{x}_T(\tau) + (L - H)\hat{x}_T(H + 1)$$

where H is the largest integer smaller than L.

When the forecast errors are independent, the standard deviation of the forecast error over the lead time (σ_{e_L}) is easily calculated. This is

$$\sigma_{e_L} = L^{0.5}\sigma_e$$

where σ_e is the one-period-ahead forecast error. This relation is seldom valid, however, since in most cases the errors are somewhat correlated.

Brown [3] has conducted a study and finds the relation

$$\sigma_{e_L} = (0.659 + 0.341L)\sigma_e$$

He reports that the approximation seems quite good for L ranging from 1 to 12 time periods.

The sample of 20 items (Table 14-1) was investigated to determine how the standard deviation over a lead time of two and three time periods is related to the one-period-ahead standard deviation. A listing of the standard deviations, sorted from low to high, is shown in Table 14-2. The average value of σ_{e_2}/σ_e is 1.51 and for σ_{e_3}/σ_e it is 1.82.

The results can be used to seek a value of b for the relation of the form

$$\sigma_{e_L} = L^b\sigma_e$$

where L is the lead time. Since

$$\sigma_{e_2} = 2^b\sigma_e = 1.51\sigma_e$$

then for $L = 2$,

$$2^b = 1.51$$

Table 14-2. SORTED VALUES OF $\sigma e_2/\sigma_e$ AND $\sigma e_3/\sigma_e$ FOR A SAMPLE OF 20 ITEMS

	$\sigma e_2/\sigma e$	$\sigma e_3/\sigma e$
	1.08	1.15
	1.14	1.23
	1.18	1.31
	1.19	1.38
	1.23	1.39
	1.26	1.43
	1.31	1.56
	1.33	1.57
	1.33	1.63
	1.39	1.78
	1.42	1.87
	1.51	1.89
	1.57	1.93
	1.63	1.94
	1.68	2.08
	1.75	2.27
	1.89	2.38
	1.93	2.47
	2.15	2.53
	2.28	2.70
Average	1.51	1.82

gives $b = 0.59$. For $L = 3$,

$$\sigma_{e_3} = 3^b\sigma_e = 1.82\sigma_e$$

or

$$3^b = 1.82$$

whereby $b = 0.55$. In this simple analysis it seems that

$$\sigma_{e_L} = L^{0.57}\sigma_e$$

is a fair relation for L limited to 2 and 3.

A plot of the results is shown in Figure 14-2. For the sample of 20 items, a wide scatter seems evident. Again, the results shown here should not be applied universally and are merely meant to demonstrate a procedure of seeking the appropriate relation for a given inventory. As before, a much broader sample is needed to gain statistical validity.

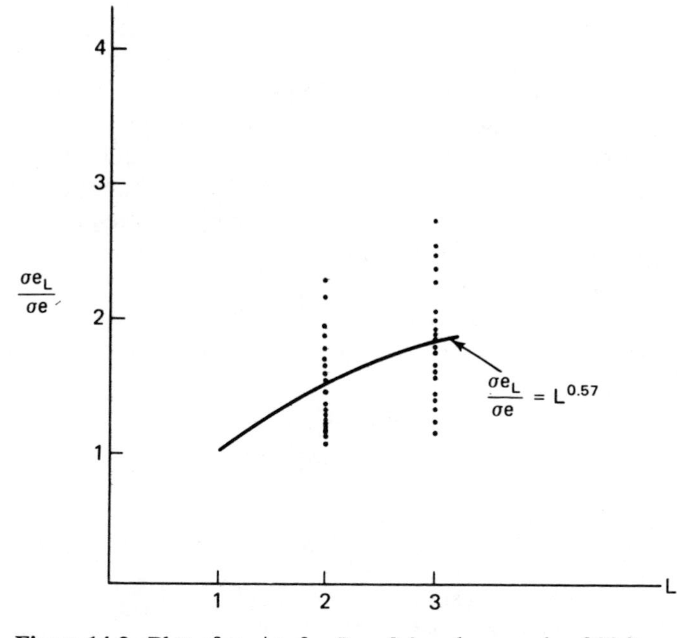

Figure 14-2. Plot of σ_{e_L}/σ_e for $L = 2.3$ and a sample of 20 items.

Example 14.4

Consider an item with a lead time of $L = 2.5$ and $\sigma_e = 10$. Using the relation $\sigma_{e_L} = L^{0.57}\sigma_e$, find an estimate for the lead-time standard deviation of the forecast error.

The result is

$$\sigma_{e_{2.5}} = (2.5)^{0.57}(10) = 16.86$$

Example 14.5

Suppose that an item has $L = 3$, $\hat{x}_T(\tau) = 20$ and $\sigma_e = 10$ for all τ. Find estimates of the average and standard deviation of the lead time demands using $\sigma_{e_L} = L^{0.57}\sigma_e$.

The average is

$$\hat{X}_T(3) = 3(20) = 60$$

and the standard deviation becomes

$$\sigma_{e_3} = 3^{0.57}(10) = 18.71$$

14-4 THE TRACKING SIGNAL

More and more companies today are using forecasting systems monitored by some type of tracking signal. The forecasting system may have a set of forecasting models and/or parameters available for use with each item. One combination of forecast model and/or parameter is selected for each item to generate the forecasts. The accuracy of the forecasts is measured at every time period through the use of a tracking signal. When the forecasts are unbiased, the tracking signal will lie within certain limits with a preset probability; otherwise (for biased forecasts), the signal will fall outside the limits.

When the tracking signal lies within the limits, the forecasting model and parameters are not changed and a new forecast is generated. If outside the limits, a change in the forecasting model and/or the parameters is made before generating the new forecasts.

A flow of how a typical system may operate is shown in Figure 14-3. The data available to the system at the current time period T are the following:

$$TS_{T-1} = \text{tracking signal as of } T - 1$$
$$\hat{x}_{T-1}(1) = \text{one-period-ahead forecast at } T - 1$$
$$x_T = \text{current demand}$$

Now the following steps are followed as shown in the diagram:

1. The current forecast error is measured.
2. The tracking signal is updated.
3. Low and high limits (L, H) are established to see if the current tracking signal lies within the limits. If within the limits, then steps 4, 5, and 6 are followed; otherwise, steps 7, 8, and 9 are used.
4. The current forecast model is used to generate new forecasts.
5. The current one-period-ahead forecast is saved for the next time period.
6. The current tracking signal is saved for the next time period.
   (This completes the steps when the tracking signal lies within the limits.)
7. A new forecast model and/or parameters are selected and the forecasts are generated.
8. The current one-period-ahead forecast is saved for the next time period.
9. The tracking signal is set to zero.
   (This completes the steps when the tracking signal lies outside the limits.)

14-5 TRACKING SIGNAL USING THE CUMULATIVE SUM OF FORECAST ERRORS

Brown [4] developed a tracking signal that is used along with single, double, or triple smoothing models. The tracking signal is set to zero at the outset and as time progresses, the signal gyrates relative to the flow of the forecast errors. For an unbiased forecasting model, the tracking signal is structured

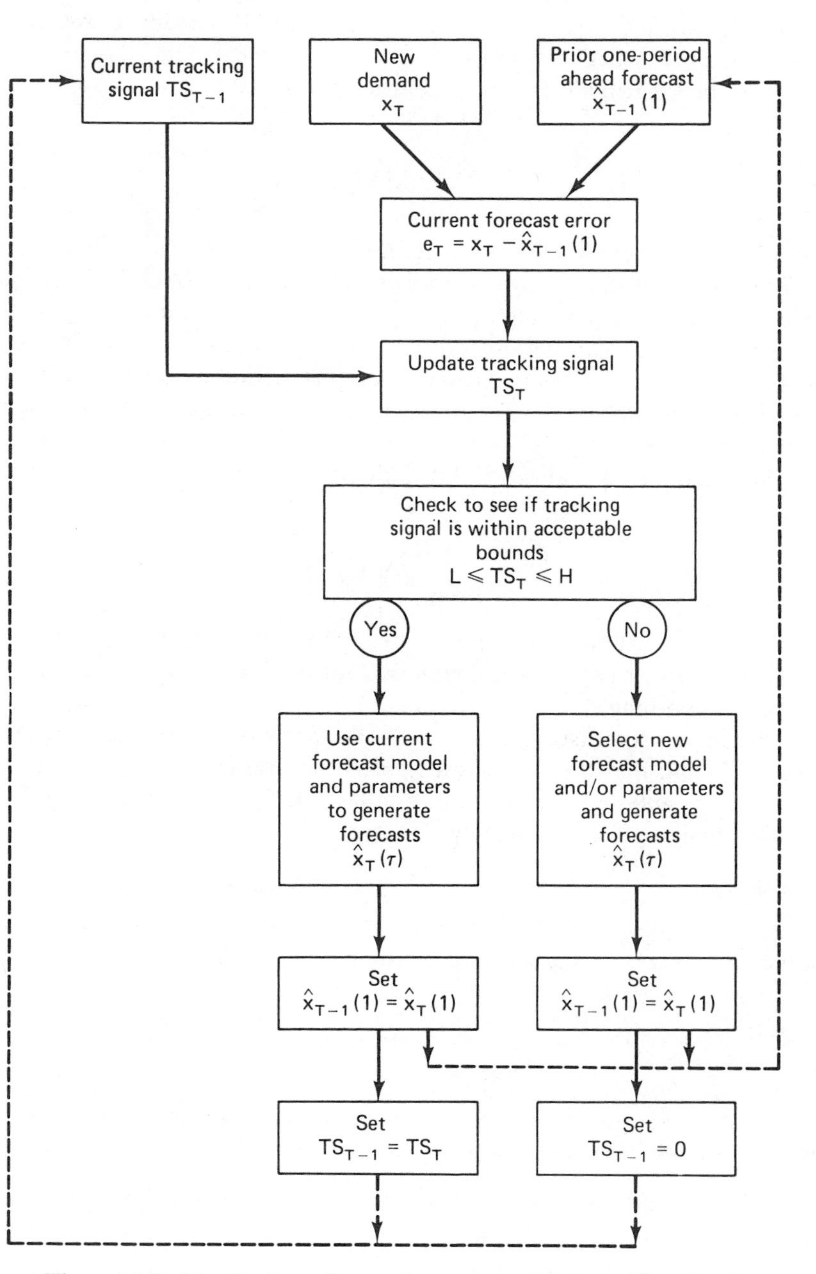

Figure 14-3. Monitoring a forecasting system with a tracking signal.

to fluctuate around zero, and for biased forecasts it steadily moves away from zero, either positive or negative.

The tracking signal at time T is measured by

$$TS_T = \frac{Y_T}{\text{MAD}_T}$$

where Y_T is the cumulative sum of forecast errors that have occurred since the forecast model and/or parameters were selected; and MAD_T is the current mean absolute deviation. It is Y_T that is set to zero whenever a change in the forecast model and/or parameters occurs. The mean absolute deviation (MAD_T) is not reset and is updated at each time period in the usual manner. In this way, should Y_T be set to zero, the tracking signal automatically becomes zero.

The standard deviation of this limit is

$$\sigma_{TS} \approx 0.884\sqrt{\frac{1+\beta}{1-\beta^{2n}}}$$

where $\beta = (1 - \alpha)$ and n represents the number of terms in the smoothing model ($n = 1$ for single smoothing, $n = 2$ for double smoothing, and $n = 3$ for triple smoothing).

When the forecasts are unbiased, the mean of the tracking signal is zero. Assuming that the signal is a random occurrence from a normal distribution, low and high limits can be established. The 95% limits are $\pm 2\sigma_{TS}$, and 99% limits are approximately $\pm 3\sigma_{TS}$.

Example 14.6

An example is shown in Table 14-3, where the single smoothing forecast model is in use with $\alpha = 0.1$. Here, the standard deviation of the tracking signal is

$$\sigma_{TS} = 0.884\sqrt{\frac{1.9}{1-0.81}} = 2.80$$

Assuming that 95% limits are needed, the limits become ± 5.60. In the table it is seen that the signal is within the limits until $t = 16$.

Mathematical Basis. When single smoothing is in use, the forecast error is

$$e_T = x_T - \hat{x}_{T-1}(1)$$

$$= x_T - \alpha \sum_{j=0}^{\infty} (1-\alpha)^j x_{T-j}$$

Table 14-3. WORKSHEET FOR EXAMPLE 14.6 USING THE TRACKING SIGNAL WITH CUMULATIVE SUMS

t	x_t	$\hat{x}_{t-1}(1)$	e_t	Y_t	MAD_t	TS_t
1	10	—	—	0.0	—	—
2	14	10.0	4.0	4.0	4.0	1.00
3	12	10.4	1.6	5.6	3.8	1.47
4	8	10.6	−2.6	3.0	3.6	0.83
5	16	10.3	5.7	8.7	3.8	2.29
6	15	10.9	4.1	12.8	3.9	3.28
7	9	11.3	−2.3	10.5	3.7	2.84
8	17	11.1	5.9	16.4	3.9	4.21
9	6	11.7	−5.7	10.7	4.1	2.61
10	8	11.1	−3.1	7.6	4.0	1.90
11	13	10.8	2.2	9.8	3.8	2.58
12	2	11.0	−9.0	0.8	4.3	0.19
13	6	10.1	−4.1	−3.3	4.3	−0.77
14	30	9.7	20.3	17.0	5.9	2.88
15	28	11.7	16.3	33.3	7.0	4.76
16	35	13.3	21.7	55.0	8.4	6.55

Now the sum of the errors is

$$\begin{aligned} Y_T &= \sum_{j=0}^{\infty} [x_{T-j} - \alpha \sum_{k=0}^{\infty} (1-\alpha)^k x_{T-j-k}] \\ &= \sum_{j=0}^{\infty} (1-\alpha)^j x_{T-j} \end{aligned}$$

Hence, the variance of Y_T becomes

$$\begin{aligned} \sigma_{Y_T}^2 &= \sum_{j=0}^{\infty} (1-\alpha)^{2j} \sigma_x^2 \\ &= \frac{\sigma_x^2}{1-\beta^2} \end{aligned}$$

where $\beta = 1 - \alpha$.

The corresponding results for double and triple smoothing are given by Brown [4] and are

$$\sigma_{Y_T}^2 \sim \frac{\sigma_x^2}{1-\beta^4} \qquad \text{(for double smoothing)}$$

$$\sigma_{Y_T}^2 \sim \frac{\sigma_x^2}{1-\beta^6} \qquad \text{(for triple smoothing)}$$

Hence, letting n designate the number of terms,

$$\sigma_{Y_T}^2 \sim \frac{\sigma_x^2}{1 - \beta^{2n}}$$

Using the relation between σ_e and σ_x, the variance of Y_T is found as it relates to σ_e. The single smoothing relation

$$\sigma_e^2 = \frac{2}{1 + \beta}\sigma_x^2$$

is used for $n = 1, 2,$ and 3. Hence,

$$\sigma_{Y_T}^2 \sim \frac{1 + \beta}{2(1 - \beta^{2n})}\sigma_e^2$$

Now since

$$\sigma_e = 1.25\text{MAD}_T$$

then

$$\sigma_{Y_T}^2 = \frac{1 + \beta}{2(1 - \beta^{2n})}(1.25\text{MAD}_T)^2$$

or

$$\sigma_{Y_T} = 0.884\sqrt{\frac{1 + \beta}{1 - \beta^{2n}}}\,\text{MAD}_T$$

Also, since $TS_T = Y_T/\text{MAD}_T$, then

$$\sigma_{TS} \sim \frac{\sigma_Y}{\text{MAD}_T} = 0.884\sqrt{\frac{1 + \beta}{1 - \beta^{2n}}}$$

14-6 TRACKING SIGNAL USING THE SMOOTHED AVERAGE FORECAST ERROR

Trigg [5] gives another tracking signal method which is based on the smoothed average of the forecast error. This system applies to the smoothing and adaptive smoothing models.

At the current time period (T), the smoothed error is updated by

$$E_T = \alpha e_T + (1 - \alpha)E_{T-1}$$

where α is the smoothing parameter that is in use in the forecast model. For adaptive smoothing models, $\alpha = 1 - \beta$. The tracking signal becomes

$$TS_T = \frac{E_T}{\text{MAD}_T}$$

where MAD_T is again the current mean absolute deviation and is updated in the standard manner using α. At the outset, E_T is set to zero as it also is with the selection of a new model and/or parameter.

The tracking signal above will never lie outside the range $(-1, 1)$. For unbiased forecasts, the tracking signal above should fluctuate near zero. Whereas, for biased forecasts, the tracking signal will approach one of the limits (-1 or $+1$).

Brown [8] shows that the standard deviation of this tracking signal is approximately

$$\sigma_{TS} = 0.55\sqrt{\alpha}$$

The relation above is primarily based on empirical results. The 95% limits on TS_T are $\pm 2\sigma_{TS}$, and 99% limits are approximately $\pm 3\sigma_{TS}$.

Johnson and Montgomery [7] point out that this tracking signal does appear to have an advantage over the method using the cumulative sum of errors. Suppose that at a particular time period there is a large error in the forecast but not one large enough to cause a trip to the tracking signal. Also, assume that several periods go by with small errors averaging to zero. The cumulative error (Y_t) will remain at its large value while the mean absolute deviation (MAD_t) will decrease toward zero. At some time in the future, if a second large random error with the same sign as the previous one occurs, then Y_t will increase and the tracking signal will incorrectly generate an out-of-control condition. With the same circumstances, the tracking signal with the smoothed error will not give a false signal.

Another example occurs when the cumulative forecast error-tracking signal is just below one of its limits. Now suppose that an almost perfect forecast comes in and reduces MAD_t but not Y_t. This circumstance could cause the tracking signal to falsely exceed the limit. On the other hand, using the smoothed average error (E_t), both E_t and MAD_t will drop when a good forecast comes in and a small change in the tracking signal will take place.

Example 14.7

An example is followed in Table 14-4. The item is the same as used in Example 14.4, where single smoothing with $\alpha = 0.1$ is applied. In this situation

$$\sigma_{TS} = 0.55\sqrt{0.1} = 0.174$$

Table 14-4. WORKSHEET FOR EXAMPLE 14.7 USING THE TRACKING SIGNAL WITH SMOOTHED ERRORS

t	x_t	$\hat{x}_{t-1}(1)$	e_t	E_t	MAD_t	TS_t
1	10	—	—	0.00		
2	14	10.0	4.0	0.40	4.00	0.100
3	12	10.4	1.6	0.52	3.76	0.138
4	8	10.6	−2.6	0.21	3.64	0.058
5	16	10.3	5.7	0.76	3.85	0.197
6	15	10.9	4.1	1.09	3.87	0.282
7	9	11.3	−2.3	0.75	3.72	0.202
8	17	11.1	5.9	1.27	3.94	0.322
9	6	11.7	−5.7	0.57	4.11	0.139
10	8	11.1	−3.1	0.21	4.01	0.052
11	13	10.8	2.2	0.41	3.83	0.107
12	2	11.0	−9.0	−0.53	4.35	−0.122
13	6	10.1	−4.1	−0.89	4.32	−0.206
14	30	9.7	20.3	1.23	5.92	0.208
15	28	11.7	16.3	2.74	6.96	0.394

and the 95% limits are ± 0.348. At $t = 1$, E_1 is set to zero, and at $t = 2$, MAD_2 is set to $e_2 = 4.0$. This is to avoid a false tracking signal trip at $t = 2$. The results show that the signal is within the limits until $t = 15$.

MATHEMATICAL BASIS. The smoothed error at time T is

$$E_T = \alpha e_T + (1 - \alpha)E_{T-1}$$

$$= \alpha \sum_{j=0}^{\infty} (1 - \alpha)^j e_{T-j}$$

For the single smoothing model, the variance of E_T is

$$\sigma_{E_T}^2 = \alpha^2 \left[\sum_{j=0}^{\infty} (1 - \alpha)^{2j} \sigma_{e_{T-j}}^2 + 2 \sum_{j=0}^{\infty} \sum_{k=j+1}^{\infty} (1 - \alpha)^{j+k} \sigma_{jk} \right]$$

where

$$\sigma_{jk} = \text{Cov}\,(e_{T-j}, e_{T-k})$$

Now

$$\sigma_{e_{T-j}}^2 = \left(\frac{2}{1 + \beta}\right) \sigma_x^2$$

$$\sigma_{j,j+l} = \left(\frac{-\alpha\beta^{l-1}}{1 + \beta}\right) \sigma_x^2$$

for $l = k - j$. Hence, with tedious calculations,

$$\sigma_{E_T}^2 = \frac{2\alpha}{(1+\beta)^3}\sigma_x^2$$

$$= \frac{2\alpha}{(1+\beta)^3}\left(\frac{1+\beta}{2}\right)\sigma_e^2$$

$$= \frac{\alpha}{(1+\beta)^2}(1.25)^2\text{MAD}_T^2$$

For β approximately 1,

$$\sigma_{E_T}^2 = (0.39)\alpha\text{MAD}_T^2$$

Now when

$$TS_T = \frac{E_T}{\text{MAD}_T}$$

then

$$\sigma_{TS} = 0.625\sqrt{\alpha}$$

which is close to the empirical result reported by Brown [8].

PROBLEMS

14-1. Consider Table 4-3, where the moving-average model with $N = 6$ is in use.

a. Calculate the mean absolute deviation of the one-period-ahead forecast error starting at $t = 7$ with $N = 1$. At $t = 8$, use $N = 2$ and continue in this fashion until $t = 12$, whereupon N will remain at 6. Continue until $t = 15$.

b. Do the same as in part a but now find the standard deviation of the forecast errors for $t = 7$ to 15.

14-2. Consider Table 4-6, where single smoothing is in use with $\alpha = 0.1$ and use only the entries for $t = 1$ to 12.

a. Calculate the mean absolute deviation of the one-period-ahead forecast error starting at $t = 2$. Use $\alpha = 1$ at $t = 2$ and use $\alpha = 0.1$ thereafter.

b. Do the same as in part a but now find the standard deviation of the forecast error.

c. From the results of part a, estimate the standard deviation for each time period and compare these with the results found in part b.

14-3. Consider Table 5-6, where double smoothing with $\alpha = 0.1$ is in use and consider only the entries for $t = 1$ to 12.

a. Calculate the mean absolute deviation of the one-period-ahead forecast error. Begin at $t = 2$ using $\text{MAD}_1 = 10$, which is an initial estimate set by the forecaster.

b. Do the same as in part a, but now find the standard deviation of the forecast errors. Use $\sigma_{e1} = 12$.

14-4. Consider Table 9-2, where the horizontal seasonal model is in use. Starting at $t = 26$, find the standard deviation of the one-period-ahead forecast error using $\alpha = 0.1$ and $\sigma_{e25} = 2$ as an initial estimate set by the forecaster.

14-5. Consider Table 8-4, where the four-term model is in use with $\beta = 0.974$. Find the standard deviation of the one-period-ahead forecast error for each time period starting at $t = 2$ and using $\sigma_{e1} = 10$ (an initial estimate set by the forecaster). Continue until $t = 12$.

14-6. Assume that three new items (A, B, C) are added to the inventory. Suppose that the estimates of the demands per period for these items are $\bar{x}_A = 2000$, $\bar{x}_B = 5$, and $\bar{x}_C = 60$.

a. Estimate the mean absolute deviation of the forecast error for each item, assuming that

$$\text{MAD} = \bar{x}^{0.80}$$

b. Now estimate the standard deviation of the forecast errors for each item.

14-7. Suppose that the policy of a company is to keep safety stock (SS) with each item, where the safety stock is

$$SS = (1.5)\sigma_e$$

Estimate the total investment in safety stock needed for the three items in Problem 14-6 when the cost per unit is \$0.10 for item A, \$200 for item B, and \$30 for item C.

14-8. Suppose that an item with a lead time of 4.5 months is forecast by the relation

$$\hat{x}_T(\tau) = 100 + 5\tau$$

where the standard deviation of the 1-month-ahead forecast error is $\sigma_e = 10$.

a. Find the lead-time forecast.

b. Estimate the lead-time standard deviation assuming that the forecast errors are independent.

c. Estimate the lead-time standard deviation using $\sigma_{eL} = L^{0.57}\sigma_e$.

d. Estimate the lead-time standard deviation using

$$\sigma_{E_L} = (0.659 + 0.341L)\sigma_e$$

14-9. Consider Table 5-2, where the double-moving-average model is in use.

a. List the one-period-ahead forecast errors for $t = 12$ to 24 and calculate the standard deviation of these errors.

b. List the two-period-ahead forecast errors and calculate the standard deviation of these errors for $t = 13$ to 24.

c. List the cumulative two-period ahead forecast errors for $t = 13$ to 24 and calculate the standard deviation of these errors.

d. Does any relation seem to hold with the results above?

14-10. Carry out the steps in Table 14-3, but now use $\alpha = 0.2$.

14-11. Consider Table 5-6, where double smoothing with $\alpha = 0.1$ is in use. Set up a worksheet that gives the tracking signal using the cumulative sum of forecast errors and follow the results for the 24 time periods.

14-12. Carry out the steps in Table 14-4, but now use $\alpha = 0.2$.

14-13. Redo Problem 14-11 using $\alpha = 0.1$ and the tracking signal that uses the smoothed average forecast error.

REFERENCES

[1] BROWN, R. G., *Smoothing, Forecasting and Prediction of Discrete Time Series.* Englewood Cliffs, N.J.: Prentice-Hall, Inc., 1962, pp. 271–287.

[2] HOLT, C. C., F. MODIGLIANI, J. F. MUTH, AND H. A. SIMON, *Planning Production, Inventories and Work Force.* Englewood Cliffs, N.J.: Prentice-Hall, Inc., 1960, Chap. 15.

[3] BROWN, R. G., *Decision Rules for Inventory Management.* New York: Holt, Rinehart and Winston, Inc., 1967, pp. 140–144.

[4] BROWN, R. G., *Smoothing, Forecasting and Prediction of Discrete Time Series.* Englewood Cliffs, N.J.: Prentice-Hall, Inc., 1962, pp. 287–290.

[5] TRIGG, D. W., "Monitoring a Forecasting System." *Operational Research Quarterly*, Vol. 15, 1964, pp. 271–274.

[6] BROWN, R. G., *Decision Rules for Inventory Management.* New York: Holt, Rinehart and Winston, Inc., 1967, pp. 152–155, 164–166.

[7] JOHNSON, L. A., AND D. C. MONTGOMERY, *Operations Research in Production Planning, Scheduling, and Inventory Control.* New York: John Wiley & Sons, Inc., 1974, p. 448.

15

CUSTOMER SERVICE AND SAFETY STOCK

One of the principal functions of forecasting is to estimate the future requirements for the items in the inventory so that management can plan to have available an adequate amount of stock to meet customer demands as they occur. Too much stock will yield the high customer service that is sought, but at the expense of high inventory costs, while too little stock will result in poor customer service and a potential loss of sales.

Customer service may be defined in many ways, depending on the function by which the inventory is used. Perhaps the most common interpretation of customer service is the percent of the total demand that is filled directly from the available stock. For the purposes of this chapter, this definition is used and the measure of this percentage is called the *service level.* Hence, the service level (SL) is

$$SL = \frac{\text{demand filled}}{\text{total demand}}$$

Generally a service level of 90 to 95% is deemed adequate by most managements, although it may be higher or lower for particular items, depending on their value or use. In order to achieve the service level sought, the stock level must be continually in review and replenished when it becomes dangerously low. A good inventory system is one that fulfills the desired service level at the minimum investment in inventory.

It is common to refer to the inventory for an item as the total stock (TS) which is partitioned into two components called the cycle stock (CS) and safety stock (SS), i.e.,

$$TS = CS + SS$$

The cycle stock is set at a level that is consistent with an economical order quantity. The safety stock is planned only in a precautionary manner and is an excess amount of stock that is available to meet higher than average demands.

Safety stock plays a key role in yielding the service level sought. It is the purpose of this chapter to explore the relationships between the safety stock and the service level for various situations. The chapter shows how these relations are used to find the proper level of safety stock for each item.

15-1 SAFETY STOCKS AND SERVICE LEVELS USING THE NORMAL DISTRIBUTION

Perhaps the most common relation between safety stocks and service levels in use today is when the demands are assumed to follow a normal probability distribution. Brown [1] shows that with this assumption, the safety stock for an item can be established in such a way where a preselected level of service is achieved in the long run.

The method is applicable for items that are subject to a continual review. When the on-hand (OH) plus on-order (OO) inventory drops to an order point (OP) level, an order is placed to the supplier for a specified order quantity (Q). The elapsed time until the order quantity is received is called the lead time (L). The demand over the lead time (x_L) is assumed to be normally distributed with a mean of $\hat{X}_L$ (the forecast for the lead time period) and a standard deviation of $\hat{\sigma}_L$ (the standard deviation of the forecast error over the lead time).

To illustrate some of the concepts, consider the plot in Figure 15-1, which shows the flow of inventory over time. At time t_1 the on-hand plus on-order ($OH + OO$) inventory reaches the order-point level, whereupon the quantity Q is ordered from the supplier. This quantity is not received until time t_2, which is a lead time beyond t_1, i.e., $t_2 = t_1 + L$. This process repeats itself as shown in the diagram, where the next order is placed at t_3 and the quantity is received at t_4. An *order cycle* represents the elapsed time between the receipts of two successive orders, such as between t_2 and t_4.

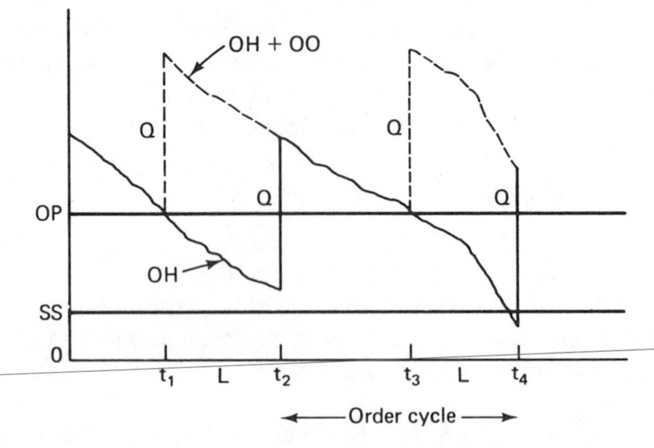

Figure 15-1. On-hand and on-order relations for a continuous review order-point system.

Using statistical methods, the safety stock and order-point values can be found to satisfy a particular service level. The safety stock becomes

$$SS = k\hat{\sigma}_L$$

where k is called a *safety factor* and the order point is

$$OP = \hat{X}_L + SS$$

The objective is to find the appropriate value of k that yields the specified service level (SL).

Using Q, $\hat{\sigma}_L$, and SL, it is possible to find the safety factor k. Consider the following reasoning: over an order cycle, the expected number of pieces short is

$$\begin{aligned} E(\text{pieces short}) &= E(x_L > OP) \\ &= \hat{\sigma}_L E(z > k) \end{aligned}$$

where $E(x_L > OP)$ represents the expected value of x_L beyond OP and $E(z > k)$ gives the expected value of z beyond k for a standard normal variable.[1] The service level becomes

$$\begin{aligned} SL &= 1 - \frac{E(\text{pieces short})}{Q} \\ &= 1 - \frac{\hat{\sigma}_L E(z > k)}{Q} \end{aligned}$$

[1]This quantity $E(z > k)$ is commonly called the *partial expectation* of z beyond k.

whereby

$$E(z > k) = \frac{(1 - SL)Q}{\hat{\sigma}_L}$$

Now with $E(z > k)$ known, a unique value of k can be found from Appendix B. Hence, the safety stock ($SS = k\hat{\sigma}_L$) and the order point ($OP = \hat{X}_L + SS$) are derived for the item.

It is noted that k could be less than zero, resulting in a negative safety stock. Although a negative safety stock has no physical meaning, the order point ($\hat{X}_L + SS$) can still be measured as a quantity larger than zero. The concept of a negative safety stock is confusing, so the usual practice is to set $k = 0$ whenever the safety factor is negative. In this case the order point is the same as the lead-time forecast ($\hat{X}_L$), and the service level that will result in the long run is higher than is otherwise planned.

Example 15.1

Consider an item where the lead-time forecast is $\hat{X}_L = 100$ and the associated standard deviation of the forecast error is $\hat{\sigma}_L = 60$. Also, assume that the order quantity is $Q = 200$ and that a 95% service level is desired. Find the safety stock and order point for this item.

Since

$$E(z > k) = \frac{(1 - 0.95)200}{60} = 0.167$$

then, using Appendix B, the safety factor is $k = 0.60$. Hence,

$$SS = 0.60(60) = 36$$

and

$$OP = 100 + 36 = 136$$

Note that the expected number of pieces short becomes

$$E(x_L > OP) = 60(0.167) = 10$$

Now since the average demand during an order cycle is $Q = 200$, the expected service level gives

$$SL = 1 - \frac{10}{200} = 0.95$$

which is the result requested at the outset.

Example 15.2

Table 15-1 lists the values of the safety stock for 10 items that are derived using a service level of 95%. For these items, the safety stock is set to zero whenever the calculated safety factor becomes less than zero.

Table 15-1. A 10 ITEM EXAMPLE GIVING SAFETY STOCKS WITH A 95% SERVICE LEVEL

Item	Q	$\hat{\sigma}_L$	k	SS
1	200	60	0.60	36
2	100	60	1.00	60
3	400	50	0.00	0
4	30	5	0.20	1
5	50	40	1.15	46
6	10	3	0.60	2
7	90	10	−0.10†	0
8	1000	250	0.50	125
9	20	2	−0.20†	0
10	150	100	1.00	100

†The safety stock is set to zero for items with negative values of k.

15-2 SAFETY STOCKS AND SERVICE LEVEL USING THE TRUNCATED NORMAL DISTRIBUTION

This section shows how safety stocks and service levels are related when the demand over the lead time is assumed to follow a truncated normal probability distribution. The fault with the normal distribution is that demands less than zero are allowed as a possibility and are taken into consideration in the calculations. The truncated normal can be structured so that only demands greater or equal to zero are permitted in the probability distribution.

The shape of the distribution depends on the relation between the standard deviation of forecast errors over the lead time (σ_L) and the average lead-time demand (μ_L). The ratio

$$c = \frac{\sigma_L}{\mu_L}$$

is called the *coefficient of variation* and is used to identify the shape of the truncated normal distribution. When $c \leq 0.33$, the truncated normal looks much like the normal distribution, since the average (μ_L) is at least three times as large as the standard deviation. Figure 15-2 shows examples of the truncated normal distribution for various values of c. Note that the shape of the distribution is exactly the same as the normal distribution for the area where demands are greater or equal to zero.

The procedure on seeking the appropriate safety stocks and order points for the items in the inventory is very similar to the method shown in

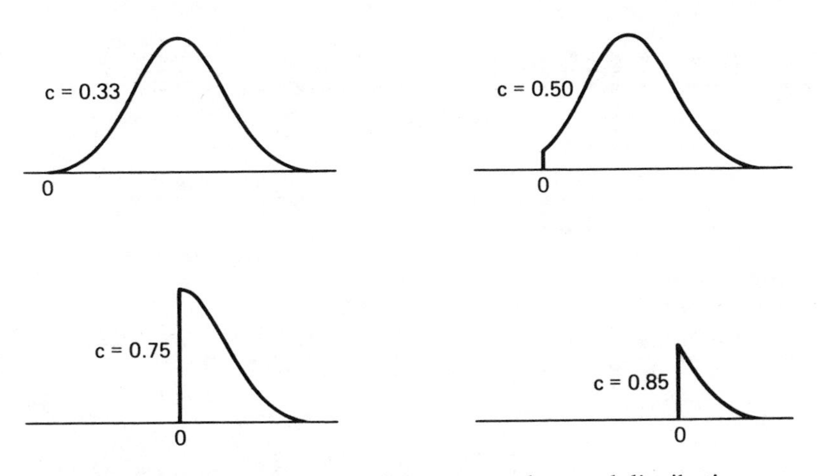

Figure 15-2. Some examples of the truncated normal distribution.

Section 15-1. As before, the data required are

$\hat{X}_L =$ lead-time forecast

$\hat{\sigma}_L =$ standard deviation of the forecast error over the lead time

$Q =$ order quantity

$SL =$ desired service level

The following four steps are now carried out. First, the quantity $E(w \geq w_o)$ is found by

$$E(w \geq w_o) = (1 - SL)\frac{Q}{\hat{\sigma}_L}$$

In this situation w represents a standardized truncated normal variable and $E(w \geq w_o)$ is the expected value of w greater than a preselected value w_o. Second, the ratio

$$c = \frac{\hat{\sigma}_L}{\hat{X}_L}$$

is calculated in order to estimate the shape of the truncated normal distribution. Using c and $E(w \geq w_o)$, the appropriate value of w_o is found by way of Table 15-2. In this table are values of $E(w \geq w_o)$ for eight representative truncated normal distributions (defined by $c = 0.33, 0.4, \ldots, 1.0$). Having found w_o, the third step is to calculate the safety stock using the relation

$$SS = w_o\hat{\sigma}_L$$

Table 15-2. VALUES OF $E(w \geq w_o)$ FOR THE TRUNCATED NORMAL PROBABILITY DISTRIBUTION WITH $c = 0.33$ TO 1.00

w_o \ c	.33	.40	.50	.60	.70	.80	.90	1.00
-3.0	3.0000	***	***	***	***	***	***	***
-2.9	2.9001	***	***	***	***	***	***	***
-2.8	2.8002	***	***	***	***	***	***	***
-2.7	2.7004	***	***	***	***	***	***	***
-2.6	2.6007	***	***	***	***	***	***	***
-2.5	2.5011	2.4999	***	***	***	***	***	***
-2.4	2.4017	2.4001	***	***	***	***	***	***
-2.3	2.3026	2.3004	***	***	***	***	***	***
-2.2	2.2037	2.2012	***	***	***	***	***	***
-2.1	2.1053	2.1023	***	***	***	***	***	***
-2.0	2.0072	2.0040	2.0016	***	***	***	***	***
-1.9	1.9098	1.9062	1.9020	***	***	***	***	***
-1.8	1.8130	1.8092	1.8035	***	***	***	***	***
-1.7	1.7170	1.7131	1.7061	***	***	***	***	***
-1.6	1.6220	1.6180	1.6100	1.5979	***	***	***	***
-1.5	1.5281	1.5242	1.5154	1.5005	***	***	***	***
-1.4	1.4356	1.4317	1.4224	1.4053	1.3966	***	***	***
-1.3	1.3445	1.3408	1.3314	1.3127	1.2996	***	***	***
-1.2	1.2552	1.2517	1.2425	1.2229	1.2065	1.2028	***	***
-1.1	1.1679	1.1647	1.1558	1.1359	1.1173	1.1085	1.0951	***
-1.0	1.0827	1.0799	1.0717	1.0520	1.0321	1.0196	0.9998	1.0013
-0.9	1.0000	0.9976	0.9902	0.9713	0.9510	0.9362	0.9119	0.8995
-0.8	0.9199	0.9180	0.9117	0.8939	0.8739	0.8579	0.8309	0.8157
-0.7	0.8428	0.8414	0.8363	0.8200	0.8010	0.7847	0.7564	0.7410
-0.6	0.7688	0.7679	0.7641	0.7497	0.7321	0.7164	0.6878	0.6665
-0.5	0.6981	0.6977	0.6954	0.6830	0.6673	0.6527	0.6248	0.5997
-0.4	0.6309	0.6311	0.6302	0.6200	0.6065	0.5935	0.5670	0.5408
-0.3	0.5674	0.5681	0.5686	0.5607	0.5497	0.5387	0.5141	0.4907
-0.2	0.5077	0.5089	0.5108	0.5052	0.4967	0.4879	0.4656	0.4429
-0.1	0.4518	0.4535	0.4568	0.4535	0.4475	0.4410	0.4212	0.3993

Table 15-2. *(continued)*

0.0	0.4000	0.4020	0.4065	0.4054	0.4019	0.3977	0.3807	0.3541
0.1	0.3521	0.3545	0.3600	0.3610	0.3599	0.3580	0.3437	0.3182
0.2	0.3081	0.3108	0.3173	0.3201	0.3212	0.3216	0.3100	0.2867
0.3	0.2680	0.2709	0.2782	0.2826	0.2858	0.2883	0.2793	0.2552
0.4	0.2317	0.2348	0.2426	0.2484	0.2535	0.2579	0.2514	0.2276
0.5	0.1991	0.2023	0.2104	0.2175	0.2241	0.2302	0.2260	0.2083
0.6	0.1699	0.1731	0.1816	0.1895	0.1974	0.2050	0.2030	0.1854
0.7	0.1441	0.1473	0.1558	0.1644	0.1733	0.1822	0.1822	0.1640
0.8	0.1214	0.1245	0.1329	0.1419	0.1517	0.1616	0.1633	0.1493
0.9	0.1015	0.1045	0.1127	0.1220	0.1323	0.1430	0.1462	0.1301
1.0	0.0844	0.0872	0.0950	0.1043	0.1150	0.1262	0.1308	0.1207
1.1	0.0696	0.0722	0.0796	0.0888	0.0996	0.1112	0.1168	0.1017
1.2	0.0570	0.0594	0.0663	0.0752	0.0859	0.0977	0.1043	0.0965
1.3	0.0463	0.0485	0.0549	0.0634	0.0739	0.0857	0.0930	0.0860
1.4	0.0374	0.0394	0.0452	0.0532	0.0633	0.0750	0.0828	0.0769
1.5	0.0299	0.0317	0.0369	0.0444	0.0540	0.0655	0.0736	0.0673
1.6	0.0238	0.0253	0.0300	0.0368	0.0460	0.0570	0.0655	0.0558
1.7	0.0188	0.0201	0.0242	0.0304	0.0389	0.0496	0.0581	0.0497
1.8	0.0147	0.0158	0.0194	0.0250	0.0329	0.0430	0.0515	0.0477
1.9	0.0114	0.0124	0.0155	0.0204	0.0277	0.0372	0.0456	0.0396
2.0	0.0088	0.0096	0.0122	0.0166	0.0232	0.0321	0.0404	0.0332
2.1	0.0067	0.0074	0.0096	0.0134	0.0193	0.0276	0.0357	0.0276
2.2	0.0051	0.0056	0.0075	0.0108	0.0161	0.0237	0.0315	0.0217
2.3	0.0038	0.0043	0.0058	0.0086	0.0133	0.0203	0.0278	0.0240
2.4	0.0028	0.0032	0.0045	0.0069	0.0110	0.0174	0.0245	0.0151
2.5	0.0021	0.0024	0.0034	0.0054	0.0090	0.0148	0.0215	0.0129
2.6	0.0015	0.0018	0.0026	0.0043	0.0074	0.0126	0.0190	0.0167
2.7	0.0011	0.0013	0.0020	0.0033	0.0060	0.0107	0.0166	0.0167
2.8	0.0008	0.0009	0.0015	0.0026	0.0049	0.0091	0.0146	0.0122
2.9	0.0006	0.0007	0.0011	0.0020	0.0040	0.0076	0.0128	0.0122
3.0	0.0004	0.0005	0.0008	0.0016	0.0032	0.0064	0.0112	0.0066

As in Section 15-1, if w_o is less than zero, the inventory manager may elect to set $w_o = 0$ ($SS = 0$). Now in the fourth step, the order point is calculated as before, i.e.,

$$OP = \hat{X}_L + SS$$

Example 15.3

Consider an item where $\hat{X}_L = 100$, $\hat{\sigma}_L = 60$, $Q = 200$, and $SL = 0.95$. Find the safety stock and order point for the item using the truncated normal distribution.

First,

$$E(w \geq w_o) = (1 - 0.95)\frac{200}{60} = 0.167$$

Now since

$$c = \frac{60}{100} = 0.60$$

then, using Table 15-2, w_o is approximately equal to 0.70. This gives

$$SS = 0.70(60) = 42$$

and

$$OP = 100 + 42 = 142$$

Example 15.4

The safety stocks with 95% service levels using the truncated normal probability distribution are calculated for the 10 items listed in Table 15-1 (Example 15.2). The results are given in Table 15-3. For comparative purposes, the safety stocks (SS_n) using the normal distribution are also listed.

Table 15-3. Worksheet to Find Safety Stocks Using the Truncated Normal with 95% Service Level

Item	$\hat{X}_L$	Q	$\hat{\sigma}_L$	c	$E(w \geq w_o)$	w_o	SS	SS_n
1	100	200	60	0.60	0.167	0.7	42	36
2	150	100	60	0.40	0.083	1.1	66	60
3	100	400	50	0.50	0.400	0.0	0	0
4	20	30	5	0.25	0.300	0.2	1	1
5	50	50	40	0.80	0.063	1.5	60	46
6	5	10	3	0.60	0.167	0.7	2	2
7	25	90	10	0.40	0.450	−0.1†	0	0
8	625	1000	250	0.40	0.200	0.5	125	125
9	8	20	2	0.25	0.500	−0.2†	0	0
10	333	150	100	0.30	0.075	1.0	100	100

†The safety stock is set to zero for negative values of w_o.

Mathematical Basis. Let $t = z - k$ for $z \geq k$, where z is the standard normal variate and where the probability of t is zero when $t < 0$. Hence, the probability density of t is

$$g(t) = \frac{f(z)}{H(k)}$$

where

$$H(k) = \int_k^\infty f(z)\,dz$$

Now

$$E(t)_k = \int_0^\infty t g(t)\,dt = \frac{1}{H(k)} \int_k^\infty (z - k) f(z)\,dz$$

$$= \frac{1}{H(k)}[f(k) - kH(k)]$$

$$E(t^2)_k = \int_0^\infty t^2 g(t)\,dt = \frac{1}{H(k)} \int_k^\infty (z - k)^2 f(z)\,dz$$

$$= \frac{1}{H(k)}[H(k)(1 + k^2) - kf(k)]$$

$$V(t)_k = E(t^2)_k - E(t)_k^2$$

For a particular value of k, there exists a unique coefficient of variation

$$c = \frac{\sigma_k}{\mu_k}$$

where $\sigma_k^2 = V(t)_k$ and $\mu_k = E(t)_k$.

For a given c, the unique value of k is found, and for $t_0 > 0$,

$$E(t > t_0)_k = \int_{t_0}^\infty (t - t_0) g(t)\,dt$$

$$= \frac{1}{H(k)} \int_{z_0}^\infty (z - z_0) f(z)\,dz$$

$$= \frac{E(z > z_0)}{H(k)}$$

where $z_0 = t_0 + k$.

For $t = \mu_k + w\sigma_k$

$$E(t > t_0)_k = E(\mu_k + w\sigma_k > \mu_k + w_0\sigma_k)_k = E(w > w_0)_k \sigma_k$$

or

$$E(w > w_0)_k = E(t > t_0)_k/\sigma_k$$
$$= \frac{E(z > z_0)}{H(k)\sigma_k}$$

with $z_0 = \mu_k + w_0\sigma_k + k$.

In Table 15-2, the entries for a given c are $E(w > w_o) = E(w > w_o)_k$, where k is found using Appendix B.

15-3 SETTING SAFETY STOCKS FOR A GROUP OF ITEMS USING AN EXCHANGE CURVE

A common need in industry is to control at a group level the total investment in safety stock (for the group) and the overall service level (for the group). The group may consist of all items in a line, a stocking location, or an entire division of the company. The objective is to spread the safety stock to all items in the group such that the minimum investment in safety stock (for the group) is needed to yield a desired service level for the group in total.

Brown [2] shows how the concept of an exchange curve can be employed to find the proper setting of safety stocks for each item in the group. Using exchange curves, the inventory management selects a policy value λ to control the amount and spread of safety stock. The value of λ is initially sought by an iterative search procedure for a particular group. Once found, the value generally does not have to be changed for several time periods to come. A description of this procedure will be given shortly.

The method is developed assuming that the probability distribution of demands is normal, as in Section 15-1. This is for convenience, however, and the method is equally applicable when the truncated normal is in use.

Suppose that the group consists of N items and for the ith item the following information is known:

$\hat{X}_i$ = forecast of the demand for 1 year

$\hat{\sigma}_i$ = standard deviation of the forecast error over the lead time

Q_i = order quantity

v_i = dollar investment per piece

Now the average investment in safety stock for the group is

$$\$SS = \sum_{i=1}^{N} v_i k_i \hat{\sigma}_i$$

where k_i is the safety factor for item i. The service level for the group becomes

$$SL = 1 - \frac{\sum_{i=1}^{N} S_i}{\sum_{i=1}^{N} \hat{X}_i}$$

where

$$S_i = \text{expected number of pieces short over a year for item } i$$
$$= \frac{\hat{X}_i}{Q_i} \sigma_i E(z > k_i)$$

Note that $(\hat{X}_i/Q_i)$ is the number of order cycles per year for item i and that $\sigma_i E(z > k_i)$ gives the expected number of pieces short per cycle. Hence, S_i is an estimate of the number of pieces short for the item over the entire year.

Once a particular value of λ is selected for the group, the safety factors (k_i $i = 1, 2, \ldots, N$) can be calculated for each item in the group. This is by first finding

$$F(k_i) = 1 - \lambda \frac{v_i Q_i}{\hat{X}_i} \qquad i = 1, 2, \ldots, N$$

Then, with the use of Appendix B,[2] the associated values k_i ($i = 1, 2, \ldots, N$) are obtained. Thereupon the safety stock for item i is

$$SS_i = k_i \hat{\sigma}_i$$

Note that $F(k_i)$ represents the probability that the ith item will not be short. The associated values of k_i are obtained by

$$k_i = \begin{cases} 0 & \text{when } F(k_i) < 0.5 \\ 4 & \text{when } F(k_i) \geq 1.0 \\ \text{Table values of Appendix B} & \text{when } 0.5 \leq F(k_i) < 1.0 \end{cases}$$

The dollar investment in inventory for the group can now be derived. This is

$$\$SS = \sum_{i=1}^{N} v_i k_i \hat{\sigma}_i$$

Also the service level for the group is easily found. For this purpose the values

---

[2]In Appendix B, $H(k) = 1 - F(k)$ is given, so use $F(k) = 1 - H(k)$.

$E(z > k_i)$ are first derived for each item with use of Appendix B. Now the shortage per year for item i is found by

$$S_i = \hat{\sigma}_i E(z > k_i)\frac{\hat{X}_i}{Q_i}$$

Hence, the service level for the group is calculated as shown earlier, i.e.,

$$SL = 1 - \frac{\sum_{i=1}^{N} S_i}{\sum_{i=1}^{N} \hat{X}_i}$$

At this point the pair $\$SS$ and SL is known for the particular value of λ selected. Since it is not possible a priori to predict these values for a particular setting of λ, a search technique is used to find the value of λ that is satisfactory to the inventory management.

In this endeavor another value of λ is selected and the procedure is carried out again. A smaller setting of λ will yield a larger $\$SS$ and a larger SL. This process continues until an appropriate relation is formed between $\$SS$ and SL. Interpolation procedures can then be used to find the proper λ for the group.

Generally, once a λ is found, it is used for several time periods to come. Afterward, should the search be carried out again, the neighborhood of the best value is already established and the search will require fewer steps than is needed at the outset.

Example 15.5

Table 15-4 shows how the safety stock for each of ten items is found when $\lambda = 0.05$. Illustrating with item 1,

$$F(k_1) = 1 - (0.05)2\left(\frac{200}{400}\right) = 0.95$$

Using Appendix B, the corresponding safety factor becomes $k_1 = 1.65$ and $E(z > 1.65) = 0.021$. Hence, the safety stock is

$$SS_1 = 1.65(60) = 99$$

It is noted here that the investment in safety stock is

$$\$SS_1 = \$2(99) = \$198$$

Table 15-4. WORKSHEET TO FIND SAFETY STOCKS USING THE EXCHANGE CURVE WITH $\lambda = 0.05$ (EXAMPLE 15.5)

i	$\$v_i$	Q_i	$\hat{X}_i$	$\hat{\sigma}_i$	$F(k_i)$	k_i	$E(z > k_i)$	SS_i	$\$SS_i$	S_i
1	2	200	400	60	0.950	1.65	0.021	99	198	2.52
2	4	100	900	60	0.978	2.02	0.007	121	484	3.78
3	2	400	300	50	0.867	1.12	0.066	56	112	2.48
4	80	30	240	5	0.500	0.00	0.399	0	0	16.00
5	1	50	250	40	0.990	2.32	0.003	93	93	0.60
6	30	10	15	3	−9.000†	0.00	0.399	0	0	1.80
7	30	90	50	10	−1.700†	0.00	0.399	0	0	2.21
8	1	1000	625	250	0.920	1.40	0.037	350	350	5.78
9	200	20	48	2	−3.167†	0.00	0.399	0	0	1.90
10	3	150	300	100	0.925	1.44	0.033	144	432	6.60
Sums			3128						\$1669	43.67

†k_i is set to zero when $F(k_i) \leq 0.50$.

and the expected number of pieces short over the year becomes

$$S_1 = (0.021)(60)\frac{400}{200} = 2.52$$

For the entire group, the dollar investment in safety stock is

$$\$SS = \$1669$$

Now since the expected number of pieces short over the year is

$$\sum_{i=1}^{10} S_i = 43.67$$

and the associated expected demand is

$$\sum_{i=1}^{10} \hat{X}_i = 3128$$

the service level for the group becomes

$$SL = 1 - \frac{43.67}{3128} = 98.6\%$$

The reader should note that the service level of 98.6% is the highest possible for the group with $\$SS = \$1669$. That is, there is no other way of spreading \$1669 of safety stock to the 10 items and achieving a higher service level. Conversely, the $\$SS = \$1669$ is the minimum investment in safety stock that is possible to achieve a 98.6% service level.

Table 15-5. WORKSHEET TO FIND SAFETY STOCK FOR EXAMPLE 15.6 WITH $\lambda = 0.10$ AND $\lambda = 0.20$

	$\lambda = 0.10$						$\lambda = 0.20$					
i	$F(k_i)$	k_i	$E(z > k_i)$	SS_i	$\$SS_i$	S_i	$F(k_i)$	k_i	$E(z > k_i)$	SS_i	$\$SS_i$	S_i
1	0.900	1.28	0.047	77	$154	5.64	0.800	0.84	0.111	50	100	13.32
2	0.956	1.71	0.018	103	412	9.72	0.912	1.36	0.040	82	328	21.60
3	0.733	0.62	0.161	31	62	6.04	0.466†	0.00	0.399	0	0	14.96
4	0.000†	0.00	0.399	0	0	16.00	−1.000†	0.00	0.399	0	0	16.00
5	0.980	2.05	0.007	82	82	1.40	0.960	1.75	0.016	70	70	3.20
6	−19.000†	0.00	0.399	0	0	1.80	−39.000†	0.00	0.399	0	0	1.80
7	−4.400†	0.00	0.399	0	0	2.21	−9.800†	0.00	0.399	0	0	2.21
8	0.840	0.99	0.084	247	247	13.13	0.680	0.47	0.210	118	118	32.80
9	−7.333†	0.00	0.399	0	0	1.92	−15.667†	0.00	0.399	0	0	1.92
10	0.850	1.04	0.077	104	312	15.40	0.700	0.53	0.190	53	159	38.00
Sums					$1269	73.26					$775	145.81

†k_i is set to zero when $F(k_i) \leq 0.50$.

Example 15.6

Table 15-5 shows comparable results to Table 15-4 for $\lambda = 0.10$ and 0.20. When $\lambda = 0.10$,

$$\$SS = \$1269$$

and

$$SL = 1 - \frac{73.26}{3128} = 97.7\%$$

Also when $\lambda = 0.20$, then

$$\$SS = \$775$$

and

$$SL = 1 - \frac{145.81}{3128} = 95.3\%$$

A plot of the results for the three values of λ is given in Figure 15-3. This plot represents an *exchange curve* for the group of 10 items.

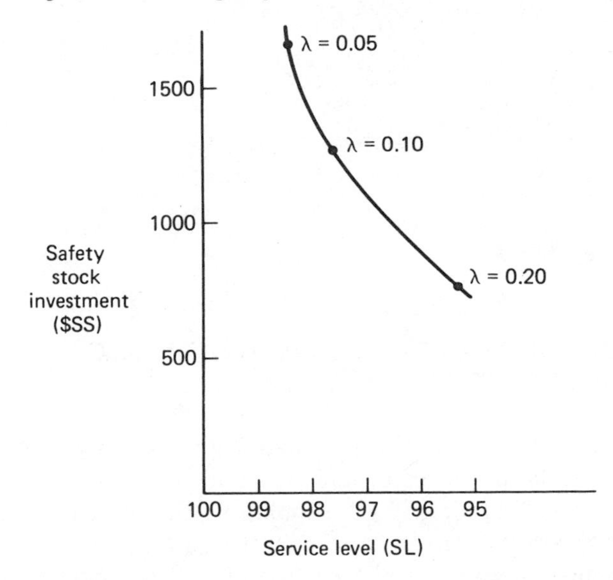

Figure 15-3. Exchange curve for Example 15.6.

If the inventory management desires a service level for the group of 97%, then with interpolation of the curve, the value of λ will be approximately $\lambda = 0.13$ and the dollar investment will be $\$SS = \$1100$. Should the desired service level be 95%, λ will be set slightly larger than 0.20 (perhaps $\lambda = 0.22$).

Now suppose that management is willing to invest $1000 in safety stock and wishes the highest service level possible with this investment. Interpolation shows that with $\$SS = \$1000$, then the service level attainable is about 96.5% and the policy variable to achieve these results will be approximately $\lambda = 0.15$.

Mathematical Basis. Recall that

$$\$SS = \sum_{i=1}^{N} k_i \sigma_i v_i$$

and

$$\sum_{i=1}^{N} S_i = \sum_{i=1}^{N} \sigma_i E(z > k_i) \frac{\hat{X}}{Q_i}$$

Now for $\$SS$ equal to a fixed amount A, the Lagrangian function with $\lambda > 0$ is

$$L(k_i, \ldots, k_N, \lambda) = \sum_{i=1}^{N} S_i - \lambda(\$SS - A)$$

The minimum total shortage is found when

$$\frac{\partial L}{\partial k_i} = \sigma_i[1 - F(k_i)]\frac{\hat{X}_i}{Q_i} - \lambda \sigma_i v_i = 0 \qquad i = 1, \ldots, N$$

Hence,

$$F(k_i) = 1 - \frac{\lambda v_i Q_i}{\hat{X}_i} \qquad i = 1, \ldots, N$$

In this respect λ is like a shadow price, and when λ increases, $\$SS$ decreases. At the extreme when $\lambda = 0$, $\$SS$ is infinity.

15-4 EFFECTS OF LEAD TIMES AND ORDER SIZES ON SAFETY STOCK

Some useful insights can be gained by studying the sensitivity of the safety stock with respect to the elements that have an influence on its size. Assuming that the conditions of Section 15-1 are applicable, the elements that influence the safety stock are the service level (SL), the order quantity (Q), and the standard deviation over the lead time (σ_L). This section shows how these elements are related to the safety stock for the particular case when the horizontal demand is applicable.

Recall that for a horizontal item, the average demand is the same for all time periods. Letting μ and σ represent the average demand and the standard deviation of the demand for each time period, the coefficient of variation for the item is the ratio of σ over μ, i.e.,

$$c = \frac{\sigma}{\mu}$$

Consider an item with a lead time of L time periods. The mean (μ_L) and standard deviation (σ_L) over the lead time become

$$\mu_L = L\mu$$
$$\sigma_L = \sqrt{L}\sigma$$

Note also that since $c = \sigma/\mu$, then $\sigma = c\mu$ and

$$\sigma_L = \sqrt{L}c\mu$$

Now the order quantity (Q), regardless of how it is derived, can be expressed by its relative size with respect to μ, i.e.,

$$Q = N\mu$$

Hence, N gives the number of time periods for which the order quantity will last.

The safety stock can be found for an item with conditions c, L, N, and μ. First, the partial expectation becomes

$$E(z > k) = \frac{(1 - SL)Q}{\sigma_L}$$
$$= \frac{(1 - SL)N\mu}{\sqrt{L}c\mu}$$
$$= \frac{(1 - SL)N}{\sqrt{L}c}$$

With use of Appendix B, the safety factor k is obtained, and then the safety stock is found from

$$SS = k\sigma_L$$
$$= k\sqrt{L}c\mu$$

Using the relations above, it is seen that the safety stock depends on μ, c, N, L, and SL.

Note that the safety stock is expressed in terms of μ, where

$k\sqrt{L}c$ = number of time periods of demand represented in the safety stock

Table 15-6 shows the number of time periods of safety stock for various conditions of c, L, N, and SL. With this table the inventory management can easily see how the elements influence the safety stock. SS increases with c, L, and SL, and SS decreases as N increases.

Table 15-6. TIME PERIODS OF SAFETY STOCK

SL= .900

C= .3

L \ N	1	2	3	4	5	6	7	8	9	10	11	12
1	.04	.00	.00	.00	.00	.00	.00	.00	.00	.00	.00	.00
2	.16	.00	.00	.00	.00	.00	.00	.00	.00	.00	.00	.00
3	.27	.02	.00	.00	.00	.00	.00	.00	.00	.00	.00	.00
4	.37	.08	.00	.00	.00	.00	.00	.00	.00	.00	.00	.00
5	.45	.15	.00	.00	.00	.00	.00	.00	.00	.00	.00	.00
6	.54	.21	.00	.00	.00	.00	.00	.00	.00	.00	.00	.00
7	.61	.27	.03	.00	.00	.00	.00	.00	.00	.00	.00	.00
8	.69	.32	.08	.00	.00	.00	.00	.00	.00	.00	.00	.00
9	.76	.38	.13	.00	.00	.00	.00	.00	.00	.00	.00	.00
10	.83	.44	.17	.00	.00	.00	.00	.00	.00	.00	.00	.00
11	.90	.49	.21	.00	.00	.00	.00	.00	.00	.00	.00	.00
12	.96	.54	.25	.03	.00	.00	.00	.00	.00	.00	.00	.00

C= .4

L \ N	1	2	3	4	5	6	7	8	9	10	11	12
1	.14	.00	.00	.00	.00	.00	.00	.00	.00	.00	.00	.00
2	.32	.05	.00	.00	.00	.00	.00	.00	.00	.00	.00	.00
3	.48	.17	.00	.00	.00	.00	.00	.00	.00	.00	.00	.00
4	.62	.27	.04	.00	.00	.00	.00	.00	.00	.00	.00	.00
5	.75	.38	.12	.00	.00	.00	.00	.00	.00	.00	.00	.00
6	.87	.47	.20	.00	.00	.00	.00	.00	.00	.00	.00	.00
7	.98	.56	.28	.04	.00	.00	.00	.00	.00	.00	.00	.00
8	1.10	.64	.34	.10	.00	.00	.00	.00	.00	.00	.00	.00
9	1.20	.73	.41	.17	.00	.00	.00	.00	.00	.00	.00	.00
10	1.30	.81	.48	.23	.01	.00	.00	.00	.00	.00	.00	.00
11	1.39	.89	.54	.28	.05	.00	.00	.00	.00	.00	.00	.00
12	1.48	.96	.61	.33	.11	.00	.00	.00	.00	.00	.00	.00

C= .5

L \ N	1	2	3	4	5	6	7	8	9	10	11	12
1	.24	.00	.00	.00	.00	.00	.00	.00	.00	.00	.00	.00
2	.50	.18	.00	.00	.00	.00	.00	.00	.00	.00	.00	.00
3	.71	.35	.10	.00	.00	.00	.00	.00	.00	.00	.00	.00
4	.90	.49	.22	.00	.00	.00	.00	.00	.00	.00	.00	.00
5	1.07	.63	.34	.10	.00	.00	.00	.00	.00	.00	.00	.00
6	1.24	.76	.44	.18	.00	.00	.00	.00	.00	.00	.00	.00
7	1.39	.89	.54	.28	.05	.00	.00	.00	.00	.00	.00	.00
8	1.53	1.00	.64	.37	.13	.00	.00	.00	.00	.00	.00	.00
9	1.66	1.11	.73	.45	.21	.00	.00	.00	.00	.00	.00	.00
10	1.80	1.22	.84	.54	.28	.06	.00	.00	.00	.00	.00	.00
11	1.92	1.33	.93	.61	.35	.13	.00	.00	.00	.00	.00	.00
12	2.06	1.42	1.00	.69	.42	.19	.00	.00	.00	.00	.00	.00

Table 15-6. (*continued*)

SL= .925

C= .3

L \ N	1	2	3	4	5	6	7	8	9	10	11	12
1	.10	.00	.00	.00	.00	.00	.00	.00	.00	.00	.00	.00
2	.24	.04	.00	.00	.00	.00	.00	.00	.00	.00	.00	.00
3	.36	.12	.00	.00	.00	.00	.00	.00	.00	.00	.00	.00
4	.47	.20	.03	.00	.00	.00	.00	.00	.00	.00	.00	.00
5	.56	.28	.09	.00	.00	.00	.00	.00	.00	.00	.00	.00
6	.65	.35	.15	.00	.00	.00	.00	.00	.00	.00	.00	.00
7	.74	.42	.21	.03	.00	.00	.00	.00	.00	.00	.00	.00
8	.82	.48	.25	.08	.00	.00	.00	.00	.00	.00	.00	.00
9	.90	.55	.31	.13	.00	.00	.00	.00	.00	.00	.00	.00
10	.98	.61	.36	.17	.01	.00	.00	.00	.00	.00	.00	.00
11	1.04	.67	.41	.21	.04	.00	.00	.00	.00	.00	.00	.00
12	1.11	.72	.46	.25	.08	.00	.00	.00	.00	.00	.00	.00

C= .4

L \ N	1	2	3	4	5	6	7	8	9	10	11	12
1	.21	.02	.00	.00	.00	.00	.00	.00	.00	.00	.00	.00
2	.42	.17	.00	.00	.00	.00	.00	.00	.00	.00	.00	.00
3	.60	.30	.11	.00	.00	.00	.00	.00	.00	.00	.00	.00
4	.75	.42	.21	.04	.00	.00	.00	.00	.00	.00	.00	.00
5	.89	.54	.30	.12	.00	.00	.00	.00	.00	.00	.00	.00
6	1.02	.65	.39	.20	.03	.00	.00	.00	.00	.00	.00	.00
7	1.14	.74	.48	.28	.10	.00	.00	.00	.00	.00	.00	.00
8	1.27	.84	.57	.34	.16	.00	.00	.00	.00	.00	.00	.00
9	1.38	.94	.64	.41	.23	.06	.00	.00	.00	.00	.00	.00
10	1.48	1.02	.72	.48	.29	.11	.00	.00	.00	.00	.00	.00
11	1.59	1.10	.80	.54	.34	.17	.01	.00	.00	.00	.00	.00
12	1.69	1.19	.86	.61	.40	.22	.06	.00	.00	.00	.00	.00

C= .5

L \ N	1	2	3	4	5	6	7	8	9	10	11	12
1	.33	.11	.00	.00	.00	.00	.00	.00	.00	.00	.00	.00
2	.62	.32	.12	.00	.00	.00	.00	.00	.00	.00	.00	.00
3	.85	.50	.28	.10	.00	.00	.00	.00	.00	.00	.00	.00
4	1.05	.67	.42	.22	.05	.00	.00	.00	.00	.00	.00	.00
5	1.24	.83	.55	.34	.15	.00	.00	.00	.00	.00	.00	.00
6	1.42	.97	.67	.44	.24	.07	.00	.00	.00	.00	.00	.00
7	1.57	1.10	.78	.54	.34	.16	.00	.00	.00	.00	.00	.00
8	1.74	1.23	.91	.64	.42	.24	.08	.00	.00	.00	.00	.00
9	1.89	1.35	1.00	.73	.51	.33	.15	.00	.00	.00	.00	.00
10	2.02	1.47	1.11	.84	.60	.40	.22	.06	.00	.00	.00	.00
11	2.16	1.59	1.21	.93	.68	.48	.30	.13	.00	.00	.00	.00
12	2.29	1.70	1.32	1.00	.76	.55	.36	.19	.03	.00	.00	.00

Table 15-6. *(continued)*

SL= .950

C= .3

L \ N	1	2	3	4	5	6	7	8	9	10	11	12
1	.18	.04	.00	.00	.00	.00	.00	.00	.00	.00	.00	.00
2	.34	.16	.04	.00	.00	.00	.00	.00	.00	.00	.00	.00
3	.48	.27	.12	.02	.00	.00	.00	.00	.00	.00	.00	.00
4	.60	.37	.20	.08	.00	.00	.00	.00	.00	.00	.00	.00
5	.71	.45	.28	.15	.03	.00	.00	.00	.00	.00	.00	.00
6	.81	.54	.35	.21	.09	.00	.00	.00	.00	.00	.00	.00
7	.90	.61	.42	.27	.14	.03	.00	.00	.00	.00	.00	.00
8	1.00	.69	.48	.32	.20	.08	.00	.00	.00	.00	.00	.00
9	1.08	.76	.55	.38	.24	.13	.02	.00	.00	.00	.00	.00
10	1.17	.83	.61	.44	.29	.17	.06	.00	.00	.00	.00	.00
11	1.24	.90	.67	.49	.34	.21	.10	.00	.00	.00	.00	.00
12	1.32	.96	.72	.54	.38	.25	.14	.03	.00	.00	.00	.00

C= .4

L \ N	1	2	3	4	5	6	7	8	9	10	11	12
1	.31	.14	.02	.00	.00	.00	.00	.00	.00	.00	.00	.00
2	.55	.32	.17	.05	.00	.00	.00	.00	.00	.00	.00	.00
3	.74	.48	.30	.17	.06	.00	.00	.00	.00	.00	.00	.00
4	.92	.62	.42	.27	.15	.04	.00	.00	.00	.00	.00	.00
5	1.07	.75	.54	.38	.24	.12	.02	.00	.00	.00	.00	.00
6	1.22	.87	.65	.47	.32	.20	.09	.00	.00	.00	.00	.00
7	1.35	.98	.74	.56	.40	.28	.15	.04	.00	.00	.00	.00
8	1.48	1.10	.84	.64	.49	.34	.21	.10	.00	.00	.00	.00
9	1.61	1.20	.94	.73	.56	.41	.29	.17	.06	.00	.00	.00
10	1.73	1.30	1.02	.81	.63	.48	.34	.23	.11	.01	.00	.00
11	1.84	1.39	1.10	.89	.70	.54	.41	.28	.17	.05	.00	.00
12	1.95	1.48	1.19	.96	.78	.61	.47	.33	.22	.11	.00	.00

C= .5

L \ N	1	2	3	4	5	6	7	8	9	10	11	12
1	.45	.24	.11	.00	.00	.00	.00	.00	.00	.00	.00	.00
2	.76	.50	.32	.18	.06	.00	.00	.00	.00	.00	.00	.00
3	1.03	.71	.50	.35	.21	.10	.00	.00	.00	.00	.00	.00
4	1.26	.90	.67	.49	.34	.22	.10	.00	.00	.00	.00	.00
5	1.46	1.07	.83	.63	.47	.34	.21	.10	.00	.00	.00	.00
6	1.65	1.24	.97	.76	.59	.44	.31	.18	.07	.00	.00	.00
7	1.84	1.39	1.10	.89	.70	.54	.41	.28	.16	.05	.00	.00
8	2.01	1.53	1.23	1.00	.81	.64	.49	.37	.24	.13	.03	.00
9	2.16	1.66	1.35	1.11	.91	.73	.58	.45	.33	.21	.10	.00
10	2.32	1.80	1.47	1.22	1.01	.84	.68	.54	.40	.28	.17	.06
11	2.47	1.92	1.59	1.33	1.11	.93	.76	.61	.48	.35	.23	.13
12	2.62	2.06	1.70	1.42	1.20	1.00	.85	.69	.55	.42	.29	.19

Table 15-6. *(continued)*

SL= .975

C= .3

L \ N	1	2	3	4	5	6	7	8	9	10	11	12
1	.30	.18	.10	.04	.00	.00	.00	.00	.00	.00	.00	.00
2	.50	.34	.24	.16	.10	.04	.00	.00	.00	.00	.00	.00
3	.66	.48	.36	.27	.19	.12	.07	.02	.00	.00	.00	.00
4	.80	.60	.47	.37	.28	.20	.14	.08	.03	.00	.00	.00
5	.93	.71	.56	.45	.36	.28	.21	.15	.09	.03	.00	.00
6	1.05	.81	.65	.54	.43	.35	.28	.21	.15	.09	.04	.00
7	1.17	.90	.74	.61	.51	.42	.34	.27	.21	.14	.09	.03
8	1.27	1.00	.82	.69	.58	.48	.40	.32	.25	.20	.14	.08
9	1.37	1.08	.90	.76	.65	.55	.46	.38	.31	.24	.18	.13
10	1.47	1.17	.98	.83	.71	.61	.51	.44	.36	.29	.23	.17
11	1.56	1.24	1.04	.90	.77	.67	.57	.49	.41	.34	.28	.21
12	1.65	1.32	1.11	.96	.83	.72	.62	.54	.4[illegible]	.38	.32	.25

C= .4

L \ N	1	2	3	4	5	6	7	8	9	10	11	12
1	.46	.31	.21	.14	.08	.02	.00	.00	.00	.00	.00	.00
2	.74	.55	.42	.32	.24	.17	.11	.05	.00	.00	.00	.00
3	.98	.74	.60	.48	.39	.30	.24	.17	.11	.06	.00	.00
4	1.18	.92	.75	.62	.52	.42	.34	.27	.21	.15	.10	.04
5	1.36	1.07	.89	.75	.64	.54	.46	.38	.30	.24	.18	.12
6	1.53	1.22	1.02	.87	.75	.65	.55	.47	.39	.32	.25	.20
7	1.68	1.35	1.14	.98	.86	.74	.65	.56	.48	.40	.34	.28
8	1.83	1.48	1.27	1.10	.96	.84	.74	.64	.57	.49	.41	.34
9	1.98	1.61	1.38	1.20	1.06	.94	.83	.73	.64	.56	.48	.41
10	2.11	1.73	1.48	1.30	1.15	1.02	.91	.81	.72	.63	.56	.48
11	2.24	1.84	1.59	1.39	1.23	1.10	.99	.89	.80	.70	.62	.54
12	2.37	1.95	1.69	1.48	1.33	1.19	1.07	.96	.86	.78	.69	.61

C= .5

L \ N	1	2	3	4	5	6	7	8	9	10	11	12
1	.63	.45	.33	.24	.17	.11	.05	.00	.00	.00	.00	.00
2	1.00	.76	.62	.50	.40	.32	.25	.18	.12	.06	.01	.00
3	1.31	1.03	.85	.71	.60	.50	.42	.35	.28	.21	.15	.10
4	1.57	1.26	1.05	.90	.78	.67	.58	.49	.42	.34	.28	.22
5	1.81	1.46	1.24	1.07	.94	.83	.73	.63	.55	.47	.40	.34
6	2.02	1.65	1.42	1.24	1.09	.97	.86	.76	.67	.59	.51	.44
7	2.24	1.84	1.57	1.39	1.23	1.10	.99	.89	.78	.70	.62	.54
8	2.42	2.01	1.74	1.53	1.37	1.23	1.10	1.00	.91	.81	.72	.64
9	2.61	2.16	1.89	1.66	1.50	1.35	1.23	1.11	1.00	.91	.82	.73
10	2.78	2.32	2.02	1.80	1.63	1.47	1.34	1.22	1.11	1.01	.92	.84
11	2.95	2.47	2.16	1.92	1.74	1.59	1.44	1.33	1.21	1.11	1.01	.93
12	3.12	2.62	2.29	2.06	1.85	1.70	1.56	1.42	1.32	1.20	1.11	1.00

It is also interesting to determine how the total stock varies with the conditions above. In this respect the total stock (TS) is composed of two components, cycle stock (CS) and safety stock (SS), where

$$TS = CS + SS$$

Now, since the average cycle stock is

$$CS = \frac{Q}{2} = \frac{N\mu}{2}$$

the average total stock becomes

$$TS = \frac{N\mu}{2} + k\sqrt{L}\,c\mu$$

When $N = 1$, $CS = \frac{1}{2}\mu$. For $N = 2$, $CS = \mu$; and so on.

Example 15.7

Consider an item where $\sigma = 10$, $\mu = 20$, $L = 4$, and $Q = 60$. Find the number of months of safety stock when the service level is 95% and show what percent this is of the total stock.

First note that

$$c = \frac{10}{20} = 0.5$$

and

$$N = \frac{60}{20} = 3$$

Using Table 15-6, with $SL = 0.95$ and $L = 4$,

$$SS = 0.67\mu = 0.67(20) = 13.4 = 14$$

or SS is 67% of 1 month's demand. Now since

$$CS = \frac{N}{2}\mu = \frac{3}{2}(20) = 30$$

and

$$TS = 30 + 14 = 44$$

the safety stock is 14/44 = 35% of the total stock.

15-5 SERVICE LEVELS FOR THE MULTIWAREHOUSE CASE

A common practice in industry is to stock an item in two or more warehouses (or depots, regions, or stores), where each warehouse is assigned its own safety stock. The question arises as to how much safety stock is needed to satisfy a desired national service level and how this safety stock differs from the one-warehouse case. Thomopoulos and Laakso [3] have studied this problem where a certain amount of cooperation is allowed between the warehouses. In this respect, when one warehouse runs out of stock, it may satisfy the excess customer demands, up to a certain amount, from another warehouse, which still has stock available. The results are based on a computer simulation model designed to produce reference tables that link safety factors for each warehouse to the national service level.

Two situations are studied. The first assumes that W warehouses are available and that the demands at each warehouse are independent from each other. In the second situation, two warehouses are available and the demands in each warehouse are dependent.

W Independent Warehouses

Table 15-7 shows results for the W-warehouse case when the demands at each warehouse are independent from each other. The results are for $c = 0.3$ and 0.5, where c represents the national coefficient of variation, i.e.,

$$c = \frac{\sigma}{\mu}$$

Here μ is the mean demand per order cycle for the item on a national level, σ is the lead-time standard deviation of the demand, and the demands are normally distributed. The table is generated for the situation where each warehouse is equally likely to incur a demand for the item. In this respect, when $W = 2$, the mean demand for warehouses 1 and 2 are

$$\mu_1 = \mu_2 = \frac{\mu}{2}$$

and

$$\sigma_1 = \sigma_2 = \frac{\sigma}{\sqrt{2}}$$

In this way,

$$\mu = \mu_1 + \mu_2$$

and

$$\sigma^2 = \sigma_1^2 + \sigma_2^2$$

Table 15-7. SERVICE LEVELS FOR THE MULTIWAREHOUSE PROBLEM

		$c = 0.3$ γ			$c = 0.5$ γ		
W	k	0	0.5	1.0	0	0.5	1.0
2	0	0.83	0.86	0.88	0.72	0.77	0.80
	0.5	0.91	0.94	0.95	0.86	0.91	0.92
	1.0	0.96	0.98	0.99	0.94	0.97	0.98
	1.5	0.99	1.00	1.00	0.98	0.99	1.00
4	0	0.76	0.83	0.87	0.60	0.72	0.79
	0.5	0.88	0.95	0.97	0.80	0.92	0.96
	1.0	0.95	0.99	1.00	0.91	0.99	1.00
	1.5	0.98	1.00	1.00	0.97	1.00	1.00
6	0	0.70	0.81	0.87	0.51	0.69	0.78
	0.5	0.85	0.95	0.98	0.75	0.93	0.97
	1.0	0.93	1.00	1.00	0.89	0.99	1.00
	1.5	0.98	1.00	1.00	0.96	1.00	1.00
8	0	0.66	0.80	0.87	0.43	0.66	0.79
	0.5	0.83	0.96	0.99	0.71	0.94	0.98
	1.0	0.92	1.00	1.00	0.87	1.00	1.00
	1.5	0.97	1.00	1.00	0.96	1.00	1.00
10	0	0.61	0.78	0.88	0.36	0.63	0.80
	0.5	0.80	0.97	0.99	0.67	0.95	0.99
	1.0	0.91	1.00	1.00	0.86	1.00	1.00
	1.5	0.97	1.00	1.00	0.95	1.00	1.00

In general, for the W warehouse case

$$\mu_i = \frac{1}{W}\mu \qquad \text{for } i = 1, 2, \ldots, W$$

and

$$\sigma_i = \frac{1}{\sqrt{W}}\sigma \qquad \text{for } i = 1, 2, \ldots, W$$

Hence,

$$\mu = \mu_1 + \mu_2 + \ldots + \mu_w$$

and

$$\sigma^2 = \sigma_1^2 + \sigma_2^2 + \ldots + \sigma_w^2$$

The parameter γ (gamma) represents a level of cooperation between the warehouses. When $\gamma = 0$, every warehouse acts independently from each other and there is no level of cooperation. So, if one warehouse is short, it cannot seek help from any other warehouse, even though a sufficient supply is available elsewhere.

When $\gamma = 1$, there is a full level of cooperation among the warehouses, and the inventory is shared by all warehouses. In this situation an out-of-stock condition can occur only when all warehouses are out of stock.

Now in the case when $\gamma = \frac{1}{2}$, a warehouse will release up to $\frac{1}{2}$ of its excess stock to help out a warehouse that has run out of stock. For example, if $W = 2$, the opening on-hand inventory (OH) at each location is $OH_1 = OH_2 = 20$, and the demands are $x_1 = 30$ and $x_2 = 8$, then

$$\text{warehouse 1 is short by } (x_1 - OH_1) = 10$$
$$\text{warehouse 2 has an excess of } (OH - x_2) = 12$$

and warehouse 2 will fill up to $\frac{1}{2}(12) = 6$ pieces of the excess demand at warehouse 1. Hence, the national service level is

$$\frac{\text{sum filled}}{\text{sum demand}} = \frac{20 + (6) + 8}{30 + 8} = \frac{34}{38} = 89.5\%$$

Another example is shown using $W = 4$, $OH_1 = OH_2 = OH_3 = OH_4 = 20$, and $x_1 = 30$, $x_2 = 8$, $x_3 = 18$, and $x_4 = 16$. In this situation

$$\text{warehouse 1 is short by } (x_1 - OH_1) = 10$$
$$\text{warehouse 2 has an excess of } (OH_2 - x_2) = 12$$
$$\text{warehouse 3 has an excess of } (OH_3 - x_3) = 2$$
$$\text{warehouse 4 has an excess of } (OH_4 - x_4) = 4$$

With $\gamma = \frac{1}{2}$, warehouse 2 fills 6 pieces of the excess demands at warehouse 1, and warehouses 3 and 4 fill 1 and 2 pieces, respectively. Hence, the national service level is

$$\frac{\text{sum filled}}{\text{sum demand}} = \frac{20 + (6 + 1 + 2) + 8 + 18 + 16}{30 + 8 + 18 + 16} = \frac{71}{72} = 98.6\%$$

Four safety factors ($k = 0, 0.5, 1.0,$ and 1.5) have been investigated. These are used to set the safety stock at each warehouse by the relation

$$SS_i = k\sigma_i \qquad i = 1, 2, \ldots, W.$$

With these safety stocks, the opening on-hand inventory at each location is

$$OH_i = \mu_i + SS_i \qquad i = 1, 2, \ldots, W$$

Table 15-7 reflects the results of measuring the service level after 500 simulated trials for each entry. Here it is seen, for example, that a higher safety factor is needed with $W = 10$ than with $W = 2$ to yield an equivalent national service level. The results also show that a lower safety factor is needed to maintain a particular service level when γ increases from 0 to 1.

Example 15.8

Consider an item that will be stocked in $W = 4$ locations and where the national mean and standard deviation of demands are $\mu = 200$ and $\sigma = 60$, respectively. Suppose that the expected demands at each location are equal and that the level of cooperation between the warehouses is $\gamma = \frac{1}{2}$. Find the safety stock that is required at each warehouse to yield a 95% national service level.

In this situation the mean and standard deviation of demands per warehouse are

$$\mu_i = \frac{1}{4}(200) = 50 \qquad i = 1, 2, 3, 4$$

$$\sigma_i = \frac{1}{\sqrt{4}}(60) = 30 \qquad i = 1, 2, 3, 4$$

Now using Table 15-7, with $c = 60/200 = 0.3$, $\gamma = \frac{1}{2}$, and $W = 4$, it is seen that when $k_i = 0.5$, the national service level is 95%. Hence, the safety stock per warehouse is

$$SS_i = 0.5\sigma_i$$
$$= 0.5(30) = 15 \qquad i = 1, 2, 3, 4$$

Two Dependent Warehouses

Table 15-8 shows comparable results to Table 15-7 for the two-warehouse case when the demands at each location are related with a correlation of $\rho = -0.5$, 0, and 0.5. Note that when $\rho = -0.5$, the demands at ware-

Table 15-8. SERVICE LEVELS FOR THE TWO-WAREHOUSE CASE

		$c = 0.3$			$c = 0.5$		
		γ			γ		
ρ	k	0	0.5	1.0	0	0.5	1.0
−0.5	0	0.76	0.83	0.88	0.60	0.72	0.79
	0.5	0.88	0.95	0.97	0.80	0.92	0.95
	1.0	0.95	0.99	1.00	0.92	0.99	1.00
	1.5	0.98	1.00	1.00	0.97	1.00	1.00
0	0	0.83	0.86	0.88	0.72	0.77	0.80
	0.5	0.91	0.94	0.95	0.86	0.91	0.92
	1.0	0.96	0.98	0.99	0.94	0.97	0.98
	1.5	0.99	1.00	1.00	0.98	0.99	1.00
0.5	0	0.86	0.87	0.88	0.77	0.79	0.80
	0.5	0.93	0.94	0.94	0.88	0.90	0.91
	1.0	0.97	0.98	0.98	0.95	0.96	0.96
	1.5	0.99	0.99	0.99	0.98	0.99	0.99

house 1 tend to lie above the mean (μ_1) for situations when the demands at warehouse 2 are below the mean (μ_2), and vice versa. In this respect the demands at the two warehouses are indirectly related to each other.

When $\rho = 0.5$, the demands are somewhat directly related. The demands at warehouse 1 will tend to rise and fall in relation to corresponding rises and falls for warehouse 2.

Table 15-8 shows results when $\rho = 0$ for comparative purposes. This situation occurs when the demands are independent and the table entries are the same as listed in Table 15-7 with $W = 2$.

In the two-warehouse case, the average demand per location is

$$\mu_1 = \mu_2 = \tfrac{1}{2}\mu$$

and the corresponding standard deviation is

$$\sigma_1 = \sigma_2 = \frac{1}{\sqrt{3}}\sigma \qquad \text{for } \rho = 0.5$$

and

$$\sigma_1 = \sigma_2 = \sigma \qquad \text{for } \rho = -0.5$$[3]

Example 15.9

Suppose that an item is stocked in two warehouses and the national mean and standard deviation of demands are $\mu = 100$ and $\sigma = 30$, respectively. Assuming that $\rho = 0.5$ and $\gamma = 0.5$, find the safety stock that is required to give a national service level of 94%.

The standard deviation at each warehouse becomes

$$\sigma_1 = \sigma_2 = \frac{1}{\sqrt{3}}30 = 17.3$$

and the safety factor using Table 15-8 is $k = 0.5$ (for $c = 0.3$, $\gamma = 0.5$, and $\rho = 0.5$). Hence, the safety stock at each warehouse becomes

$$SS_1 = SS_2 = 0.5(17.3) = 8.7$$

Since this rounds up to nine,

$$SS_1 = SS_2 = 9$$

---

[3]In general, $\sigma_1^2 + \sigma_2^2 + 2\rho\sigma_1\sigma_2 = \sigma^2$ and

$$\sigma_i = \sqrt{\frac{1/2}{1+\rho}}\sigma \qquad \text{for } i = 1, 2$$

and for the case when $\sigma_1 = \sigma_2$.

PROBLEMS

15-1. Suppose that the following information is known for an item whose demands are normally distributed:

$$\hat{x}_L = 20 \text{ (lead-time forecast)}$$

$$\hat{\sigma}_L = 12 \text{ (lead-time standard deviation)}$$

$$Q = 30 \text{ (order quantity)}$$

a. Find the safety stock and order point that correspond with a 96% service level. What is the expected number of pieces short over a cycle?
b. Find the safety stock and order point that correspond with a 93% service level. What is the expected number of pieces short per cycle?
c. In part a, what is the average total stock for the item?
d. If the safety stock is set at 18 pieces, what is the expected number of pieces short per cycle, and what is the expected service level?
e. If the safety factor is $k = 0.5$, what is the expected service level for the item?
f. What is the expected service level when no safety stock is carried?

15-2. Consider an item with the forecast equation of $\hat{x}_T(\tau) = 50 + 2\tau$ and where the one-period-ahead forecast error is $\sigma_e = 30$. Suppose that the lead time is four time periods and the order quantity is 250 pieces. Assuming independence and the normal distribution applies, find the safety stock needed to maintain a 98% service level. What is the expected number of pieces short per cycle? Find the probability that a shortage will occur.

15-3. Consider Table 15-1 and find the safety stocks that are needed when a 97% service level is desired. Use the normal probability distribution.

15-4. Consider Table 15-1 and find the safety stocks that are needed when a 92% service level is desired. Use the normal probability distribution.

15-5. Suppose that the lead-time forecast is $\hat{x}_L = 50$, the lead standard deviation is $\sigma_L = 35$, and the order quantity is $Q = 60$. Now using the truncated normal, find the following:

a. The safety stock that is needed to give a 95% service level.
b. The expected number of pieces short over a cycle.
c. The average total stock for the item.
d. Using the safety stock from part a, show what service level you would estimate when the normal distribution is assumed.
e. What safety stock would be calculated when a 95% service level is desired and the normal distribution is assumed?

15-6. Consider Table 15-3 and show what safety stocks are needed to yield a 97% service level. Use the truncated normal.

15-7. Suppose that an item has a horizontal demand where the forecast per month is 10 pieces and the corresponding standard deviation of the forecast error is 10 pieces. Also, assume that the order quantity is 50 pieces and that a service level of 95% is desired.

a. Find the safety stock that is needed when the lead time is $L = 2$ and the monthly forecast errors are assumed independent.
b. Redo part a, but let $L = 3$.

15-8. Suppose that the following information is known for an item:

$$v = \$10 \text{ (cost per unit)}$$

$$Q = 600 \text{ (order quantity)}$$

$$\hat{x} = 1800 \text{ (forecast of demand per year)}$$

$$\sigma_L = 100 \text{ (lead-time standard deviation of the forecast errors)}$$

Assume that the normal probability distribution is applicable and that safety stocks will be generated using an exchange curve with $\lambda = 0.05$. Find:
a. The safety stock that corresponds with $\lambda = 0.05$.
b. What is the probability that the item will not be out of stock?
c. What is the expected number of pieces short?
d. What is the expected service level?

15-9. Redo Problem 15-8 but now use $\lambda = 0.10$.

15-10. Consider Table 15-4 and find the total investment in safety stock and the service level for the group of items when $\lambda = 0.15$.

15-11. Using the results of Figure 15-3 and of Problem 15-10, find the following:
a. What safety stock investment is needed to maintain a service level for the group of 97%? What λ should be used in this situation?
b. If management wishes to invest approximately $800 in safety stock, what is the maximum service level they can expect? What λ should be used in this situation?
c. If $\lambda = 0.125$ is used, what investment in safety stock and what service level can be expected?

15-12. Consider Table 15-4 and find the service levels for each item.

15-13. Consider the items in Table 15-4. Find the total safety stock investment and service level for the group when all the safety factors (k) are set to 1 (i.e., $k_i = 1$ and $SS_i = \hat{\sigma}_i$).

15-14. Consider the results from Table 15-1. Find the total investment in safety stock and the service level for the group. Compare the results with those in Figure 15-3. (*Note:* See Table 15-4 for v and $\hat{X}$.)

15-15. Assume an item whose monthly demands are independent, normally distributed with $\mu = 10$ and $\sigma = 4$. Find the approximate safety stock and the average amount of total stock that is needed under the following conditions:
a. The order quantity (Q) $= 50$, the lead time (L) $= 4$ months, and the service level desired (SL) $= 95\%$.
b. $Q = 50$, $L = 4$, and $SL = 90\%$
c. $Q = 50$, $L = 5$, and $SL = 90\%$
d. $Q = 40$, $L = 4$, and $SL = 95\%$

15-16. An item with monthly demands normal, independent, $\mu = 100$, and $\sigma = 50$ has a lead time of 2 months.

a. To maintain a 97.5% service level, how much safety stock is needed when $Q = 400$?

b. What percent of the average total stock is the safety stock from part a?

c. Using the safety stock found in part a, find the approximate service level that would be expected if the order quantity was lowered from 400 to 300.

d. With an order quantity of 300, find the average total stock that will result when the desired service level is 97.5%. Compare this result to that when $Q = 400$.

15-17. Assume that an item is stocked in four locations and the national demand for the item over the order cycle has a mean of 40 and the lead time standard deviation is 20. Also, the demand is assumed normally distributed and the warehouse's demands are independent and each is equally likely to incur a demand for the item.

a. Find the mean and standard deviation of the demand at each warehouse.

b. Approximately how much safety stock should be available at each location to yield a national service level of 95%, when no cooperation is maintained among the warehouses (i.e., $\gamma = 0$)?

c. Using the safety stocks found in part b, what service level is obtained when the level of cooperation is increased to $\gamma = 0.5$?

d. If $\gamma = 0.5$, approximately how much safety stock is needed to maintain a 95% national service level?

e. If $\gamma = 1.0$, approximately how much safety stock is needed to maintain a 95% national service level?

f. If two of the warehouses are closed and the remaining two are equally likely to share the total national demand, approximately how much safety stock is needed in each warehouse to maintain a 95% national service level? Assume here that $\gamma = 0$.

g. If, instead, two more warehouses are added ($W = 6$) and each of the six warehouses is equally likely to share the national demand, approximately how much safety stock is needed per warehouse to maintain a 95% service level? Assume here that $\gamma = 0$.

15-18. Suppose that the monthly national demands for a part are independent, normally distributed, with $\mu = 19$ and $\sigma = 20$. When the order cycle is 3 months and the lead time is $L = 2$ months, $Q = 3(19) = 57$ and $\sigma_L = \sqrt{2}(20) = 28$. So for planning purposes, $\mu = 57$ and $\sigma = 28$ for the order-cycle period.

a. To maintain a 95% service level, approximately how much safety stock is needed?

b. Suppose that the item is located in two warehouses that are equally likely to share the demands. Assume also that each warehouse will be given equally the total stock for the item at the outset of the order cycle. Now using the safety stock found in part a, find the approximate national service level that can be expected when $\gamma = 0$.

c. Redo part b, but assume now that $\gamma = 0.5$.

15-19. Assume a part with demands normally distributed with $\mu = 100$ for the order cycle and $\sigma = 50$ for the lead time. Also, suppose the part is stocked in two warehouses that are equally likely to share the total demand.

a. If the demands per warehouse are related with a correlation of $\rho = 0.5$, what is the expected mean and standard deviation of demands per warehouse?

b. If instead, $\rho = -0.5$, what are μ_i and σ_i $(i = 1, 2)$?

c. If $\rho = 0.5$ and $\gamma = 0$, what safety stock is needed at each warehouse to maintain a 95% service level?

d. If $\rho = 0.5$ and $\gamma = 0.5$, what safety stock is needed at each warehouse to maintain a 95% service level?

REFERENCES

[1] Brown, R. G., *Smoothing, Forecasting and Prediction of Discrete Time Series.* Englewood Cliffs, N.J.: Prentice-Hall, Inc., 1962, pp. 370–372.

[2] Brown, R. G., *Materials Management Systems.* New York: John Wiley & Sons, Inc., 1977, Chap. 11.

[3] Thomopoulos, N. T., and C. Laakso, "Productivity in a Multi-Location Inventory System," *Manufacturing Productivity Frontiers,* Vol. 2, No. 2, Dec. 1978, pp. 23–25.

REFERENCES FOR FURTHER STUDY

Laakso, C. A., Safety Stocks and Service Levels for Multi-Location Inventory. Unpublished M.S. Thesis, Illinois Institute of Technology, Chicago, Illinois, 1978.

Silver, E. A., "A Modified Formula for Calculating Customer Service Under Continuous Inventory Review." *American Institute of Industrial Engineers Transactions,* Vol. 2, No. 3, Sept. 1970, pp. 241–245.

APPENDICES

A

SOME USEFUL RELATIONS

A-1 FINITE SUMS

$$\sum_{k=1}^{N} 1 = N$$

$$\sum_{k=1}^{N} k = \frac{N(N+1)}{2}$$

$$\sum_{k=1}^{N} k^2 = \frac{N(N+1)(2N+1)}{6}$$

$$\sum_{k=1}^{N} k^3 = \left(\sum k\right)^2 = \frac{N^2(N+1)^2}{4}$$

$$\sum_{k=1}^{N} k^4 = \frac{N}{30}(6N^4 + 15N^3 + 10N^2 - 1)$$

$$\sum_{k=0}^{N} x^k = \frac{1 - x^{N+1}}{1 - x} \qquad x \neq 1$$

$$\sum_{k=0}^{N} kx^k = \frac{x[1 - (N+1)x^N + Nx^{N+1}]}{(1-x)^2} \qquad x \neq 1$$

$$\sum_{k=0}^{N} k^2 x^k = \frac{x^{N+1}[(2N^2 + 2N - 1)x - N^2x^2 - (N + 1)^2] + x(1 + x)}{(1 - x)^3} \qquad x \neq 1$$

$$\sum_{k=1}^{N} \sin k\omega = \frac{\sin (N\omega/2) \sin [(N + 1)/2]\omega}{\sin \omega/2}$$

$$\sum_{k=1}^{N} \cos k\omega = \frac{\cos (N\omega/2) \sin [(N + 1)/2]\omega}{\sin \omega/2} - 1$$

$$\sum_{k=1}^{N} k \sin k\omega = \frac{\sin (N + 1)\omega}{4 \sin^2 \omega/2} - \frac{(N + 1) \cos [(2N + 1)/2]\omega}{2 \sin \omega/2}$$

$$\sum_{k=1}^{N} k \cos k\omega = \frac{(N + 1) \sin [(2N + 1)/2]\omega}{2 \sin \omega/2} - \frac{1 - \cos (N + 1)\omega}{4 \sin^2 \omega/2}$$

A-2 INFINITE SUMS[1]

$$\sum \beta^k = \frac{1}{1 - \beta}$$

$$\sum k\beta^k = \frac{\beta}{(1 - \beta)^2}$$

$$\sum k^2\beta^k = \frac{\beta(1 + \beta)}{(1 - \beta)^3}$$

$$\sum k^3\beta^k = \frac{\beta(1 + 4\beta + \beta^2)}{(1 - \beta)^4}$$

$$\sum k^4\beta^4 = \frac{\beta(1 + 11\beta + 11\beta^2 + \beta^3)}{(1 - \beta)^5}$$

$$\sum k^5\beta^k = \frac{\beta(1 + 26\beta + 66\beta^2 + 26\beta^3 + \beta^4)}{(1 - \beta)^6}$$

$$\sum k^6\beta^k = \frac{\beta(1 + 57\beta + 302\beta^2 + 302\beta^3 + 57\beta^4 + \beta^5)}{(1 - \beta)^7}$$

$$\sum \beta^k \sin \omega k = \frac{\beta \sin \omega}{1 - 2\beta \cos \omega + \beta^2}$$

$$\sum \beta^k \cos \omega k = \frac{1 - \beta \cos \omega}{1 - 2\beta \cos \omega + \beta^2}$$

$$\sum k\beta^k \sin \omega k = \frac{\beta(1 - \beta^2) \sin \omega}{(1 - 2\beta \cos \omega + \beta^2)^2}$$

$$\sum k\beta^k \cos \omega k = \frac{2\beta^2 - \beta(1 + \beta^2) \cos \omega}{(1 - 2\beta \cos \omega + \beta^2)^2}$$

---

[1]The summations range from k equal zero to infinity.

$\sum \beta^k \sin \omega_1 k \sin \omega_2 k$

$$= \frac{-1}{2}\left[\frac{1 - \beta \cos(\omega_1 + \omega_2)}{1 - 2\beta \cos(\omega_1 + \omega_2) + \beta^2} - \frac{1 - \beta \cos(\omega_1 - \omega_2)}{1 - 2\beta \cos(\omega_1 - \omega_2) + \beta^2}\right]$$

$\sum \beta^k \sin \omega_1 k \cos \omega_2 k$

$$= \frac{1}{2}\left[\frac{\beta \sin(\omega_1 + \omega_2)}{1 - 2\beta \cos(\omega_1 + \omega_2) + \beta^2} - \frac{\beta \sin(\omega_1 - \omega_2)}{1 - 2\beta \cos(\omega_1 - \omega_2) + \beta^2}\right]$$

$\sum \beta^k \cos \omega_1 k \cos \omega_2 k$

$$= \frac{1}{2}\left[\frac{1 - \beta \cos(\omega_1 + \omega_2)}{1 - 2\beta \cos(\omega_1 + \omega_2) + \beta^2} + \frac{1 - \beta \cos(\omega_1 - \omega_2)}{1 - 2\beta \cos(\omega_1 - \omega_2) + \beta^2}\right]$$

A-3 TRIGONOMETRIC RELATIONS

$\sin a \sin b = -\frac{1}{2}[\cos(a + b) - \cos(a - b)]$

$\cos a \cos b = \frac{1}{2}[\cos(a + b) + \cos(a - b)]$

$\cos a \sin b = \frac{1}{2}[\sin(a + b) - \sin(a - b)]$

$\sin a \cos b = \frac{1}{2}[\sin(a + b) + \sin(a - b)]$

B

SOME PROPERTIES OF THE NORMAL AND TRUNCATED NORMAL PROBABILITY DISTRIBUTIONS

k	$f(k)$	$H(k)$	$E(z > k)$	$\sigma(z > k)$	μ_k	σ_k	[illegible]
-3.0	.0044	.9987	3.0004	.9988	3.0044	.9933	.3
-2.9	.0060	.9981	2.9005	.9983	2.9060	.9913	.3
-2.8	.0079	.9974	2.8008	.9977	2.8079	.9888	.3
-2.7	.0104	.9965	2.7011	.9968	2.7105	.9857	.3
-2.6	.0136	.9953	2.6015	.9958	2.6136	.9820	.3
-2.5	.0175	.9938	2.5020	.9944	2.5176	.9775	.3
-2.4	.0224	.9918	2.4027	.9926	2.4226	.9723	.4
-2.3	.0283	.9893	2.3037	.9904	2.3286	.9661	.4
-2.2	.0355	.9861	2.2049	.9876	2.2360	.9589	.4
-2.1	.0440	.9821	2.1065	.9841	2.1448	.9508	.4
-2.0	.0540	.9772	2.0085	.9799	2.0552	.9415	.4
-1.9	.0656	.9713	1.9111	.9748	1.9676	.9312	.4
-1.8	.0790	.9641	1.8143	.9686	1.8819	.9197	.4
-1.7	.0940	.9554	1.7183	.9613	1.7984	.9072	.5
-1.6	.1109	.9452	1.6232	.9526	1.7174	.8936	.5
-1.5	.1295	.9332	1.5293	.9425	1.6388	.8789	.5
-1.4	.1497	.9192	1.4367	.9309	1.5629	.8634	.5
-1.3	.1714	.9032	1.3455	.9176	1.4897	.8470	.5

k	$f(k)$	$H(k)$	$E(z > k)$	$\sigma(z > k)$	μ_k	σ_k	c_k
-1.2	.1942	.8849	1.2561	.9025	1.4194	.8298	.5846
-1.1	.2179	.8643	1.1686	.8855	1.3520	.8119	.6005
-1.0	.2420	.8413	1.0833	.8667	1.2876	.7935	.6163
-.9	.2661	.8159	1.0004	.8459	1.2261	.7747	.6318
-.8	.2897	.7881	.9202	.8231	1.1676	.7555	.6471
-.7	.3123	.7580	.8429	.7985	1.1119	.7362	.6621
-.6	.3332	.7257	.7687	.7721	1.0591	.7167	.6767
-.5	.3521	.6915	.6978	.7439	1.0092	.6973	.6909
-.4	.3683	.6554	.6304	.7142	.9619	.6779	.7048
-.3	.3814	.6179	.5668	.6832	.9172	.6587	.7181
-.2	.3910	.5793	.5069	.6509	.8751	.6397	.7311
-.1	.3970	.5398	.4509	.6177	.8353	.6211	.7435
.0	.3989	.5000	.3989	.5838	.7979	.6028	.7555
.1	.3970	.4602	.3509	.5495	.7626	.5849	.7670
.2	.3910	.4207	.3069	.5150	.7294	.5675	.7780
.3	.3814	.3821	.2668	.4805	.6982	.5506	.7886
.4	.3683	.3446	.2304	.4464	.6688	.5341	.7986
.5	.3521	.3085	.1978	.4129	.6411	.5182	.8082
.6	.3332	.2743	.1687	.3803	.6150	.5027	.8174
.7	.3123	.2420	.1429	.3486	.5905	.4878	.8261
.8	.2897	.2119	.1202	.3182	.5674	.4734	.8344
.9	.2661	.1841	.1004	.2891	.5456	.4596	.8422
1.0	.2420	.1587	.0833	.2615	.5251	.4462	.8497
1.1	.2179	.1357	.0686	.2355	.5058	.4333	.8568
1.2	.1942	.1151	.0561	.2112	.4876	.4210	.8635
1.3	.1714	.0968	.0455	.1885	.4703	.4091	.8698
1.4	.1497	.0808	.0367	.1676	.4541	.3977	.8758
1.5	.1295	.0668	.0293	.1483	.4387	.3867	.8815
1.6	.1109	.0548	.0232	.1307	.4241	.3762	.8869
1.7	.0940	.0446	.0183	.1146	.4104	.3661	.8920
1.8	.0790	.0359	.0143	.1002	.3973	.3564	.8969
1.9	.0656	.0287	.0111	.0871	.3849	.3470	.9016
2.0	.0540	.0228	.0085	.0755	.3732	.3381	.9060
2.1	.0440	.0179	.0065	.0651	.3621	.3295	.9101
2.2	.0355	.0139	.0049	.0559	.3515	.3213	.9140
2.3	.0283	.0107	.0037	.0478	.3414	.3133	.9176
2.4	.0224	.0082	.0027	.0408	.3319	.3056	.9208
2.5	.0175	.0062	.0020	.0346	.3228	.2981	.9235
2.6	.0136	.0047	.0015	.0292	.3141	.2908	.9258
2.7	.0104	.0035	.0011	.0245	.3059	.2838	.9278
2.8	.0079	.0026	.0008	.0206	.2980	.2770	.9293
2.9	.0060	.0019	.0005	.0171	.2905	.2703	.9302
3.0	.0044	.0013	.0004	.0142	.2833	.2641	.9323
3.1	.0033	.0010	.0003	.0118	.2764	.2584	.9350
3.2	.0024	.0007	.0002	.0097	.2697	.2536	.9405
3.3	.0017	.0005	.0001	.0080	.2631	.2505	.9522
3.4	.0012	.0003	.0001	.0066	.2565	.2488	.9699
3.5	.0009	.0002	.0001	.0054	.2501	.2495	.9977
3.6	.0006	.0002	.0000	.0045	.2431	.2560	1.0529
3.7	.0004	.0001	.0000	.0037	.2363	.2645	1.1193
3.8	.0003	.0001	.0000	.0031	.2288	.2794	1.2209
3.9	.0002	.0000	.0000	.0026	.2198	.3076	1.3999
4.0	.0001	.0000	.0000	.0022	.2106	.3366	1.5982

$$f(k) = \frac{1}{\sqrt{2\pi}} e^{-k^2/2}$$

$$H(k) = \int_k^\infty f(z)\, dz$$

$$E(z > k) = \int_k^\infty (z - k) f(z)\, dz = f(k) - kH(k)$$

$$\sigma^2(z > k) = E(z > k)^2 - [E(z > k)]^2 \quad \text{where} \quad E(z > k)^2 = \int_k^\infty (z - k)^2 f(z)\, dz$$

$$= H(k)(1 + k^2) - kf(k)$$

$$\mu_k = \frac{1}{H(k)} \int_k^\infty (z - k) f(z)\, dz = \frac{1}{H(k)} [f(k) - kH(k)]$$

$$\sigma_k^2 = E(z^2)_k - \mu_k^2 \quad \text{where} \quad E(z^2)_k = \frac{1}{H(k)} [H(k)(1 + k^2) - kf(k)]$$

$$c_k = \frac{\sigma_k}{\mu_k}$$

C

STUDENT'S t DISTRIBUTION

The table entries give values of t_α where the probability of a value of t exceeding t_α is α.

Degrees of Freedom	*Value of* α					
	0.25	0.10	0.050	0.025	0.010	0.005
1	1.000	3.078	6.314	12.706	31.821	63.657
2	0.816	1.886	2.920	4.303	6.965	9.925
3	0.765	1.638	2.353	3.182	4.541	5.841
4	0.741	1.533	2.132	2.776	3.747	4.604
5	0.727	1.476	2.015	2.571	3.365	4.032
6	0.718	1.440	1.943	2.447	3.143	3.707
7	0.711	1.415	1.895	2.365	2.998	3.499
8	0.706	1.397	1.860	2.306	2.896	3.355
9	0.703	1.383	1.833	2.262	2.821	3.250
10	0.700	1.372	1.812	2.228	2.764	3.169
11	0.697	1.363	1.796	2.201	2.718	3.106
12	0.695	1.356	1.782	2.179	2.681	3.055
13	0.694	1.350	1.771	2.160	2.650	3.012

Degrees of Freedom	*Value of α*					
	0.25	0.10	0.050	0.025	0.010	0.005
14	0.692	1.345	1.761	2.145	2.624	2.977
15	0.691	1.341	1.753	2.131	2.602	2.947
16	0.690	1.337	1.746	2.120	2.583	2.921
17	0.689	1.333	1.740	2.110	2.567	2.898
18	0.688	1.330	1.734	2.101	2.552	2.878
19	0.688	1.328	1.729	2.093	2.539	2.861
20	0.687	1.325	1.725	2.086	2.528	2.845
21	0.686	1.323	1.721	2.080	2.518	2.831
22	0.686	1.321	1.717	2.074	2.508	2.819
23	0.685	1.319	1.714	2.069	2.500	2.807
24	0.685	1.318	1.711	2.064	2.492	2.797
25	0.684	1.316	1.708	2.060	2.485	2.787
26	0.684	1.315	1.706	2.056	2.479	2.779
27	0.684	1.314	1.703	2.052	2.473	2.771
28	0.683	1.313	1.701	2.048	2.467	2.760
29	0.683	1.311	1.699	2.045	2.462	2.756
30	0.683	1.310	1.697	2.042	2.457	2.750
∞	0.674	1.282	1.645	1.960	2.326	2.576

Source: Adapted from R. A. Fisher, *Statistical Methods for Research Workers*, 14th ed. (New York: Hafner Press, 1970), Table IV, by permission of the copyright holder of record, Mr. V. A. Edgeloe, Registrar, University of Adelaide, and the publisher. (© 1970 University of Adelaide.)

BIBLIOGRAPHY

AYRES, F., *Matrices*. New York: McGraw-Hill Book Company, 1960.

BAMBER, D. J., "A Versatile Family of Forecasting Systems." *Operational Research Quarterly*, Vol. 20, April 1969, pp. 111–121.

BATES, J. M., AND C. W. J. GRANGER, "Combination of Forecast." *Operational Research Quarterly*, Vol. 20, No. 4, 1969, pp. 451–468.

BEALS, R. E., *Statistics for Economists*. Chicago: Rand McNally, 1972.

BIEGLE, J. E., *Production Control*. Englewood Cliffs, N.J.: Prentice-Hall, Inc., 1971.

BOSSONS, J., "The Effects of Parameter Misspecification and Non-stationarity on the Applicability of Adaptive Forecasts." *Management Science*, Vol. 12, No. 9, 1966, pp. 659–669.

BOX, G. E. P., AND G. M. JENKINS, "Some Recent Advances in Forecasting and Control." *Applied Statistics*, Vol. 17, 1968, pp. 91–109.

BOX, G. E. P., AND G. M. JENKINS, *Time Series Analysis, Forecasting and Control*. San Francisco: Holden-Day, Inc., 1976.

BROWN, R. G., *Statistical Forecasting for Inventory Control*. New York: McGraw-Hill Book Company, 1959.

BROWN, R. G., *Smoothing, Forecasting and Prediction of Discrete Time Series*. Englewood Cliffs, N.J.: Prentice-Hall, Inc., 1962.

BROWN, R. G., *Decision Rules for Inventory Management.* New York: Holt, Rinehart and Winston, Inc., 1967.

BROWN, R. G., "Simulations to Explore Alternative Sequencing Rules." *Naval Research Logistics Quarterly*, Vol. 15, No. 2, 1968.

BROWN, R. G., *Management Decisions for Production Operations.* Hinsdale, Ill.: Dryden Press, 1970.

BROWN, R. G., "Detection of Turning Points in a Time Series." *Decision Sciences*, Vol. 2, No. 4, 1971, pp. 383–403.

BROWN, R. G., *Materials Management Systems.* New York: John Wiley & Sons, Inc., 1977.

BROWN, R. G., AND R. F. MEYER, "The Fundamental Theorem of Exponential Smoothing." *Operations Research*, Vol. 9, 1961.

BURMAN, J. P., "Moving Seasonal Adjustments of Economic Time Series." *Journal of the Royal Statistical Society*, Ser. A, Vol. 128, 1965, pp. 534–558.

CHAMBERS, J. C., S. K. MULLICK, AND D. D. SMITH, "How to Choose the Right Forecasting Technique." *Harvard Business Review*, July–August 1971, pp. 45–74.

CHAMBERS, J. C., S. K. MULLICK, AND D. D. SMITH, *An Executive's Guide to Forecasting.* New York: John Wiley & Sons, Inc., 1974.

CHATFIELD, C., *The Analysis of Time Series.* London: Chapman & Hall Ltd., 1975.

CHATFIELD, C., AND D. L. PROTHERO, "Box–Jenkins Seasonal Forecasting: Problems in a Case-Study." *Journal of the Royal Statistical Society*, Vol. 136, Ser. A, Part 3, pp. 295–336.

CHISHOLM, R. K., AND G. R. WHITAKER, JR., *Forecasting Methods.* Homewood, Ill.: Richard D. Irwin, Inc., 1971.

CHOW, W. M., "Adaptive Control of the Exponential Smoothing Constant." *Journal of Industrial Engineering*, Vol. 16, No. 5, 1965, pp. 314–317.

COGGER, K. O., "The Optimality of General-Order Exponential Smoothing." *Operations Research*, Vol. 22, No. 4, 1974, pp. 858–867.

COHEN, G. D., "Bayesian Adjustment of Sales Forecasts in Multi-item Inventory Control Systems." *Journal of Industrial Engineering*, Vol. 17, No. 9, 1966, pp. 474–479.

COX, D. R., "Prediction by Exponentially Weighted Moving Averages and Related Methods." *Journal of the Royal Statistical Society*, Ser. B, Vol. 23, No. 2, 1961, pp. 414–422.

CRANE, D. G., AND J. R. CROTTY, "A Two-Stage Forecasting Model: Exponential Smoothing and Multiple Regression." *Management Science*, Vol. 13, No. 8, 1967, pp. 501–507.

DAUTEN, C. A., AND L. M. VALENTINE, *Business Cycles and Forecasting.* Cincinnati, Ohio: South-Western Publishing Company, 1974.

D'ESOPO, D. A., "A Note on Forecasting by the Exponential Smoothing Operator." *Operations Research*, Vol. 9, No. 5, 1961, pp. 686–687.

DOBBIE, J. M., "Forecasting Periodic Trends by Exponential Smoothing." *Operations Research*, Vol. 11, No. 6, 1963, pp. 908–918.

DRAPER, N., AND H. SMITH, *Applied Regression Analysis*. New York: John Wiley & Sons, Inc., 1966.

DURBIN, J., AND G. S. WATSON, "Testing for Serial Correlation in Least Squares Regression: I and II." *Biometrica*, Vol. 37, Dec. 1950, pp. 404–428, and Vol. 38, June 1951, pp. 159–178.

FARLEY, J. U., AND M. J. HINICH, "Detecting 'Small' Mean Shifts in Time Series." *Management Science*, Vol. 17, No. 3, November 1970, pp. 189–199.

FRANK, C. R., *Statistics and Econometrics*. New York: Holt, Rinehart and Winston, Inc., 1971.

GOODMAN, M. L., "A New Look at Higher-Order Exponential Smoothing for Forecasting." *Operations Research*, Vol. 22, No. 4, 1974, pp. 880–888.

GREEN, J. H., *Production and Inventory Control Handbook*. New York: McGraw-Hill Book Company, 1970.

GRENANDER, V., AND M. ROSENBLATT, *Statistical Analysis of Stationary Time Series*. New York: John Wiley & Sons, Inc., 1957.

GROFF, G. K., "Empirical Comparison of Models for Short-Range Forecasting." *Management Science*, Vol. 20, No. 1, September 1973, pp. 22–31.

GROSS, C. W., AND R. T. PETERSON, *Business Forecasting*. Boston: Houghton Mifflin Company, 1976.

GROSS, D., AND J. L. RAY, "A General Purpose Forecast Simulator." *Management Science*, Vol. 11, No. 6, 1965, pp. 119–135.

HANNAN, E. J., "The Estimation of Seasonal Variation in Economic Time Series." *Journal of the American Statistical Association*, Vol. 58, 1963, pp. 31–44.

HARRIS, L., "A Decision-Theoretic Approach on Deciding When a Sophisticated Forecasting Technique Is Needed." *Management Science*, Vol. 13, No. 2, 1966, pp. 66–69.

HARRISON, P. J., "Short-Term Forecasting." *Applied Statistics*, Vol. 14, 1965, pp. 102–139.

HARRISON, P. J., "Exponential Smoothing and Short-Term Sales Forecasting." *Management Science*, Vol. 13, No. 11, 1967, pp. 821–842.

HARRISON, P. J., AND O. L. DAVIES, "The Use of Cumulative Sum (CUSUM) Techniques for the Control of Routine Forecasts of Product Demand." *Operations Research*, Vol. 12, No. 2, 1964, pp. 325–333.

HARRISON, P. J., AND C. F. STEVENS, "A Bayesian Approach to Short-Term Forecasting." *Operational Research Quarterly*, Vol. 22, No. 4, 1971, pp. 341–362.

HARTUNG, P., "A Simple Style Goods Inventory Model." *Management Science*, Vol. 19, No. 2, August 1973, pp. 1452–1458.

HAUSMAN, W. H., AND R. S. G. SIDES, "Mail Order Demands for Style Goods: Theory and Data Analysis." *Management Science*, Vol. 20, No. 2, October 1973, pp. 191–202.

HERTZ, D. B., AND K. H. SCHAFFIR, "A Forecasting Model for Management of Seasonal Style Goods Inventories." *Operations Research*, Vol. 8, No. 2, 1960, pp. 45–52.

HOLT, C. C., "Forecasting Seasonal and Trends by Exponentially Weighted Moving Averages." Carnegie Institute of Technology, Pittsburgh, Pa., 1957.

HOLT, C. C., MODIGLIANI, F., MUTH, J. F., AND H. A. SIMON, *Planning Production, Inventories and Work Force*. Englewood Cliffs, N.J.: Prentice-Hall, Inc., 1960.

JOHNSON, L. A., AND D. C. MONTGOMERY, *Operations Research in Production Planning, Scheduling, and Inventory Control*. New York: John Wiley & Sons, Inc., 1974.

KELEGIAN, H. H., AND W. E. OATES, *Introduction to Econometrics*. New York: Harper & Row, Inc., 1974.

KENDALL, M. G., *Time Series*. New York: Hafner Press, 1973.

KIRBY, R. M., "A Comparison of Short and Medium Range Statistical Forecasting Methods." *Management Science*, Vol. 13, No. 2, 1966, pp. B202–B210.

LAAKSO, C. A., Safety Stocks and Service Levels for Multi-Location Inventory. Unpublished M.S. thesis, Illinois Institute of Technology, Chicago, Ill., 1978.

LEVIN, R. I., *Statistics for Management*. Englewood Cliffs, N.J.: Prentice-Hall, Inc., 1978.

MABERT, V. A., "An Introduction to Short-Term Forecasting Using the Box–Jenkins Methodology." *AIIE* (PP&G—75-1), Norcross, Ga., 1975.

MAKRIDAKIS, S., "A Survey of Time Series." *International Statistical Review*, Vol. 44, No. 1, 1976.

MAKRIDAKIS, S., AND S. C. WHEELWRIGHT, *Interactive Forecasting*. San Francisco: Holden-Day, Inc., 1977.

MAKRIDAKIS, S., AND S. WHEELRIGHT, *Forecasting: Methods and Applications*. New York: John Wiley & Sons, Inc., 1978.

MARQUARDT, D. W., "An Algorithm for Least Squares Estimation of Nonlinear Parameters." *Society for Industrial and Applied Mathematics*, Vol. 11, No. 2, June 1963.

MCLAUGHLIN, R. L., *Time-Series Forecasting. Marketing-Research Technique*, Ser. 6. New York: American Marketing Association, 1962.

MCLAUGHLIN, R. L., "A New Five-Phase Economic Forecasting System." *Business Economics*, September 1975, pp. 49–60.

MCLAUGHLIN, R. L., AND J. J. BOYLE, *Short-Term Forecasting*. New York: American Marketing Association, 1968.

MCCLAIN, J. O., AND L. J. THOMAS, "Response-Variance Tradeoffs in Adaptive Forecasting." *Operations Research*, Vol. 21, No. 2, March–April 1973, pp. 554–568.

MEYER, R. F., "An Adaptive Method for Routine Short Term Forecasting." International Federation of Operational Research Societies, Oslo, July 1963.

MICHAEL, G. C., "A Computer Simulation Model for Forecasting Catalog Sales." *Journal of Marketing Research*, May 1971, pp. 224–229.

MILLER, I., AND J. E. FREUND, *Probability and Statistics for Engineers*. Englewood Cliffs, N.J.: Prentice-Hall, Inc., 1977.

MONTGOMERY, D. C., "An Introduction to Short-Term Forecasting." *Journal of Industrial Engineering*, Vol. 19, No. 10, 1968, pp. 500–503.

MONTGOMERY, D. C., "An Application of Statistical Forecasting Techniques in an Inventory Control Policy." *Production and Inventory Management*, First Quarter, 1969, pp. 66–74.

MONTGOMERY, D. C., "Adaptive Control of Exponential Smoothing Parameters by Evolutionary Operation." *AIIE Transactions*, Vol. 2, No. 3, 1970, pp. 268–269.

MONTGOMERY, D. C., AND L. A. JOHNSON, *Forecasting and Time Series Analysis*. New York: McGraw-Hill Book Company, 1976.

MORRISON, N., *Introduction to Sequential Smoothing and Prediction*. New York: McGraw-Hill Book Company, 1969.

MUTH, J. F., "Optimal Properties of Exponentially Weighted Forecasts of Time Series with Permanent and Transitory Components." *Journal of the American Statistical Association*, Vol. 55, No. 2, 1960, pp. 299–306.

NATRELLA, M. G., *Experimental Statistics*. Washington, D.C.: National Bureau of Standards Handbook 91, 1963.

NAYLOR, T. H., AND T. G. SEAKS, "Box–Jenkins Methods: An Alternative to Econometric Models." *International Statistical Review*, Vol. 40, No. 2, 1972, pp. 123–137.

NELSON, C. R., *Applied Time Series Analysis for Managerial Forecasting*. San Francisco: Holden-Day, Inc., 1973.

NERLOVE, M., AND S. WAGE, "On the Optimality of Adaptive Forecasting." *Management Science*, Vol. 10, No. 2, 1964, pp. 207–224.

PARZEN, E., *Time Series Analysis Papers*. San Francisco: Holden-Day, Inc., 1967.

PEGELS, C. C., "A Note on Exponential Forecasting." *Management Science*, Vol. 15, No. 5, 1969, pp. 311–315.

ROBERTS, S. D., AND R. REED, "The Development of a Self-Adaptive Forecasting Technique." *AIIE Transactions*, Vol. 1, No. 4, 1969, pp. 314–322.

SCHUSSEL, G., "Sales Forecasting with the Aid of a Human Behavior Simulator." *Management Science*, Vol. 13, No. 10, 1967, pp. B593–B611.

SHISKIN, J., et al., "The X-11 Variant of the Census II Method Seasonal Adjustment Program." *Bureau of the Census, Technical Paper No. 15*, 1967.

SILVER, E. A., "A Modified Formula for Calculating Customer Service Under Continuous Inventory Review." *American Institute of Industrial Engineers Transactions*, Vol. 2, No. 3, Sept. 1970, pp. 241–245.

SIMPSON, K. F., "In-Process Inventories." *Operations Research*, Vol. 7, No. 6, 1959, pp. 797–805.

SULLIVAN, W. G., AND W. W. CLAYCOMBE, *Fundamentals of Forecasting*. Reston, Va.: Reston Publishing Company, Inc., 1977.

THEIL, H., AND S. WAGE, "Some Observations on Adaptive Filtering." *Management Science*, Vol. 10, No. 2, January 1964, pp. 198–224.

THOMOPOULOS, N. T., AND C. LAAKSO, "Productivity in a Multi-Location Inventory System," *Manufacturing Productivity Frontiers,* Vol. 2, No. 2, Dec. 1978, pp. 23–25.

THOMPSON, H. E., AND W. BERANEK, "The Efficient Use of an Imperfect Forecast." *Management Science*, Vol. 13, No. 3, 1966, 233–243.

TRIGG, D. W., "Monitoring a Forecasting System," *Operational Research Quarterly*, Vol. 15, 1964, pp. 271–274.

TRIGG, D. W., AND D. H. LEACH, "Exponential Smoothing with an Adaptive Response Rate." *Operational Research Quarterly*, Vol. 18, 1967, pp. 53–59.

VAN DOBBEN DE BRUYN, C. S., *Cumulative Sum Tests.* London: Charles Griffin & Co. Ltd., 1968.

WADE, J. B., "Determining Reorder Points When Demand Is Lumpy." *Management Science*, Vol. 24, No. 6, Feb. 1978, pp. 623–632.

WHEELWRIGHT, S. C., AND S. MAKRIDAKIS, *Forecasting Methods for Management.* New York: John Wiley & Sons, Inc., 1973.

WILCOX, J. E., "How to Forecast Lumpy Items." *American Production and Inventory Management*, First Quarter, 1970, pp. 51–54.

WILD, R., *The Techniques of Production Management.* New York: Holt, Rinehart and Winston, Inc., 1971.

WINTERS, P. R., "Forecasting Sales by Exponentially Weighted Moving Averages." *Management Science*, April 1960, pp. 324–342.

WONNACOTT, T. H., AND R. J. WONNACOTT, *Introductory Statistics.* New York: John Wiley & Sons, Inc., 1977.

WOODWARD, R. H., AND P. L. GOLDSMITH, *Cumulative Sum Techniques.* Edinburgh: Oliver & Boyd Ltd., 1964.

INDEX

C

D

E

F

G

H

I

J

L

M

N

O

P

Q

R

S

T

U

V

W